思科系列丛书

思科网络实验室 CCNP（路由技术）实验指南

（第 2 版）

梁广民　徐　磊　谢晓广　编著

李涤非　审校

电子工业出版社

Publishing House of Electronics Industry

北京·BEIJING

内 容 简 介

本书以 Cisco2911 路由器和 Catalyst3560 交换机为硬件平台，以新版 CCNP 内容为基础，以实验为依托，从行业的实际需求出发组织全部内容，全书共 9 章，主要内容包括：实验准备、IP 路由原理、EIGRP、OSPF、IS-IS、路由重分布与路径控制、BGP、分支连接和 IPv6。

本书既可以作为思科网络技术学院的配套实验教材，用来增强学生的网络知识和操作技能，也可以作为电子和计算机等专业的网络集成类课程的教材或者实验指导书，还可以作为相关企业的培训教材；同时，对于从事网络管理和维护的技术人员，也是一本很实用的技术参考书。

未经许可，不得以任何方式复制或抄袭本书之部分或全部内容。
版权所有，侵权必究。

图书在版编目（CIP）数据

思科网络实验室 CCNP（路由技术）实验指南 / 梁广民，徐磊，谢晓广编著. —2 版. —北京：电子工业出版社，2019.3
（思科系列丛书）
ISBN 978-7-121-35957-6

Ⅰ. ①思… Ⅱ. ①梁… ②徐… ③谢… Ⅲ. ①计算机网络—路由选择—实验—指南 Ⅳ. ①TN915.05-33

中国版本图书馆 CIP 数据核字 (2019) 第 015336 号

策划编辑：宋 梅
责任编辑：宋 梅
印　　刷：北京七彩京通数码快印有限公司
装　　订：北京七彩京通数码快印有限公司
出版发行：电子工业出版社
　　　　　北京市海淀区万寿路 173 信箱　邮编　100036
开　　本：787×1 092　1/16　印张：21.25　字数：544 千字
版　　次：2012 年 3 月第 1 版
　　　　　2019 年 3 月第 2 版
印　　次：2024 年 1 月第 8 次印刷
定　　价：89.00 元

凡所购买电子工业出版社图书有缺损问题，请向购买书店调换。若书店售缺，请与本社发行部联系，联系及邮购电话：(010) 88254888，88258888。
质量投诉请发邮件至 zlts@phei.com.cn，盗版侵权举报请发邮件至 dbqq@phei.com.cn。
本书咨询联系方式：mariams@phei.com.cn。

前 言

作为全球领先的互联网设备供应商，思科公司的产品涉及路由、交换、安全、语音、无线和存储等诸多方面。而思科推出的系列职业认证 CCNA、CCNP 和 CCIE 无疑是 IT 领域最为成功的职业认证规划之一。本书以 CCNP（路由技术）职业认证为依托，从实际应用的角度出发，以思科网络实验室为背景设计拓扑，全面、详细地介绍了新版 CCNP 中路由技术的内容。本书的特色如下：

在目标上，以企业实际需求为向导，以培养学生的网络设计能力、对网络设备的配置和调试能力、分析和解决问题能力以及创新能力为目标，讲求实用。

在内容选取上，集先进性、科学性和实用性为一体，全面覆盖新版 CCNP（路由技术）的内容，但又不局限于 CCNP 范围，尽可能覆盖最新、最实用的技术。如本书对 IS-IS 路由协议、IPv6 技术、IPSec VPN 技术、PPPoE 技术等做了适当的扩展和有益的补充。

在内容表现形式上，把握"理论够用、技能为主"的原则，用最简单和最精练的描述讲解网络基本知识，然后通过详尽的实验现象分析来分层、分步骤地讲解网络技术；而对实验调试信息，笔者根据多年实验调试的经验加以汇总和注释，写入本书，直观、易懂。

在内容结构上，本书按照 CCNP（路由技术）新版教材的结构和布局设计为 9 章：第 1 章实验准备、第 2 章 IP 路由原理、第 3 章 EIGRP、第 4 章 OSPF、第 5 章 IS-IS、第 6 章路由重分布与路径控制、第 7 章 BGP、第 8 章分支连接和第 9 章 IPv6，从配置开始，逐渐展开，结合实验调试结果来巩固和深化所学的内容，最后达到学习知识和培养能力的目的。

本书以 Cisco2911 路由器和 Catalyst3560 交换机为硬件平台来搭建实验环境，由于各个实验室的具体情况不同，在实际使用过程中，教师可能需要做稍微的改动，以适应自己实验室不同实验设备和环境。

本书既可以作为思科网络技术学院的配套实验教材，用来增强学生的网络知识和操作技能，也可以作为电子和计算机等专业的网络集成类课程的教材或者实验指导书使用，还可以作为相关企业的培训教材；同时，对于从事网络管理和维护的技术人员，也是一本很实用的技术参考书。

本书由梁广民（CCIE#14496 R/S，Security）、徐磊和谢晓广组织编写并统稿，参加编写的还有王隆杰、张喜生、石淑华、杨旭、刘平、张立涓、石光华、邹润生、杨名川、成荣、周鸣琦、韦凯和齐治文等。从复杂和庞大的 Cisco 网络技术中，编写一本简明的、适合实验室使用的实验教材确实不是一件容易的事情，衷心感谢思科大中华区网络技术学院技术经理李涤非老师在百忙之中审校全书。感谢沃尔夫网络实验室（www.wolf-lab.com）对本书中的关键技术给予的指导和帮助。如果没有他们的帮助，本书是不可能在很短的时间内高质量完成的。

由于时间仓促，加上作者水平有限，书中难免有不妥和错误之处，恳请同行专家指正。
E-mail：gmliang@szpt.edu.cn。

编 著 者
2019 年 3 月于深圳

目 录

第 1 章 实验准备 ... 1
1.1 实验拓扑搭建 ... 1
1.1.1 网络设备之间的连接 ... 1
1.1.2 终端访问服务器的连接 ... 2
1.1.3 终端访问服务器的配置 ... 3
1.2 准备实验软件 ... 5
1.2.1 准备操作系统软件 ... 5
1.2.2 工具软件 ... 6
本章小结 ... 7

第 2 章 IP 路由原理 ... 8
2.1 IP 路由概述 ... 8
2.1.1 静态路由特征 ... 8
2.1.2 动态路由协议特征 ... 8
2.1.3 填充路由表 ... 10
2.1.4 查找路由表 ... 12
2.2 RIP 概述 ... 13
2.2.1 RIP 特征 ... 13
2.2.2 RIP 数据包格式 ... 14
2.3 配置静态路由和 RIPv2 ... 15
2.3.1 实验 1：配置静态路由 ... 15
2.3.2 实验 2：配置 RIPv2 ... 19
本章小结 ... 24

第 3 章 EIGRP ... 25
3.1 EIGRP 概述 ... 25
3.1.1 EIGRP 特征 ... 25
3.1.2 DUAL 算法 ... 25
3.1.3 EIGRP 数据包类型 ... 26
3.1.4 EIGRP 数据包格式 ... 27
3.1.5 EIGRP 的 SIA 及查询范围的限定 ... 30

3.2 配置 EIGRP ... 31
3.2.1 实验 1：配置基本 EIGRP ... 31
3.2.2 实验 2：配置高级 EIGRP ... 39
3.2.3 实验 3：配置 EIGRP stub ... 46
3.2.4 实验 4：配置命名 EIGRP ... 50
本章小结 ... 54

第 4 章 OSPF ... 55
4.1 OSPF 概述 ... 55
4.1.1 OSPF 特征 ... 55
4.1.2 OSPF 术语 ... 55
4.1.3 OSPF 路由器类型 ... 56
4.1.4 OSPF 网络类型 ... 57
4.1.5 OSPF 区域类型 ... 57
4.1.6 OSPF LSA 类型 ... 58
4.1.7 OSPF 数据包格式 ... 58
4.1.8 OSPF 邻居关系建立 ... 63
4.1.9 OSPF 运行步骤 ... 64
4.2 配置单区域 OSPF ... 65
4.2.1 实验 1：配置单区域 OSPF ... 65
4.2.2 实验 2：配置 OSPF 验证 ... 74
4.3 配置多区域 OSPF ... 81
4.3.1 实验 3：配置多区域 OSPF ... 81
4.3.2 实验 4：配置 OSPF 路由手工汇总 ... 86
4.3.3 实验 5：配置 OSPF 末节区域和完全末节区域 ... 89
4.3.4 实验 6：配置 OSPF NSSA 区域 ... 91
4.3.5 实验 7：配置虚链路 ... 95
本章小结 ... 99

第 5 章 IS-IS ... 100
5.1 IS-IS 概述 ... 100
5.1.1 IS-IS 特征 ... 100
5.1.2 IS-IS 术语 ... 100
5.1.3 IS-IS 路由器类型 ... 102
5.1.4 IS-IS 数据包格式 ... 102
5.2 配置集成 IS-IS ... 108
5.2.1 实验 1：配置单区域集成 IS-IS ... 108
5.2.2 实验 2：配置多区域集成 IS-IS ... 114
5.2.3 实验 3：配置集成 IS-IS 验证 ... 119

本章小结 ... 123

第 6 章 路由重分布与路径控制 ... 124

6.1 路由重分布概述 ... 124
6.1.1 路由重分布种子度量值 ... 124
6.1.2 路由重分布存在的问题 ... 124

6.2 路径控制概述 ... 125
6.2.1 路由映射表（Route Map）... 125
6.2.2 分布列表、前缀列表和偏移列表 ... 126
6.2.3 IP SLA ... 126
6.2.4 策略路由（PBR）... 127
6.2.5 VRF ... 127

6.3 路由重分布 ... 128
6.3.1 实验 1：路由重分布基本配置 ... 128
6.3.2 实验 2：路由重分布中次优路由和路由环路问题及其解决方案 ... 133

6.4 控制路由更新 ... 139
6.4.1 实验 3：配置被动接口和分布列表控制路由更新 ... 139
6.4.2 实验 4：配置前缀列表和路由映射表控制路由更新 ... 142
6.4.3 实验 5：配置 Cisco IP SLA 控制路径选择 ... 145

6.5 策略路由 ... 150
6.5.1 实验 6：配置基于源 IP 地址的策略路由 ... 150
6.5.2 实验 7：配置基于数据包长度的策略路由 ... 152
6.5.3 实验 8：配置基于应用的策略路由 ... 154

6.6 VRF Lite ... 155
6.6.1 实验 9：配置 VRF Lite ... 155

本章小结 ... 160

第 7 章 BGP ... 161

7.1 BGP 概述 ... 161
7.1.1 BGP 特征 ... 161
7.1.2 BGP 术语 ... 161
7.1.3 BGP 属性 ... 162
7.1.4 BGP 消息类型及格式 ... 163
7.1.5 BGP 路由决策 ... 166
7.1.6 BGP 路由抑制 ... 167
7.1.7 BGP 邻居状态 ... 168

7.2 配置基本 BGP ... 169
7.2.1 实验 1：配置 IBGP 和 EBGP ... 169
7.2.2 实验 2：配置 BGP 验证、路由抑制和 EBGP 多跳 ... 177

  7.3 配置高级 BGP 180
    7.3.1 实验 3：配置 BGP 地址聚合 180
    7.3.2 实验 4：配置路由反射器（RR）...185
    7.3.3 实验 5：配置 BGP 联邦...188
    7.3.4 实验 6：配置 BGP 团体...191
  7.4 配置 BGP 属性控制选路...195
    7.4.1 实验 7：配置 BGP ORIGIN 属性控制选路...195
    7.4.2 实验 8：配置 BGP AS-PATH 属性控制选路...198
    7.4.3 实验 9：配置 BGP LOCAL_PREF 属性控制选路....................................199
    7.4.4 实验 10：配置 BGP Weight 属性控制选路..200
    7.4.5 实验 11：用 MED 属性控制选路...201
  本章小结..205

第 8 章 分支连接 206

  8.1 分支连接概述..206
    8.1.1 公共 WAN 基础设施远程连接 206
    8.1.2 专用 WAN 基础设施远程连接 207
  8.2 PPPoE 概述...208
    8.2.1 PPPoE 简介..208
    8.2.2 PPPoE 数据包类型..208
    8.2.3 PPPoE 会话建立过程..208
  8.3 隧道技术概述..210
    8.3.1 GRE 简介...210
    8.3.2 IPSec VPN 简介..212
    8.3.3 AH 和 ESP...212
    8.3.4 安全关联和 IKE 214
    8.3.5 IPSec 操作步骤...215
  8.4 配置 PPPoE 216
    8.4.1 实验 1：配置 ADSL...216
    8.4.2 实验 2：配置 PPPoE 服务器和客户端..219
  8.5 配置隧道..221
    8.5.1 实验 3：配置 GRE 隧道..221
    8.5.2 实验 4：配置 Site To Site VPN..225
    8.5.3 实验 5：配置 Remote VPN...233
    8.5.4 实验 6：配置 GRE Over IPSec...239
    8.5.5 实验 7：配置 Redundancy VPN...244
    8.5.6 实验 8：配置 DMVPN...248
  本章小结..257

第 9 章 IPv6 258

9.1 IPv6 概述 258
   9.1.1 IPv6 特点 258
   9.1.2 IPv6 地址与基本包头格式 259
   9.1.3 IPv6 扩展包头 260
   9.1.4 IPv6 地址类型 262
   9.1.5 IPv6 邻居发现协议（NDP）.. 264
   9.1.6 IPv6 过渡技术 264
9.2 配置 IPv6 地址 266
   9.2.1 实验 1：手工配置 IPv6 单播地址 266
   9.2.2 实验 2：通过有状态 DHCPv6 获得 IPv6 地址 272
9.3 配置 IPv6 路由 275
   9.3.1 实验 3：配置 IPv6 静态路由 275
   9.3.2 实验 4：配置 RIPng 278
   9.3.3 实验 5：配置 OSPFv3 283
   9.3.4 实验 6：配置 IPv6 EIGRP 291
   9.3.5 实验 7：配置 IPv6 集成 IS-IS 296
   9.3.6 实验 8：配置 MBGP 300
9.4 IPv6 路由重分布和策略路由 305
   9.4.1 实验 9：配置 OSPFv3、IS-IS 和 MBGP 路由重分布 305
   9.4.2 实验 10：配置 IPv6 策略路由 310
9.5 配置 IPv4 向 IPv6 过渡 312
   9.5.1 实验 11：配置手工隧道 312
   9.5.2 实验 12：配置 GRE 隧道 316
   9.5.3 实验 13：配置 6to4 隧道 318
   9.5.4 实验 14：配置 ISATAP 隧道 320
   9.5.5 实验 15：配置 IPv6 静态 NAT-PT 322
   9.5.6 实验 16：配置 IPv6 动态 NAT-PT 324
本章小结 326

参考文献 327

第 1 章 实 验 准 备

要顺利完成本书各个章节的实验，必须具备相应的网络设备（路由器、交换机和服务器等）、软件（IOS 和相关工具软件）以及合理的网络连接，避免每次实验都要花费大量的时间来搭建网络拓扑。本章介绍本书使用的网络设备的选型、拓扑搭建以及相关软件的选择，力求完全满足 Cisco 的 CCNP 认证（路由技术）的所有实验需要。当然，本书中涉及的实验也可以通过 GNS3 模拟器完成，某种意义上讲，用模拟器搭建实验环境更加方便。

1.1 实验拓扑搭建

为了完成本书中的实验内容，需要构建不同的网络拓扑，为此，笔者设计了一个功能强大的网络拓扑，可以满足 CCNP 课程中和路由技术相关的实验需要。

1.1.1 网络设备之间的连接

笔者设计的实验拓扑（以太网连接部分）如图 1-1 所示（注意：图中不包含终端访问服务器和各设备的连接）。该网络拓扑中的路由器和交换机均通过终端访问服务器来进行访问控制。当然，如果实验室没有搭建终端服务器所需要的设备和模块，也可以通过计算机串行通信端口（COM 口）或者 USB 端口和网络设备的控制台（Console）端口连接，但是需要经常插拔连接网路设备的 Console 线缆，非常不方便。

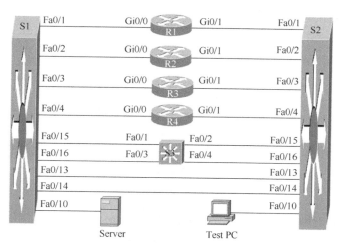

图 1-1 实验拓扑（以太网连接部分）

图 1-1 中包括 4 台（设备名称为 R1、R2、R3、R4）Cisco2911 路由器（安装 1~2 块 HWIC-2T 模块）和 3 台（设备名称为 S1、S2、S3）3560V2 交换机（24 个百兆位和 2 个千兆

位以太网接口)。读者也可以根据实验室中实验设备的具体情况选择合适的设备搭建网络拓扑。路由器也可以采用 Cisco ISR 的 4300、4400、1900、3900 系列路由器或者早期的 1800、2800 和 3800 系列路由器,不同的路由器支持的模块数量和模块类型可能不同。当然,操作系统软件也需要匹配。交换机也可以采用 2960、3650、3850 系列以及早期的 3750 系列的设备。路由器 R1～R4 的 Gi0/0 以太网接口与交换机 S1 的 Fa0/1～Fa0/4 接口相连接;Gi0/1 以太网接口与交换机 S2 的 Fa/1～Fa0/4 接口相连接。交换机 S1 和 S2 之间通过 Fa0/13 和 Fa0/14 接口进行连接;交换机 S3 的 Fa0/1 和 Fa0/3 接口连接到交换机 S1 的 Fa0/15 和 Fa0/16 接口上,交换机 S3 交换机的 Fa0/2 和 Fa0/4 接口连接到交换机 S2 的 Fa0/15 和 Fa0/16 接口上。交换机 S1 的 Fa0/10 接口连接 Server 网卡,交换机 S2 的 Fa0/10 接口连接 Test PC 网卡,读者可以根据实验的实际需要灵活地连接 Test PC 到交换机的相应以太网接口。

实验拓扑(串行连接部分)如图 1-2 所示。路由器 R1 的 Se0/0/0 和 Se0/0/1 串行口和路由器 R2 的 Se0/0/0 和 Se0/1/1 串行口连接,路由器 R2 的 Se0/0/1 串行口和路由器 R3 的 Se0/0/1 串行口连接,路由器 R2 的 Se0/1/0 串行口和路由器 R4 的 Se0/0/1 串行口连接,路由器 R3 的 Se0/0/0 串行口和路由器 R4 的 Se0/0/0 串行口连接。

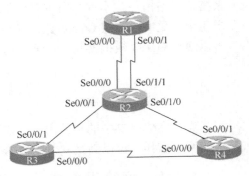

图 1-2 实验拓扑(串行连接部分)

1.1.2 终端访问服务器的连接

在实验过程中,综合和复杂的实验会用到多台路由器或者交换机,如果通过计算机串行通信端口(COM 端口)和网络设备的控制台(Console)端口连接,就需要多台计算机或者经常性插拔连接网路设备的 Console 线缆,非常不方便,而且如果带电插拔线缆,也可能把网络设备的 Console 端口烧掉,造成设备损坏。终端访问服务器可以解决这个问题,终端访问服务器和网络设备的连接如图 1-3 所示。

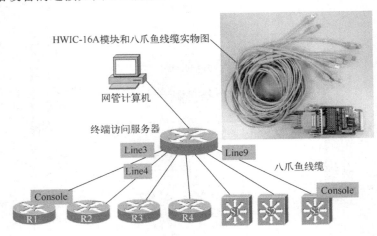

图 1-3 终端访问服务器和网络设备的连接

终端访问服务器通常由一台配置了 HWIC-8A 模块或者 HWIC-16A 模块的路由器来充当,从它引出多条连接线到各个被控设备的 Console 端口。使用时,用户首先通过计算机 COM 端口或者 Telnet 连接到终端访问服务器，然后再从终端访问服务器访问各个网络设备，这样就能在一台计算机上同时控制对多台网络设备的访问，而不用频繁插拔 Console 线缆。

1.1.3 终端访问服务器的配置

在本书设计的实验拓扑中，终端访问服务器和网络设备的物理连接如图 1-4 所示。

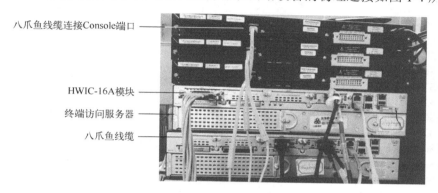

图 1-4 终端访问服务器和网络设备的物理连接

为了方便用户使用终端访问服务器，可以制作一个简单的菜单供用户使用，这样用户可以清楚地知道如何登录到相应的网络设备。本节给出终端访问服务器的配置以及使用 SecureCRT 软件同时访问多台网络设备的方法。

（1）完成终端访问服务器的基本配置

```
Router(config)#hostname TS
TS(config)#enable secret Cisco123@ccnp
TS(config)#line vty 0 15
TS(config-line)#no login            //登录时不进行密码检查
TS(config-line)#logging synchronous  //日志同步
TS(config-line)#exec-timeout 0 0    //超时时间为 0
TS(config-line)#exit
TS(config)#interface gigabitEthernet 0/0
TS(config-if)#ip address 10.3.24.15 255.255.255.0
TS(config-if)#no shutdown
TS(config-if)#exit
TS(config)#no ip routing            //关闭终端访问服务器路由功能，相当于一台计算机
TS(config)#ip default-gateway 10.3.24.254  //配置网关，允许从外网访问该设备
```

（2）配置线路和制作简易使用菜单

```
TS#show line
   Tty    Line   Typ    Tx/Rx       A  Modem  Roty AccO AccI  Uses  Noise Overruns  Int
*  0      0      CTY                -  -      -    -    -     0     0     0/0       -
   1      1      AUX    9600/9600   -  -      -    -    -     0     0     0/0       -
   2      2      TTY    9600/9600   -  -      -    -    -     0     0     0/0       -
   0/0/0  3      TTY    9600/9600   -  -      -    -    -     0     0     0/0       -
   0/0/1  4      TTY    9600/9600   -  -      -    -    -     0     0     0/0       -
   0/0/2  5      TTY    9600/9600   -  -      -    -    -     0     0     0/0       -
```

（此处省略部分输出）

以上输出给出了终端访问服务器上异步模块的各异步口所在的线路编号，含有 Tty 列的输出显示异步模块接口和所对应的线路编号，该终端访问服务器模块有 16 个接口，线路编号为 3～18，本书实验中只使用了线路 3～9。

```
TS#configure terminal
TS(config)#line 3 9
TS(config-line)#transport input telnet
//默认情况下线路允许所有输入，本配置只允许 Telnet 输入
TS(config-line)#no exec              //不允许 line 接受 exec 会话
TS(config-line)#exec-timeout 0 0
TS(config-line)#logging synchronous
TS(config-line)#exit
TS(config)#interface loopback0
TS(config-if)#ip address 1.1.1.1 255.255.255.255   //创建环回接口 Loopback0 并配置 IP 地址
TS(config)#ip host R1 2003 1.1.1.1                 //定义主机名及反向 Telnet 的端口号
TS(config)#ip host R2 2004 1.1.1.1
TS(config)#ip host R3 2005 1.1.1.1
TS(config)#ip host R4 2006 1.1.1.1
TS(config)#ip host S1 2007 1.1.1.1
TS(config)#ip host S2 2008 1.1.1.1
TS(config)#ip host S3 2009 1.1.1.1
TS(config)#alias exec cr1 clear line 3             //定义命令别名
TS(config)#alias exec cr2 clear line 4
TS(config)#alias exec cr3 clear line 5
TS(config)#alias exec cr4 clear line 6
TS(config)#alias exec cs1 clear line 7
TS(config)#alias exec cs2 clear line 8
TS(config)#alias exec cs3 clear line 9
TS(config)#privilege exec level 0 clear line       //配置命令授权
TS(config)#privilege exec level 0 clear
//通过以上两行命令授权，使得用户在用户模式下可以执行 clear line 命令
TS(config)#banner motd #
Enter TEXT message.  End with the character '#'.
        *******************************************
        R1-------R1      cr1------clear line 3
        R2-------R2      cr2------clear line 4
        R3-------R3      cr3------clear line 5
        R4-------R4      cr4------clear line 6
        S1-------S1      cs1------clear line 7
        S2-------S2      cs2------clear line 8
        S3-------S3      cs3------clear line 9
        *******************************************
#
```

以上是制作一个简易的操作菜单，提醒用户：要控制路由器 R1，可以使用 **R1** 命令（大小写不敏感）；要清除路由器 R1 所在的线路，可以使用 **cr1** 命令。

```
TS#copy   running-config startup-config    //保存配置文件
```

（3）使用 SecureCRT 软件建立多个会话，可以同时访问多台网络设备

开启 SecureCRT 软件，为本书实验中用到的 7 台设备分别创建一个会话，这样后续的实验就不用每次都建立新的会话进行连接了。在 SecureCRT 窗口中分别双击每台设备，这样就可以在同一个 SecureCRT 窗口打开不同设备的访问窗口。使用 SecureCRT 软件访问多个网络设备如图 1-5 所示。

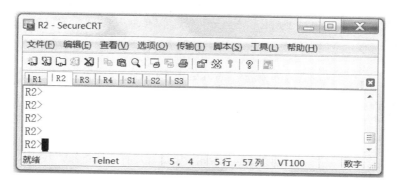

图 1-5　使用 SecureCRT 软件访问多个网络设备

1.2　准备实验软件

完成网络拓扑搭建后，接下来准备本书实验所需要的相关软件，主要包括 Cisco 路由器和交换机操作系统软件（Internetwork Operating System，IOS）和相关工具软件。

1.2.1　准备操作系统软件

不同系列和不同型号的路由器和交换机需要的 IOS 是不同的，请读者选择适合自己实验设备的 IOS。如果需要较新的 IOS，可以从 Cisco 官网（www.cisco.com）下载，并且对设备进行 IOS 升级。下载时请确认自己的网络设备是否满足 IOS 运行所需要的内存和 Flash 空间。

在本书实验环境中，路由器的型号选择 Cisco 的 ISR 2911 路由器，相应的 IOS 选择 c2900-universalk9-mz.SPA.157-3.M.bin。Cisco 官网路由器 IOS 下载页面如图 1-6 所示。

图 1-6　路由器 IOS 下载页面

在本书实验环境中，交换机的型号选择 Cisco 的 WS-C3560V2-24PS-S 三层交换机，相应的 IOS 选择 c3560-ipservicesk9-mz.150-2.SE11.bin。Cisco 官网交换机 IOS 下载页面如图 1-7 所示。

图 1-7　交换机 IOS 下载页面

1.2.2　工具软件

为了确保实验顺利进行并完成相应的功能，本书中使用了如下工具软件。读者也可以在学习相应内容时再下载、准备和安装相应软件。

（1）Wireshark

Wireshark 是网络数据包协议分析工具，它可以捕获网络数据，并显示数据包的尽可能详细的信息，对于读者深入理解网络技术非常有帮助。下载地址：www.wireshark.org。

（2）SecureCRT

SecureCRT 是最常用的终端仿真程序，支持通过串行通信、Telnet 或者 SSH 配置和管理路由器和交换机。下载地址：www.vandyke.com。

（3）TFTPD

请读者根据自己的操作系统是 32 位或 64 位系统选择 TFTPD32 或者 TFTPD64，两者的功能完全一样。TFTPD 是一款集成多种服务的袖珍网络服务器包，包括 SYSLOG 服务器、SNTP 服务器、DHCP 服务器、DNS 服务器、日志查看器以及 TFTP 服务器端和客户端。选择相应的服务器完成相应的实验内容，比如当完成 IOS 的升级或者恢复及配置文件备份时，需要选择 TFTP 服务器，当模拟 DHCP 服务时选择 DHCP 服务器。下载地址：tftpd32.jounin.net。

（4）Cisco Console 转 USB 驱动程序

新款的 Cisco 路由器都配置了 Mini-B USB Console 端口，以方便对设备进行网络管理，使用前需要安装此驱动程序。下载地址：www.cisco.com，或者通过搜索引擎选择下载地址。

（5）USB 转串口驱动程序

如果计算机没有 COM 端口，需要使用 USB 转串口数据线连接计算机和网络设备的控制台端口，此时需要下载相应的 USB 转串口驱动程序。

本 章 小 结

本章介绍了贯穿本书的网络拓扑的搭建以及如何配置终端访问服务器，方便我们在一台计算机上同时访问多台路由器或者交换机。本章还介绍了网络设备的选型以及 IOS 下载和工具软件的准备，为后续各个章节实验的顺利完成做好准备。

第 2 章 IP 路由原理

网络互联的核心任务是解决路由问题，路由器的作用就是将各个网络彼此连接起来，负责不同网络之间的数据包传送。而路由器工作的核心就是路由表，路由器使用路由表来确定转发数据包的最佳路径。路由器构建路由表的来源包括直连路由、静态路由和动态路由。在许多情况下，路由器结合使用动态路由和静态路由来构建路由表。

2.1 IP 路由概述

2.1.1 静态路由特征

1. 静态路由优点

① 占用的 CPU 和内存资源较少。
② 可控性强，便于管理员了解整个网络路由信息。
③ 不需要动态更新路由，可以减少对网络带宽的占用。
④ 简单和易于配置。

2. 静态路由缺点

① 配置和维护耗费管理员大量时间。
② 配置时容易出错，尤其对于大型网络。
③ 当网络拓扑发生变化时，需要管理员维护变化的路由信息。
④ 随着网络规模的增长和配置的扩展，维护越来越麻烦。
⑤ 需要管理员对整个网络的情况完全了解后才能进行恰当的操作和配置。

3. 静态路由使用场合

① 网络中仅包含几台路由器。在这种情况下，使用动态路由协议可能会增加额外的管理负担。
② 网络仅通过单个 ISP 接入 Internet。因为该 ISP 就是唯一的 Internet 出口点，所以不需要在此链路间运行动态路由协议。
③ 路由器没有足够的 CPU 和内存来运行动态路由协议。
④ 通过浮动静态路由为动态路由提供备份。
⑤ 链路的带宽较小。因为动态路由更新和维护会带来额外的链路负担。

2.1.2 动态路由协议特征

动态路由是路由器之间通过路由协议（如 RIP、EIGRP、OSPF、IS-IS 和 BGP 等）动态交换路由信息来构建路由表的。使用动态路由协议最大的好处是，当网络拓扑结构发生变化时，路由器会自动地相互交换路由信息，因此路由器不仅能够自动获知新增加的网络信息，还可以在当前网络连接失败时找出备用路径。

1. 动态路由协议功能

① 发现远程网络信息。
② 动态维护最新的路由信息。
③ 自动计算并选择通往目的网络的最佳路径。
④ 在当前路径无法使用时找出新的最佳路径。

2. 动态路由协议优点

① 当增加或删除网络时，管理员维护路由配置的工作量较少。
② 当网络拓扑结构发生变化时，路由协议可以自动做出调整来更新路由表。
③ 配置不容易出错。
④ 扩展性好，网络规模越大，越能体现出动态路由协议的优势。

3. 动态路由协议缺点

① 需要占用额外的资源，如路由器 CPU 和内存以及链路带宽等。
② 需要掌握更多的网络知识才能进行配置、验证和故障排除等工作，特别是一些复杂的动态路由协议对管理员的要求相对较高。

4. 常见动态路由协议

在路由 IP 数据包时常用的动态路由协议如下。
① RIP（Routing Information Protocol）：路由信息协议。
② EIGRP（Enhanced Interior Gateway Routing Protocol）：增强型内部网关路由协议。
③ OSPF（Open Shortest Path First）：开放最短路径优先。
④ IS-IS（Intermediate System-to-Intermediate System）：中间系统-中间系统。
⑤ BGP（Border Gateway Protocol）：边界网关协议。

5. 动态路由协议分类

（1）IGP 和 EGP

动态路由协议按照作用的 AS（Autonomous System，自治系统）来划分，分为 IGP（Interior Gateway Protocols，内部网关协议）和 EGP（Exterior Gateway Protocols，外部网关协议）。IGP 用于自治系统内部，包括 RIP、EIGRP、OSPF 和 IS-IS 等。而 EGP 用于不同机构管控下的不同自治系统之间的路由。BGP 是目前唯一使用的 EGP 协议，也是 Internet 使用的主要路由协议。

（2）距离矢量路由协议和链路状态路由协议

根据路由协议的工作原理，IGP 还可以进一步分为距离矢量路由协议和链路状态路由协议。距离矢量路由协议主要有 RIP 和 EIGRP。链路状态路由协议主要有 OSPF 和 IS-IS。距离矢量路由协议和链路状态路由协议的区别如表 2-1 所示。

表 2-1　距离矢量路由协议和链路状态路由协议的区别

距离矢量（Distance Vector）	链路状态（Link State）
从网络邻居的角度了解网络拓扑信息	自己有整个网络的拓扑信息
频繁、定期发送路由信息，数据包多，收敛速度慢，EIGRP 支持触发更新	通过事件触发来发送路由信息，数据包少，收敛速度快
复制完整路由表发送给邻居路由器，EIGRP 支持部分更新	仅将链路状态的变化部分传送到其他路由器
简单、占用较少的 CPU 和内存资源	复杂、占用较多的 CPU 和内存资源

距离矢量路由协议适用场合：
① 网络结构简单和扁平化，不需要特殊的分层设计。
② 管理员没有足够的知识来配置链路状态协议和排查故障。
③ 无须关注网络最差情况下的收敛时间。

链路状态路由协议适用场合：
① 网络需要进行分层设计。
② 管理员对于网络中采用的链路状态路由协议非常熟悉。
③ 网络对收敛速度的要求极高。

（3）有类（Classful）路由协议和无类（Classless）路由协议

路由协议按照所支持的 IP 地址类别可划分为有类路由协议和无类路由协议。有类路由协议在路由信息更新过程中不发送子网掩码信息，RIPv1 属于有类路由协议。而无类路由协议在路由更新信息中携带子网掩码，同时支持 VLSM 和 CIDR 等。RIPv2、EIGRP、OSPF、IS-IS 和 BGP 属于无类路由协议。

6. 动态路由协议运行过程

所有路由协议的用途都是获知远程网络，在拓扑发生变化时快速做出调整。所用的方式由该协议所使用的算法及其运行特点决定。一般来说，动态路由协议的运行过程如下。
① 路由器通过其接口发送和接收路由消息。
② 路由器与使用同一路由协议的其他路由器共享路由信息。
③ 路由器通过交换路由信息来了解远程网络。
④ 如果路由器检测到拓扑变化，路由协议可以将这一变化告知其他路由器。

2.1.3　填充路由表

1. 管理距离（Administrative Distance，AD）

管理距离用来定义路由信息来源的可信程度，范围是 0~255 的整数值，值越低表示路由信息来源的优先级别越高。管理距离值为 0 表示优先级别最高。默认情况下，只有直连路由的管理距离为 0，而且这个值不能更改。而静态路由和动态路由协议的管理距离是可以修改的。表 2-2 列出了直连路由、静态路由以及常见动态路由协议的默认管理距离。

表 2-2 直连路由、静态路由以及常见动态路由协议的默认管理距离

路由或路由协议	管理距离（AD）
直连路由	0
静态路由	1
EIGRP 汇总路由	5
外部 BGP（EBGP）	20
内部 EIGRP	90
OSPF	110
IS-IS	115
RIP	120
外部 EIGRP	170
内部 BGP（IBGP）	200

2. 度量值（Metric）

度量值是指路由协议计算到达远程网络的路由开销的值。对于同一种路由协议，当有多条路径通往同一目的网络时，路由协议使用度量值来确定最佳路径。度量值越小，路径越优先。每一种路由协议都有自己的度量标准，所以同一个网络中不同的路由协议选择出的最佳路径可能是不一样的。IP 路由协议中经常使用的度量标准如下。

① 跳数：数据包经过的路由器台数。
② 带宽：链路的数据承载能力。
③ 负载：链路的通信使用情况。
④ 延迟：数据包从源到达目的需要的时间。
⑤ 可靠性：通过接口错误计数或以往链路故障次数来估计出现链路故障的可能性。
⑥ 开销：链路上的费用，OSPF 中的开销值是根据接口带宽计算的。

3. 将路由加入路由表应遵循的原则

路由表存储了与直连网络以及远程网络相关的路径信息。路由表包含网络与下一跳的关联信息。这些关联信息告知路由器，要以最佳方式到达某一目的地，可以将数据包发送到特定路由器（即在到达最终目的地的途中的下一跳）。下一跳也可以关联到通向最终目的地的送出接口。路由器在查找路由表的过程中通常采用"递归查询"。路由器通常用以下三种途径构建路由表。

① 直连网络：就是直连到路由器某一接口的网络，只要该接口处于活动（Up）状态，路由器就会自动添加和自己直接连接的网络到路由表中。
② 静态路由：通过网络管理员手工配置添加到路由表中。
③ 动态路由：由动态路由协议（如 RIP、EIGRP、OSPF、IS-IS 和 BGP 等）通告，路由器自动学习来构建路由表。

当路由器添加路由条目到路由表中时，遵循如下原则。
① 有效的下一跳地址。
② 如果下一跳地址有效，路由器通过不同的路由协议学到多条去往同一目的网络的路

由，路由器会将管理距离最小的路由条目放入路由表中。

③ 如果下一跳地址有效，路由器通过同一种路由协议学到多条去往同一目的网络的路由，路由器会将度量值最小的路由条目放入路由表中。

路由表的工作原理如下。

① 每台路由器根据其自身路由表中的信息独立做出转发决定。

② 一台路由器的路由表中包含某些信息并不表示其他路由器也包含相同的信息。

③ 数据包从一个网络能够到达另一个网络并不意味着数据包一定可以返回，也就是说路由信息必须双向可达，才能确保网络可以双向通信，所以静态路由一般都需要双向配置。

2.1.4 查找路由表

1. 相关术语

为了深入理解路由查找过程，下面首先介绍一下相关术语。

（1）1 级路由

1 级路由指子网掩码长度等于或小于网络地址有类掩码长度的路由。192.168.1.0/24 属于 1 级网络路由，因为它的子网掩码长度等于网络有类掩码长度。1 级路由可以是：

① 默认路由——指地址为 0.0.0.0/0 的静态路由，或者路由代码后紧跟*的路由条目。

② 超网路由——指掩码长度小于有类掩码长度的网络地址。

③ 网络路由——指子网掩码长度等于有类掩码长度的路由。网络路由也可以是父路由。

（2）最终路由

最终路由指路由条目中包含下一跳 IP 地址或送出接口的路由。

（3）1 级父路由

1 级父路由指路由条目中不包含网络的下一跳 IP 地址或送出接口的网络路由。父路由实际上是表示存在 2 级路由的一个标题。只要向路由表中添加一个子网，就会在路由表中自动创建 1 级父路由。

（4）2 级路由

2 级路由指有类网络地址的子网路由，2 级路由也称为子路由，2 级路由的来源可以是直连路由、静态路由或动态路由。2 级路由也属于最终路由，因为 2 级路由包含下一跳 IP 地址或送出接口。

2. 路由查找原则

路由查找过程遵循最长匹配原则，即最精确匹配。假设路由表中有两条静态路由条目：

```
S 172.16.1.0/24 is directly connected, Serial0/0/0
S 172.16.0.0/16 is directly connected, Serial0/0/1
```

当有去往目的 IP 地址为 172.16.1.85 的数据包到达路由器时，IP 地址同时与这两条路由条目匹配，但是与 172.16.1.0/24 路由条目有 24 位匹配，而与 172.16.0.0/16 路由条目仅有 16

位匹配，所以路由器将使用有 24 位匹配的静态路由转发数据包，即最长匹配。

3. 路由器查找路由表的具体过程

路由器查找路由表的具体过程如下：

① 路由器会检查 1 级路由（包括网络路由和超网路由），查找与 IP 数据包的目的地址最佳匹配的路由。

② 如果最佳匹配的路由是 1 级最终路由，则会使用该路由转发数据包。

③ 如果最佳匹配的路由是 1 级父路由，则路由器检查该父路由的子路由，以找到最佳匹配的路由。

④ 如果在 2 级路由中存在匹配的路由，则会使用该子路由转发数据包。

⑤ 如果所有的 2 级路由都不符合匹配条件，则判断路由器当前执行的是有类路由行为还是无类路由行为。通过全局命令 **ip classless** 来配置无类路由行为，或者通过全局命令 **no ip classless** 来配置有类路由行为。Cisco 路由器默认配置是无类路由行为。

⑥ 如果执行的是有类路由行为，则会终止查找过程并丢弃数据包。

⑦ 如果执行的是无类路由行为，则继续在路由表中搜索 1 级超网路由或默认路由以寻找匹配条目。

⑧ 如果此时存在匹配位数相对较少的 1 级超网路由或默认路由，那么路由器会使用该路由转发数据包。

⑨ 如果路由表中没有匹配的路由，则路由器会丢弃数据包。

2.2 RIP 概述

2.2.1 RIP 特征

RIP（Routing Information Protocols，路由信息协议）是由 Xerox 公司在 20 世纪 70 年代开发的，作为典型的距离矢量路由协议，运行 RIP 的路由器不知道网络的全局情况，如果路由更新信息在网络上传播慢，将会导致网络收敛较慢，可能造成路由环路。为了避免路由环路，RIP 采用水平分割、毒化反转、定义最大跳数、触发更新和抑制计时等机制来避免路由环路。

RIP 路由协议有版本 1 和版本 2 两个版本，不论是版本 1 或版本 2，都具备下面的特征：

① 是距离矢量路由协议。

② 使用跳数（Hop Count）作为度量值。

③ 默认周期性进行路由更新，更新周期为 30 秒。

④ 管理距离（AD）为 120。

⑤ 度量值的最大跳数为 15 跳。

⑥ 源端口和目的端口都使用 UDP 520 端口进行操作，在没有验证的情况下，一个更新数据包最大可以包含 25 个路由条目，数据包最大为 512 字节（UDP 包头 8 字节+RIP 包头 4 字节+路由条目 25×20 字节）。

RIPv1 和 RIPv2 的区别如表 2-3 所示。

表 2-3 RIPv1 和 RIPv2 的区别

RIPv1	RIPv2
在路由更新过程中不携带子网信息	在路由更新过程中携带子网信息
不提供验证	提供明文和 MD5 验证
不支持 VLSM 和 CIDR	支持 VLSM 和 CIDR
采用广播方式更新路由	采用组播（224.0.0.9）方式更新路由
有类（Classful）路由协议	无类（Classless）路由协议

2.2.2 RIP 数据包格式

RIPv1 数据包格式如图 2-1 所示，RIPv2 数据包格式如图 2-2 所示。

```
⊞ User Datagram Protocol, Src Port: router (520), Dst Port: router (520)
⊟ Routing Information Protocol
    Command: Response (2)
    Version: RIPv1 (1)
  ⊟ IP Address: 1.0.0.0, Metric: 1
      Address Family: IP (2)
      IP Address: 1.0.0.0 (1.0.0.0)
      Metric: 1
```

图 2-1 RIPv1 数据包格式

```
⊞ User Datagram Protocol, Src Port: router (520), Dst Port: router (520)
⊟ Routing Information Protocol
    Command: Response (2)
    Version: RIPv2 (2)
    Routing Domain: 0
  ⊟ IP Address: 2.2.2.0, Metric: 1
      Address Family: IP (2)
      Route Tag: 0
      IP Address: 2.2.2.0 (2.2.2.0)
      Netmask: 255.255.255.0 (255.255.255.0)
      Next Hop: 0.0.0.0 (0.0.0.0)
      Metric: 1
```

图 2-2 RIPv2 数据包格式

RIPv2 与 RIPv1 数据包格式基本相同，但 RIPv2 添加了三项重要扩展，分别为子网掩码、路由标记和下一跳，各个字段含义如下。

① 命令：8 比特，取值为 1 或 2，1 表示 RIP 请求消息，2 表示 RIP 响应消息。

② 版本：8 比特，对于 RIPv1，该字段值为 1；对于 RIPv2，该字段值为 2。

③ 地址类型标识符：16 比特，对于 IP，该字段设置为 2；当数据包向路由器请求整个路由选择表时，该字段设置为 0。

④ 路由标记：16 比特，该字段用于标记外部路由或重分布到 RIPv2 中的路由。

⑤ IP 地址：32 比特，表示路由条目，可以是主类网络地址、子网地址或主机路由。

⑥ 子网掩码：32 比特，用来确定 IP 地址的网络或子网部分。

⑦ 下一跳：32 比特，如果存在的话，表示路由条目有更好的一下跳地址，也就是说，它指出的下一跳地址，其度量值比同一个子网上的通告路由器更靠近目的地。如果该字段设置为全 0（0.0.0.0），说明通告路由器的地址就是最好的下一跳地址。

⑧ 度量：32 比特，是一个 1~16 之间的整数。

2.3 配置静态路由和 RIPv2

2.3.1 实验 1：配置静态路由

1. 实验目的

通过本实验可以掌握：
① 带送出接口的静态路由的配置。
② 带下一跳地址的静态路由的配置。
③ 带送出接口和带下一跳地址配置静态路由的不同点。
④ 代理 ARP 的作用。
⑤ 静态路由汇总配置。
⑥ 静态默认路由配置。
⑦ 路由表的含义。

2. 实验拓扑

配置静态路由实验拓扑如图 2-3 所示。

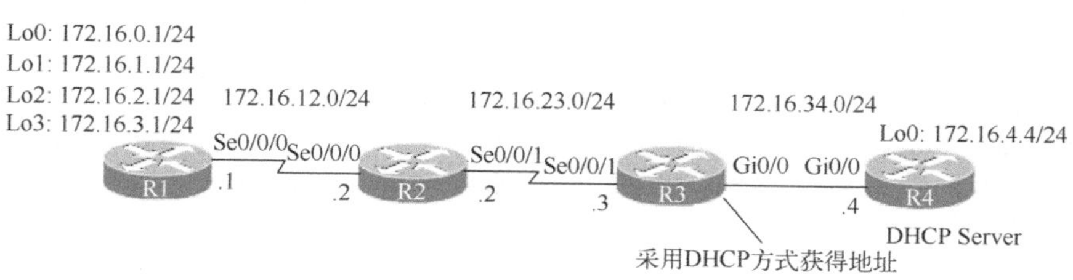

图 2-3 配置静态路由实验拓扑

3. 实验步骤

（1）配置路由器 R1

> R1(config)#**ip route 172.16.4.0 255.255.255.0 Serial0/0/0**
> //配置带送出接口的静态路由
> R1(config)#**ip route 172.16.23.0 255.255.255.0 Serial0/0/0**
> R1(config)#**ip route 172.16.34.0 255.255.255.0 Serial0/0/0**

✓【技术要点】

配置静态路由的命令是：**ip route** *prefix mask* {*address* | *interface* [*address*]} [**dhcp**] [*distance*] [**name** *next-hop-name*] [**permanent**| **track** *number*] [**tag** *tag*]，命令参数含义如下。

① *prefix*：目的网络地址。
② *mask*：目的网络的子网掩码，可对此子网掩码进行修改，实现路由汇总。
③ *address*：将数据包转发到目的网络时使用的下一跳 IP 地址。
④ *interface*：将数据包转发到目的网络时使用的本地送出接口。
⑤ **dhcp**：指定一条前往 DHCP 配置的默认网关的静态路由。
⑥ *distance*：静态路由条目的管理距离，默认为 1。
⑦ **name**：静态路由名称。
⑧ **permanent**：和路由条目相关联的接口进入 **down** 状态后路由条目也不会从路由表中消失。
⑨ **track**：将一个跟踪对象和路由关联起来。
⑩ *tag*：标记，可以在 route-map 中匹配该值。
⑪ *next-hop-name*：下一跳名字。

（2）配置路由器 R2

> R2(config)# **ip route 172.16.0.0 255.255.252.0 Serial0/0/0 name toR1**
> //配置静态路由汇总，将四条明细路由汇总成一条路由
> R2(config)#**ip route 172.16.4.0 255.255.255.0 Serial0/0/1 name toR3**
> //配置带 **name** 参数的静态路由
> R2(config)#**ip route 172.16.34.0 255.255.255.0 Serial0/0/1 name toR3**

（3）配置路由器 R3

> R3(config)#**interface gigabitEthernet0/0**
> R3(config-if)#**ip address dhcp**   //接口的 IP 地址通过 DHCP 方式获得
> R3(config-if)#**no shutdown**
> R3(config)#**ip route 172.16.0.0 255.255.252.0   Serial0/0/1 tag 110**
> R3(config)#**ip route 172.16.12.0 255.255.255.0 Serial0/0/1 tag 100**
> //以上两条配置带 **tag** 参数的静态路由
> R3(config)#**ip route 172.16.4.0 255.255.255.0 gigabitEthernet0/0   dhcp**
> //配置指向 DHCP 服务器定义的默认网关的默认路由

✓【技术要点】

① 带送出接口的静态路由在路由表中显示的是直连（**directly connected**），而带下一跳

地址的静态路由在路由表中显示的是[1/0]，但是管理距离默认情况下都是1。

② 带送出接口的静态路由条目后面直接跟送出接口，路由器只需要查找一次路由表，便能将数据包转发到送出接口，从这点来讲，查找路由表效率比带下一跳地址的效率要高。但是如果开启了CEF功能，则不存在此问题。

③ 使用送出接口而不是下一跳IP地址配置的静态路由是大多数串行点对点网络（如HDLC和PPP封装）的理想选择。

④ 对于带送出接口的静态路由配置，如果出站接口为以太网接口，为了解决ARP的问题（Cisco默认情况下以太网接口启用了ARP代理功能），如果接口关闭ARP代理（命令 **no ip proxy-arp**），同时采用送出接口配置静态路由，会造成数据包封装失败，路由器会显示 **encapsulation failed** 的日志消息，因此，最好同时使用下一跳地址和送出接口来配置，如下所示：

```
R3(config)#ip route 172.16.4.0 255.255.255.0 gigabitEthernet0/0 172.16.34.4
```

（4）配置路由器R4

```
R4(config)#ip dhcp excluded-address 172.16.34.4
R4(config)#ip dhcp pool CCNP
R4(dhcp-config)#network 172.16.34.0 255.255.255.0
R4(dhcp-config)#default-router 172.16.34.4
R4(config)#track 100 interface Loopback0 ip routing
//定义跟踪的接口和路由状态
R4(config)#ip route 0.0.0.0 0.0.0.0 gigabitEthernet0/0 track 100
//将此默认静态路由条目和跟踪对象关联起来
```

【技术要点】

Cisco IOS提供了一种跟踪（track）特性，可以跟踪接口不同的状态，在全局模式下配置，命令的格式为

track *object-number* **interface** *type number* **{line-protocol | ip routing}**，各参数含义如下。

① **object-number**：范围为1～500。

② **line-protocol:** 跟踪接口的线性协议状态。

③ **ip routing:** 跟踪接口的IP路由状态。

当然，也可以跟踪IP路由的状态，命令格式为

track *object-number* **ip route** *ip-address/prefix-length* **{reachability | metric threshold}**

例如：

```
R4(config)#track 10 ip route 172.16.12.0 255.255.255.0 reachability
```

通过全局命令 **track timer {interface | ip route}** *seconds* 配置查询被跟踪对象状态的时间间隔，范围为1～3000秒，默认对接口状态的查询间隔为1秒，对路由条目状态的查询间隔为15秒。

可以通过下面的命令验证track的配置和引用情况：

```
R4#show track 100
Track 100                                   //跟踪的号码
  Interface Loopback0 ip routing            //跟踪的接口和对象参数
  IP routing is Up                          //跟踪的状态
    1 change, last change 01:04:21          //跟踪的状态变化次数和最近一次状态变化以来的时间
  Tracked by:
    STATIC-IP-ROUTING 0                     //跟踪结果和静态路由关联和联动
```

4. 实验调试

查看路由表：

```
① R1#show ip route static
     172.16.0.0/24 is subnetted, 8 subnets              //父路由
S       172.16.34.0 is directly connected, Serial0/0/0   //二级子路由
S       172.16.23.0 is directly connected, Serial0/0/0
S       172.16.4.0 is directly connected, Serial0/0/0
② R2#show ip route static
     172.16.0.0/16 is variably subnetted, 5 subnets, 2 masks
S       172.16.34.0/24 is directly connected, Serial0/0/1
S       172.16.4.0/24 is directly connected, Serial0/0/1
S       172.16.0.0/22 is directly connected, Serial0/0/0   //汇总路由
③ R3#show ip route static
     172.16.0.0/16 is variably subnetted, 4 subnets, 2 masks
S       172.16.12.0/24 is directly connected, Serial0/0/1
S       172.16.0.0/22 is directly connected, Serial0/0/1
S       172.16.4.0 [1/0] via 172.16.34.4, GigabitEthernet0/0   //AD 值为 1，度量值为 0
S*      0.0.0.0/0 [254/0] via 172.16.34.4
//该路由条目是接口下配置 ip address dhcp 产生的默认静态路由，*表示默认，下一跳地址是 R4
上的 DHCP 服务器指定的网关地址 172.16.34.4，但是该默认路由条目的管理距离是 254，这一点要注意。同
时也看到路由条目 172.16.4.0 的下一跳是 DHCP 配置的默认网关地址。如果路由器 R4 配置 DHCP 的时候没
有指定默认网关，则上述的两条路由都不会出现在路由器 R3 的路由表中
```

可以通过如下命令来查看路由条目更详细的信息：

```
R3#show ip route 172.16.12.0 255.255.255.0
Routing entry for 172.16.12.0/24
  Known via "static", distance 1, metric 0 (connected)
//通过静态路由添加到路由表中，管理距离为 1，度量值为 0
  Tag 100                                    //路由标记
  Routing Descriptor Blocks:                 //路由描述区块
  * directly connected, via Serial0/0/1      //显示直连以及出接口
      Route metric is 0, traffic share count is 1    //路由度量值及负载分担数
      Route tag 100                          //路由标记
④ R4#show ip route static
S*    0.0.0.0/0 is directly connected, GigabitEthernet0/0    //静态默认路由
```

2.3.2 实验 2：配置 RIPv2

1. 实验目的

通过本实验可以掌握：
① RIPv1 和 RIPv2 的区别。
② 在路由器上启动 RIPv2 路由进程。
③ 激活参与 RIPv2 路由协议的路由器接口的方法。
④ 自动汇总的开启和关闭的方法。
⑤ 被动接口的含义、配置和应用场合。
⑥ RIPv2 路由的手工汇总配置方法。
⑦ RIPv2 验证的配置。
⑧ RIPv2 触发更新的含义和配置
⑨ 理解 RIP 路由表的含义。
⑩ 查看和调试 RIPv2 路由协议相关信息的方法。

2. 实验拓扑

配置 RIPv2 实验拓扑如图 2-4 所示。

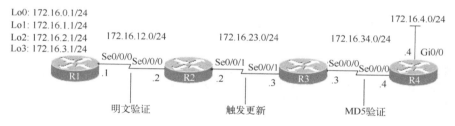

图 2-4　配置 RIPv2 实验拓扑

3. 实验步骤

（1）配置路由器 R1

```
R1(config)#key chain ccna                              //配置钥匙链
R1(config-keychain)#key 1                              //配置 key id
R1(config-keychain-key)#key-string cisco               //配置 key id 的密钥
R1(config)#interface Serial0/0/0
R1(config-if)#ip rip authentication mode text
//启用 RIPv2 明文验证模式，默认验证模式就是明文，可以不用配置
R1(config-if)#ip rip authentication key-chain ccna     //在接口下调用钥匙链
R1(config-if)#ip summary-address rip 172.16.0.0 255.255.252.0
//接口下配置 RIPv2 手工路由汇总
R1(config)#router rip                                  //启动 RIP 进程
R1(config-router)#version 2                            //配置 RIP 版本 2
```

```
R1(config-router)#no auto-summary          //关闭 RIP 路由自动汇总，默认开启
R1(config-router)#network 172.16.0.0
//配置参与 RIPv2 的接口的匹配范围，使之能够发送和接收 RIPv2 更新信息
```

【技术要点】

network 命令的作用如下：
① 匹配指定范围的所有接口下启用 RIP，相关接口将开始发送和接收 RIP 更新信息。
② 路由器通告运行 RIP 协议的接口下的网络和掩码。

（2）配置路由器 R2

```
R2(config)#key chain ccna
R2(config-keychain)#key 1
R2(config-keychain-key)#key-string cisco
R2(config)#interface Serial0/0/0
R1(config-if)#ip rip authentication mode text
R2(config-if)#ip rip authentication key-chain ccna
R2(config)#interface Serial0/0/1
R2(config-if)#ip rip triggered     //配置触发更新
```

【技术要点】

① 在以太网接口下，不支持触发更新 RIP。
② 触发更新需要协商，同一链路的两端都需要配置。

```
R2(config)#router rip
R2(config-router)#version 2
R2(config-router)#no auto-summary
R2(config-router)#network 172.16.0.0
```

（3）配置路由器 R3

```
R3(config)#key chain ccnp
R3(config-keychain)#key 1
R3(config-keychain-key)#key-string cisco
R3(config)#interface Serial0/0/0
R3(config-if)#ip rip authentication mode md5        //启用 RIPv2 MD5 验证模式
R3(config-if)#ip rip authentication key-chain ccnp  //在接口下调用钥匙链
R3(config)#interface Serial0/0/1
R3(config-if)#ip rip triggered
R3(config)#router rip
R3(config-router)#version 2
R3(config-router)#no auto-summary
R3(config-router)#network 172.16.0.0
```

（4）配置路由器 R4

```
R4(config)#key chain ccnp
```

```
R4(config-keychain)#key 1
R4(config-keychain-key)#key-string cisco
R4(config)#interface Serial0/0/0
R4(config-if)#ip rip authentication mode md5
R4(config-if)#ip rip authentication key-chain ccnp
R4(config)#router rip
R4(config-router)#version 2
R4(config-router)#no auto-summary
R4(config-router)#network 172.16.0.0
R4(config-router)#passive-interface gigabitEthernet0/0         //配置被动接口
```

4. 实验调试

（1）查看路由表

① R1#show ip route rip
```
         172.16.0.0/24 is subnetted, 8 subnets
R        172.16.34.0 [120/2] via 172.16.12.2, 00:00:02, Serial0/0/0
R        172.16.23.0 [120/1] via 172.16.12.2, 00:00:02, Serial0/0/0
R        172.16.4.0 [120/3] via 172.16.12.2, 00:00:02, Serial0/0/0
```

以上输出表明路由器 R1 学到了 3 条 RIP 路由，而且携带了子网信息，RIP 的管理距离为 120。其中路由条目 **R   172.16.4.0 [120/3] via 172.16.12.2, 00:00:2, Serial0/0/0** 的含义如下：**R** 表示路由条目是通过 RIP 路由协议学习来的；**172.16.4.0** 表示目的网络；**120** 表示 RIP 路由协议的默认管理距离；**3** 表示度量值，从路由器 R1 到达网络 **172.16.4.0** 的度量值为 3 跳；**172.16.12.2** 表示路由条目的下一跳 IP 地址；**00:00:02** 表示自上次更新以来已经过了 2 秒；**Serial0/0/0** 表示接收该路由条目的本地路由器的接口。

② R2#show ip route rip
```
         172.16.0.0/16 is variably subnetted, 5 subnets, 2 masks
R        172.16.34.0/24 [120/1] via 172.16.23.3, 00:11:27, Serial0/0/1
R        172.16.4.0/24 [120/2] via 172.16.23.3, 00:11:27, Serial0/0/1
R        172.16.0.0/22 [120/1] via 172.16.12.1, 00:00:23, Serial0/0/0
```

以上输出表明路由器 R2 的路由表包含了 3 条 RIP 路由条目，其中包括 R1 发送的 **172.16.0.0/22** 手工汇总路由。

③ R3#show ip route rip
```
         172.16.0.0/16 is variably subnetted, 5 subnets, 2 masks
R        172.16.12.0/24 [120/1] via 172.16.23.2, 00:12:48, Serial0/0/1
R        172.16.4.0/24 [120/1] via 172.16.34.4, 00:00:01, Serial0/0/0
R        172.16.0.0/22 [120/2] via 172.16.23.2, 00:12:26, Serial0/0/1
```

④ R4#show ip route rip
```
         172.16.0.0/16 is variably subnetted, 5 subnets, 2 masks
R        172.16.23.0/24 [120/1] via 172.16.34.3, 00:00:04, Serial0/0/0
R        172.16.12.0/24 [120/2] via 172.16.34.3, 00:00:04, Serial0/0/0
R        172.16.0.0/22 [120/3] via 172.16.34.3, 00:00:04, Serial0/0/0
```

（2）查看 IP 路由协议配置和统计信息

```
R3#show ip protocols
 *** IP Routing is NSF aware ***
```
// IP 路由的 NSF 感知（NSF-aware）能力，帮助有 NSF 能力的路由器执行 NSF（NonStop Forwarding，不间断转发，主要用于配置了 2 个引擎的高端路由器切换时不中断用户的数据业务，更大程度地提高网络的可靠性和稳定性
```
Routing Protocol is "application"
  Sending updates every 0 seconds
  Invalid after 0 seconds, hold down 0, flushed after 0
  Outgoing update filter list for all interfaces is not set
  Incoming update filter list for all interfaces is not set
  Maximum path: 32
  Routing for Networks:
  Routing Information Sources:
    Gateway          Distance       Last Update
  Distance: (default is 4)
```
//以上称为应用的路由协议是为 Cisco 软件定义网络（Software Defined Network，SDN）服务的，路由是通过控制器和应用程序安装进路由表的，默认管理距离为 4
```
Routing Protocol is "rip"      //路由器上运行的路由协议是 RIP
  Outgoing update filter list for all interfaces is not set
```
//在出方向上没有配置分布列表（distribute-list）
```
  Incoming update filter list for all interfaces is not set
```
//在入方向上没有配置分布列表（distribute-list）
```
  Sending updates every 30 seconds, next due in 25 seconds
```
//更新周期是 30 秒，距离下次更新还有 25 秒

【注意】

为了防止更新同步，RIP 会以 15%的误差发送更新信息，即实际发送更新信息的周期的范围是 25.5～30 秒。

```
  Invalid after 180 seconds, hold down 0, flushed after 240
```
//计时器相关的几个参数，其中 invalid 计时器表示路由器针对某条路由条目如果在 180 秒还没有收到更新信息，则被标记为无效；hold down 计时器表示路由条目抑制计时器的时间为 180 秒，用来防止路由环路；flushed 计时器表示路由器针对某条路由条目如果在 240 秒还没有收到更新信息，则从路由表中删除该路由条目

【提示】

可以通过下面的命令来调整以上四个计时器参数：

```
R3(config-router)#timers basic update invalid holddown flushed
```

```
  Redistributing: rip          //只运行 RIP 协议，没有其他协议重分布进来
  Default version control: send version 2, receive version 2
```
//默认发送 RIP 版本 2 的路由更新信息，接收 RIP 版本 2 的路由更新信息

Interface	Send	Recv	Triggered RIP	Key-chain
Serial0/0/0	2	2		ccnp
Serial0/0/1	2	2	Yes	

//以上三行显示了运行 RIP 协议的接口、接收和发送的 RIP 路由更新的版本，同时在接口 Se0/0/1
配置了触发更新，在接口 Se0/0/0 启用了验证并给出了调用钥匙链的名称
Automatic network summarization is not in effect　　//RIP 路由协议自动汇总功能被关闭
Maximum path: 4　　//RIP 路由协议默认可以支持 4 条等价路径

【注意】

默认情况下，RIP 最多只能自动在 4 条开销相同的路径上实施负载均衡。不同的 IOS 版本，RIP 能够支持的最大等价路径的条数可能也不同。可以通过下面的命令来修改 RIP 路由协议支持等价路径的条数：

R3(config-router)#**maximum-paths** *number-paths*

```
Routing for Networks:
    172.16.0.0
//以上两行表明在 RIP 路由模式下 network 命令的配置
Routing Information Sources:
    Gateway         Distance    Last Update
    172.16.34.4     120         00:00:18
    172.16.23.2     120         00:14:05      //超过 30 秒，是由于 R2 与 R3 之间配置了触发更新
//以上四行表明 RIP 路由信息源，即从哪些邻居接收 RIP 路由更新信息，其中 gateway 表示学到
路由信息的邻居路由器的接口地址，也就是下一跳地址；Distance 表示接收邻居发送的路由更新信息使用的
管理距离；Last Update 表示距离上次路由更新经过的时间
    Distance: (default is 120)      //RIP 路由协议默认管理距离是 120
```

（3）查看 RIP 路由协议的动态更新过程

```
R4#debug ip rip
R4#clear ip route *                //清除路由表
03:50:16: RIP: sending request on Serial0/0/0 to 224.0.0.9
03:50:16: RIP: sending request on Serial0/0/0 to 224.0.0.9
//以上两行表明在 Se0/0/0 接口下向组播地址（224.0.0.9）发送更新请求
03:50:16: RIP: received packet with MD5 authentication    //收到 MD5 验证的更新信息
03:50:16: RIP: received v2 update from 172.16.34.3 on Serial0/0/0
03:50:16:       172.16.0.0/22 via 0.0.0.0 in 3 hops
03:50:16:       172.16.12.0/24 via 0.0.0.0 in 2 hops
03:50:16:       172.16.23.0/24 via 0.0.0.0 in 1 hops
//以上四行表明从接口 Se0/0/0 收到 3 条 RIPv2 的更新信息，该更新信息包括子网掩码信息和度量
信息
03:50:44: RIP: sending v2 update to 224.0.0.9 via Serial0/0/0 (172.16.34.4)
03:50:44: RIP: build update entries
03:50:44:       172.16.4.0/24 via 0.0.0.0, metric 1, tag 0
//以上三行表明从接口 Se0/0/0 以组播方式（224.0.0.9）发送 1 条 RIPv2 的更新信息，该更新信息
包括子网掩码信息、度量信息和路由标记
```

（4）查看 RIP 数据库

```
R2#show ip rip database
172.16.0.0/16     auto-summary     //自动汇总路由
172.16.0.0/22     //R1 发送的总结路由
```

```
           [1] via 172.16.12.1, 00:00:07, Serial0/0/0
      172.16.4.0/24
           [2] via 172.16.23.3, 00:57:52 (permanent), Serial0/0/1
      //以上两行表明学到 172.16.4.0 路由的下一跳地址、时间和接口，[2]表示到达目的网络度量值，启
      动触发更新学到的条目显示为永久的（permanent）
       * Triggered Routes:      //触发更新路由
        - [2] via 172.16.23.3, Serial0/0/1
      172.16.12.0/24      directly connected, Serial0/0/0    //直连接口名字和网络地址
      172.16.23.0/24      directly connected, Serial0/0/1
      172.16.34.0/24
           [1] via 172.16.23.3, 00:57:52 (permanent), Serial0/0/1
       * Triggered Routes:
        - [1] via 172.16.23.3, Serial0/0/1
```

本 章 小 结

　　本章介绍了 IP 路由原理，包括静态路由和动态路由协议的优缺点和使用场合、路由表填充、查找路由表过程，并通过实验演示和验证了静态路由和 RIPv2 的配置和调试。本章内容很基础，配置命令也相对简单，但是却有一些技巧性很强的知识点，读者要好好掌握。

第 3 章　EIGRP

　　EIGRP 是 Cisco 公司于 1992 年开发的一个无类别距离矢量路由协议，它融合了距离矢量和链路状态两种路由协议的优点。EIGRP 是 Cisco 的专有路由协议，是 Cisco 的 IGRP 协议的增强版。由于 TCP/IP 是当今网络中最常用的协议，因此本章重点讨论 IP 网络环境中的 EIGRP。

3.1　EIGRP 概述

3.1.1　EIGRP 特征

　　EIGRP（Enhanced Interior Gateway Routing Protocol，增强型内部网关路由协议）是一个高效的路由协议，它的特征如下：
　　① 通过发送和接收 Hello 数据包来建立和维持邻居关系。
　　② 采用组播（224.0.0.10）或单播方式进行路由更新，仅支持 MD5 验证。
　　③ EIGRP 的默认管理距离为 5、90 或 170。
　　④ 采用触发更新和部分更新，减少带宽消耗。
　　⑤ 是无类别的路由协议，支持 VLSM 和不连续子网，从 IOS 15 版本开始默认关闭路由自动汇总功能，支持在任意运行 EIGRP 协议的接口下手工路由汇总。
　　⑥ 使用协议相关模块（Protocol Dependent Module，PDM）来支持 IPv4 和 IPv6 等多种网络层协议。对每一种网络协议，EIGRP 都维持独立的邻居表、拓扑表和路由表，并且存储整个网络拓扑结构信息，以便快速适应网络变化。
　　⑦ 采用带宽、延迟、可靠性和负载计算度量值，度量值的颗粒度精细（32 位），范围为 1～4294967296。
　　⑧ EIGRP 使用扩散更新算法（Diffusing Update Algorithm，DUAL）来实现快速收敛，并确保没有路由环路。
　　⑨ EIGRP 是支持多种网络层协议的路由协议，EIGRP 数据包的发送和接收不使用 UDP 或 TCP 承载，而使用可靠传输协议（Reliable Transport Protocol，RTP）承载，保证路由信息传输的可靠性和有序性，它支持组播和单播的混合传输。
　　⑩ 支持等价（Equal-Cost）和非等价（Unequal-Cost）的负载均衡（Load Balancing）。
　　⑪ 与数据链路层协议无缝连接，EIGRP 不要求针对第二层协议做特殊配置。
　　⑫ 支持不间断转发（NonStop Forwarding，NSF），允许发生故障路由器的 EIGRP 邻居设备保留它所通告的路由信息，并继续使用此信息直到故障路由器恢复正常操作并可以交换路由信息。

3.1.2　DUAL 算法

　　DUAL 作为驱动 EIGRP 的计算引擎，是 EIGRP 路由协议的核心，它能够确保整个路由域内的无环路径和无环备用路径工作正常。通过使用 DUAL，EIGRP 会保存所有能够到达目的地的可用路由，在主路由失效时迅速切换到替代路由。学习 DUAL 需要掌握以下术语。

① 后继（Successor）：是一个直接连接的 EIGRP 邻居路由器，通过它到达目的网络的度量值最小。后继是提供主要路由的路由器，该路由被放入 EIGRP 拓扑表和路由表中。对于同一目的网络，可能存在多个后继。

② 可行后继（Feasible Successor）：是一个直接连接的 EIGRP 邻居路由器，但是通过它到达目的网络的度量值比通过后继路由器的大，而且它的通告距离小于通过后继路由器到达目的网络的可行距离。可行后继是提供备份路由的路由器，该路由仅被放入 EIGRP 拓扑表中。对于同一目的网络，可能存在多个可行后继。

③ 可行距离（Feasible Distance，FD）：本路由器到达目的网络的最小度量值。

④ 通告距离（Reported Distance，RD）：EIGRP 邻居路由器所通告的它自己到达目的网络的最小的度量值，也有的资料把 RD 称为 AD（Advertised Distance）。

⑤ 可行性条件（Feasible Condition，FC）：是 EIGRP 路由器更新路由表和拓扑表的依据。可行性条件可以有效地阻止路由环路，实现路由的快速收敛。可行性条件的公式为 RD<FD。

EIGRP 有限状态机（Finite State Machine，FSM）包含用于在 EIGRP 网络中计算和比较路由的所有逻辑，EIGRP FSM 逻辑如图 3-1 所示。

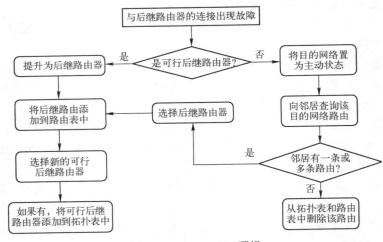

图 3-1 EIGRP FSM 逻辑

3.1.3 EIGRP 数据包类型

EIGRP 数据包类型有以下 5 种，其中查询和应答数据包会成对使用。EIGRP 数据包根据具体情况可以采用单播或者组播方式发送。

① Hello 数据包：以组播的方式定期发送，用于建立和维持 EIGRP 邻居关系。Hello 数据包的确认号始终为 0，因此不需要确认。默认情况下，在点到点链路或者带宽大于 T1 的多点链路上，EIGRP Hello 数据包每 5 秒发送一次。在带宽小于 T1 的低速链路上，EIGRP Hello 数据包每 60 秒发送一次。保持时间（Hold Time）是收到 Hello 数据包的 EIGRP 邻居在认为发出该数据包的路由器发生故障之前应该等待的最长时间。默认情况下，保持时间是 Hello 数据包发送间隔的 3 倍。到达保持时间后，EIGRP 将删除邻居以及从邻居学到的所有拓扑表中的条目。

② 更新（Update）数据包：当路由器收到某个邻居路由器的第一个 Hello 数据包时，邻居关系协商建立成功后，以单播传送方式发送包含它所知道的路由信息的更新数据包。当路由信息发生变化时，以组播的方式发送只包含变化路由信息的更新数据包。EIGRP 在路由更新过程中使用部分更新（Partial Update）和限定更新（Bounded Update）。部分更新是指路由更新信息中仅包含与变化路由相关的信息，限定更新是指部分更新信息仅发送给受变化影响的路由器。限定更新可帮助 EIGRP 最大限度减少发送 EIGRP 更新信息所需的带宽。更新数据包以可靠方式传递，因此需要邻居发送确认数据包进行确认。

③ 查询（Query）数据包：当一条链路失效并且在拓扑表中没有可行后继路由时，路由器需要重新进行路由计算，路由器就以组播方式向它的邻居发送一个查询数据包，以询问它们是否有到目的网络的路由。查询数据包通常是组播包。

④ 应答（Reply）数据包：以单播方式响应邻居的查询，应答数据包都是单播包。

⑤ 确认（ACK）数据包：以单播方式发送的没有任何数据的 Hello 数据包，包含一个不为 0 的确认号，用来确认更新数据包、查询数据包和应答数据包。确认数据包不需要确认。

3.1.4　EIGRP 数据包格式

每个 EIGRP 数据包都是由 EIGRP 数据包头部和 TLV（Type，类型 / Length，长度 / Value，值）构成的。EIGRP 数据包头部和 TLV 被封装在一个 IP 数据包中，该 IP 数据包中的协议字段为 88，用来代表 EIGRP。如果 EIGRP 数据包为组播包，则目的地址为组播地址 224.0.0.10；如果 EIGRP 数据包被封装在以太网帧内，则组播目的 MAC 地址为 01-00-5E-00-00-0A。EIGRP 数据包格式如图 3-2 所示。

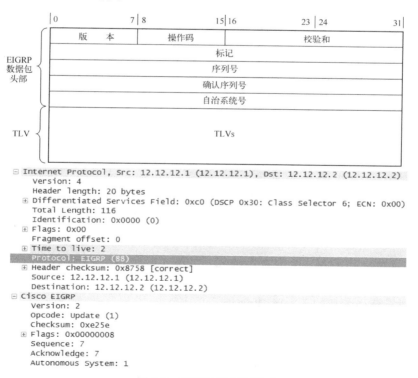

图 3-2　EIGRP 数据包格式

每个 EIGRP 数据包都包含数据包头部，它是每个 EIGRP 数据包的开始部分，各字段的含义如下。

① 版本：始发 EIGRP 进程处理的版本。
② 操作码：EIGRP 数据包的类型。
③ 校验和：基于除了 IP 头部的整个 EIGRP 数据包来计算的校验和。
④ 标记：通常设置为 0x00000001。
⑤ 序列号：用在 RTP 中的 32 位序列号。
⑥ 确认序列号：是本地路由器从邻居路由器那里收到的最新的一个 32 位序列号。
⑦ 自治系统号：EIGRP 路由进程的 ID。

EIGRP 头部的后面跟的就是多种类型的 TLV 字段，TLV 通常包括以下三种类型。

1. 带 EIGRP 参数的 TLV

带 EIGRP 参数的 TLV（类型为 0x0001）如图 3-3 所示，它用于传递度量值计算的权重和保持时间。具体的 EIGRP 度量值计算方法稍后介绍。

0　　　　　　7	8　　　　　　15	16　　　　　　23	24　　　　　　31
类型=0x0001		长　　度	
K1	K2	K3	K4
K5	保留	保持时间	

```
□ Cisco EIGRP
    Version: 2
    Opcode: Hello/Ack (5)
    Checksum: 0xeecb
  ⊞ Flags: 0x00000000
    Sequence: 0
    Acknowledge: 0
    Autonomous System: 1
  ⊟ EIGRP Parameters
    Type: EIGRP Parameters (1)
    Size: 12
    K1: 1
    K2: 0
    K3: 1
    K4: 0
    K5: 0
    Reserved: 0
    Hold Time: 15
```

图 3-3　带 EIGRP 参数的 TLV（类型为 0x0001）

2. 内部路由的 TLV

EIGRP 内部路由的 TLV（类型为 0x0102）如图 3-4 所示，各个字段的含义如下。

① 长度：数据包的长度。
② 下一跳：路由条目的下一跳 IP 地址。
③ 延迟：从源到达目的地的延迟总和，单位为 μs。
④ 带宽：链路上所有接口的最小带宽，单位为 kbps。
⑤ 最大传输单元（MTU）：路由传递方向的所有链路中最小的 MTU，某些 EIGRP 文档可能讲述 MTU 是计算 EIGRP 路由度量值的参数之一，但这是错误的。MTU 并不是 EIGRP 所用

的度量标准参数。MTU 虽然包括在路由更新信息中，但不用于计算路由度量值。

图 3-4 EIGRP 内部路由的 TLV（类型为 0x0102）

⑥ 跳数：到达目的地路由器的台数，范围为 1～255。
⑦ 可靠性：到达目的地的路径上接口的出站误码率的总和，范围为 1～255。
⑧ 负载：到达目的地的路径上接口的出站负载的总和，范围为 1～255。
⑨ 保留：保留位，总是设置为 0x0000。
⑩ 前缀长度：子网掩码的长度。
⑪ 目的地：路由的目的地址，即目标网络地址。

3. 外部路由的 TLV

当向 EIGRP 路由进程中重分布外部路由时，就会使用 EIGRP 外部路由的 TLV。EIGRP 外部路由的 TLV（类型为 0x0103）如图 3-5 所示，各字段的含义如下（这里只解释比内部路由 TLV 增加的字段）。

① 源路由器：重分布外部路由到 EIGRP 自治系统的路由器 ID，它是 IPv4 地址格式。
② 源自治系统号：始发路由的路由器所在的自治系统号。
③ 任意标记：用来携带路由映射图的标记。
④ 外部协议度量：外部协议的度量。
⑤ 外部协议 ID：表示外部路由是从哪中协议学到的。

⑥ 标志：目前仅定义了两个标志，0x01 表示外部路由，0x02 表示该路由是候选的默认路由。

0 7	8 15	16 23	24 31
类型=0x0103		长度	
下一跳			
源路由器			
源自治系统号			
任意标记			
外部协议度量			
保留	外部协议ID		标志
延迟			
带宽			
最大传输单元		跳数	
可靠性	负载	保留	
目的地			

```
□ Cisco EIGRP
    Version: 2
    Opcode: Update (1)
    Checksum: 0xe25e
  ⊞ Flags: 0x00000008
    Sequence: 7
    Acknowledge: 7
    Autonomous System: 1
  ⊞ IP internal route    =   1.1.1.0/24
  ⊟ IP external route    =   11.11.11.0/24
      Type: IP external route (259)
      Size: 48
      Next Hop: 0.0.0.0 (0.0.0.0)
      Originating router: 11.11.11.1 (11.11.11.1)
      Originating A.S.: 0
      Arbitrary tag: 0
      External protocol metric: 0
      Reserved: 0
      External protocol ID: Connected link (11)
    ⊞ Flags: 0x00
      Delay: 128000
      Bandwidth: 256
      MTU: 1514
      Hop Count: 0
      Reliability: 255
      Load: 1
      Reserved: 0
      Prefix Length: 24
      Destination = 11.11.11.0
```

图 3-5　EIGRP 外部路由的 TLV（类型为 0x0103）

3.1.5　EIGRP 的 SIA 及查询范围的限定

作为一种高级距离矢量协议，EIGRP 依靠邻居提供路由信息，当丢失路由后，EIGRP 路由器将在拓扑表中查找可行后继，如果找到，将不把原来的路由切换到主动状态，而是将可行后继提升为后继，并把路由放到路由表中，无须使用 DUAL 算法重新计算路由。如果拓扑表中没有可行后继,则该路由被置为主动状态,EIGRP 路由器向所有邻居路由器发送查询消息(除了到达后继的那个接口,水平分割限制),以便寻找一条可以替代的路由。如果被查询的路由器知道一条替代路由,它就把这条替代路由放进应答数据包中发送给发送路由查询消息的源路由器。如果接收到查询消息的路由器没有替代路由的信息,它将继续发送给自己的其他邻居,直到找到可以替代的路由为止。因为 EIGRP 使用可靠的组播方式来寻找替代路由,路由器必须要收到被查询的所有路由器的应答才能重新计算路由,如果有一个路由器的应答还没有收到,发出查询消息的源路由器就必须等待,默认如果在 3 分钟内某些路由器没有对查询消息做出响应,这条路由就进入"卡在活跃(Stuck in Active,SIA)"状态,然后路由器将重置和这个没有做出应答的路由器的邻居关系。为了避免 SIA 情形的发生或者降低 SIA 的发生概率,必须限制查询的范围,通常有以下两种方法。

1. 在路由器的出接口配置路由汇总

良好的 IP 地址规划可以方便地执行 EIGRP 路由汇总。路由汇总不仅可以减小路由表,节省 CPU 时间、内存和带宽,还可以降低网络进入 SIA 状态的概率,因为它可以减少收到 EIGRP 查询消息的数量。

2. 把远程路由器配置为 EIGRP 的 stub 路由器

当一台路由器被配置为 stub 路由器时,它的邻居就不会向它发送查询数据包,从而可以限制 EIGRP 的 SIA 查询范围。

除了上述的两种方法,还有其他的一些限制查询范围的方法,诸如路由过滤等。

3.2 配置 EIGRP

3.2.1 实验 1:配置基本 EIGRP

1. 实验目的

通过本实验可以掌握:
① 在路由器上启动 EIGRP 路由进程的方法。
② 激活参与 EIGRP 路由协议的接口的方法。
③ EIGRP 度量值的计算方法。
④ 可行距离(FD)、通告距离(RD)和可行性条件(FC)的概念。
⑤ 邻居表、拓扑表和路由表的含义。
⑥ 被动接口的含义。
⑦ EIGRP 路由自动汇总的条件。
⑧ 查看和调试 EIGRP 路由协议相关信息的方法。

2. 实验拓扑

配置基本 EIGRP 的实验拓扑如图 3-6 所示。

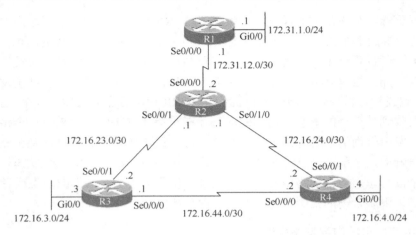

图 3-6　配置基本 EIGRP 的实验拓扑

3. 实验步骤

（1）配置路由器 R1

```
R1(config)#router eigrp 1                    //启动 EIGRP 进程，进程号为 1
R1(config-router)#auto-summary
//开启 EIGRP 有类网络边界自动汇总功能，但是只对本地产生的路由进行自动汇总，对于穿越本
路由器的路由条目，EIGRP 是不能进行自动汇总的，这一点和 RIP 不同，IOS 15 版本以后 EIGRP 路由自动
汇总功能默认关闭
R1(config-router)#eigrp router-id 1.1.1.1    //配置 EIGRP 路由器 ID
R1(config-router)#network 172.31.0.0
//匹配网络地址的接口都将被启用 EIGRP，可以发送和接收 EIGRP 更新信息
R1(config-router)#passive-interface gigabitEthernet0/0   //配置被动接口
```

【技术要点】

全局配置命令 **router eigrp *as-number*** 用来启动 EIGRP 的路由进程，其中，*as-number* 参数由网络管理员选择，取值范围为 1～65535。尽管 EIGRP 将进程 ID 称为自治系统 ID，它实际上起进程 ID 的作用，与 BGP 路由协议的自治系统号码无关。EIGRP 路由域内的所有路由器都必须使用相同的进程 ID 号。一台路由器可以启动多个 EIGRP 进程，必要的话，可以通过重分布实现不同 EIGRP 进程之间的路由信息共享。

EIGRP 确定路由器 ID 遵循如下顺序：

① 最优先的路由器 ID 是在 EIGRP 进程中用命令 **eigrp router-id** 指定的路由器 ID。

② 如果没有在 EIGRP 进程中指定路由器 ID，那么选择 IP 地址最大的环回接口的 IP 地址作为 EIGRP 路由器 ID。

③ 如果没有配置环回接口，就选择最大的活动物理接口的 IP 地址作为 EIGRP 路由器 ID。

④ 建议用命令 **eigrp router-id** 来指定路由器 ID，这样可控性比较好。

默认情况下，当在命令 **network** 中使用诸如 **172.31.0.0** 等有类网络地址时，该路由器上属于该有类网络地址的所有接口都将启用 EIGRP。然而，有时网络管理员并不想让所有接口启用 EIGRP。如果要配置 EIGRP 仅仅在某些特定的接口启用，请将 **wildcard-mask**（通配符掩码）选项与命令 **network** 一起使用。**wildcard-mask**（通配符掩码）是广播地址（**255.255.255.255**）减去子网掩码得到的。例如子网掩码是 **255.255.248.0**，则通配符掩码是 **0.0.7.255**。在高版本的 IOS 中，在配置 EIGRP 时也支持网络掩码的写法，但是系统会自动转换成通配符掩码，可以通过 **show running-config** 命令来验证。

对于 EIGRP，配置被动接口含义如下：
① 将接口配置为被动接口后，就不能建立 EIGRP 邻接关系。
② 不能通过被动接口接收或者发送路由更新信息。
③ EIGRP 进程可以通告被动接口所在的网络或子网。

（2）配置路由器 R2

```
R2(config)#router eigrp 1
R2(config-router)#eigrp router-id 2.2.2.2
R2(config-router)#auto-summary
R2(config-router)#network 172.31.0.0
R2(config-router)#network 172.16.0.0
```

（3）配置路由器 R3

```
R3(config)#router eigrp 1
R3(config-router)#eigrp router-id 3.3.3.3
R3(config-router)#auto-summary
R3(config-router)#network 172.16.0.0
R3(config-router)#passive-interface gigabitEthernet0/0
```

（4）配置路由器 R4

```
R4(config)#router eigrp 1
R4(config-router)#auto-summary
R4(config-router)#network 172.16.0.0
R4(config-router)#passive-interface gigabitEthernet0/0
```

4. 实验调试

（1）查看 EIGRP 路由表

以下各命令的输出全部省略路由代码部分。

```
① R1#show ip route eigrp
  D    172.16.0.0/16 [90/2681856] via 172.31.12.2, 00:41:04, Serial0/0/0
```

以上输出表明路由器 R1 通过 EIGRP 学到了 1 条路由，管理距离为 90。注意 EIGRP 协议路由代码用字母 **D**（DUAL 算法）表示，如果是通过重分布方式进入 EIGRP 网络的路由条目，则默认管理距离为 170，路由代码用 **D EX** 表示，也说明 EIGRP 路由协议能够区分内部路由和外部路由。

```
② R2#show ip route eigrp
```

```
                  172.16.0.0/16 is variably subnetted, 8 subnets, 4 masks
D                 172.16.0.0/16 is a summary, 00:45:34, Null0
D                 172.16.34.0/30 [90/2681856] via 172.16.24.2, 00:00:28, Serial0/1/0
                                 [90/2681856] via 172.16.23.2, 00:00:28, Serial0/0/1
//以上两行表明到达 172.16.34.0/30 目的网络有两条等价路径
D                 172.16.4.0/24 [90/2172416] via 172.16.24.2, 00:00:28, Serial0/1/0
D                 172.16.3.0/24 [90/2172416] via 172.16.23.2, 00:00:29, Serial0/0/1
                  172.31.0.0/16 is variably subnetted, 4 subnets, 4 masks
D                 172.31.1.0/24 [90/2172416] via 172.31.12.1, 00:45:34, Serial0/0/0
D                 172.31.0.0/16 is a summary, 00:45:35, Null0
```

以上输出包含 2 条出接口为 **Null0** 接口（软件意义上的直连接口）的汇总路由（称为系统路由），该路由条目管理距离为 **5**。默认情况下，EIGRP 使用 Null0 接口丢弃与主类网络路由条目匹配但与该主类网络下所有子网路由都不匹配的数据包，从而可以有效避免路由环路。

【技术要点】

EIGRP 向路由表中自动加入出接口为 Null0 的汇总路由的条件必须同时满足下面三点：
- EIGRP 进程中至少有两个不同主类网络；
- 每个主类网络通过 EIGRP 至少发现了一个子网；
- EIGRP 进程中启用了自动汇总，如果关闭了自动汇总，则出接口为 Null0 的汇总路由将被删除，除非接口下执行手工路由汇总。

```
③ R3#show ip route eigrp
                 172.16.0.0/16 is variably subnetted, 8 subnets, 3 masks
D                172.16.24.0/30 [90/2681856] via 172.16.34.2, 00:01:05, Serial0/0/0
                                [90/2681856] via 172.16.23.1, 00:01:05, Serial0/0/1
D                172.16.4.0/24 [90/2172416] via 172.16.34.2, 00:01:05, Serial0/0/0
D                172.31.0.0/16 [90/2681856] via 172.16.23.1, 00:01:06, Serial0/0/1
```

以上输出中并没有看到出接口为 **Null0** 的汇总路由，这是因为路由器 R3 上所有运行 EIGRP 的接口都属于相同的主类网络（172.16.0.0/16），因此不会执行路由自动汇总，即使开启了路由自动汇总功能，这也是为什么会学到 **172.16.24.0/30** 和 **172.16.4.0/24** 明细路由的原因。EIGRP 度量值的计算采用比较复杂的度量方法，接下来以路由器 R3 路由表中的 **D 172.16.4.0/24 [90/2172416] via 172.16.34.2, 00:01:05, Serial0/0/0** 路由条目为例来说明。

EIGRP 度量值的计算公式 = [K1×Bandwidth + (K2×Bandwidth)/(256-Load) + K3×Delay]×[K5/(Reliability + K4)]×256

默认情况下，K1 = K3 = 1，K2 = K4 = K5 = 0。

Bandwidth = [10^7 / 传递路由条目所经由链路中入接口带宽（单位为 kbps）的最小值]×256

Delay = [传递路由条目所经由链路中入接口的延迟之和（单位为 μs）/10]×256

接下来看一下在路由器 R3 中的 **172.16.4.0/24** 路由条目的度量值是如何计算的？

首先，带宽应该是从路由器 R4 的 Gi0/0 到路由器 R3 经过的所有接口最小带宽（学习到路由方向的接口），所以应该是 R3 的 Se0/0/0 接口的带宽，即 1544 kbps，而延迟是路由器 R4 的 Gi0/0 接口的延迟（100 μs）和路由器 R3 的 Se0/0/0 接口的延迟（20000 μs）之和，所以最后的度量值应该是$[10^7/1544+(100+20000)/10]\times256=$**2172416**，这和路由器计算的结果是一致的。

【提示】

接口的带宽和延迟可以通过 **show interface** 命令来查看。可以通过下面的命令修改接口的带宽和延迟：

> R3(config-if)#**bandwidth** *bandwidth*
> R3(config-if)#**delay** *delay*

需要注意的是，命令 **Bandwidth** 并不能更改链路的物理带宽，只会影响 EIGRP 或 OSPF 等路由协议选路。有时，网络管理员可能会出于加强对出接口控制的目的而更改带宽值。默认情况下，EIGRP 会使用不超过 50%的接口带宽来传输 EIGRP 数据。这可避免因 EIGRP 进程过度占用链路带宽而使得路由正常数据流量所需的带宽不足。配置接口下可供 EIGRP 使用的带宽百分比的命令为 **ip bandwidth-percent eigrp** *as-number percent*。

④ R4#**show ip route eigrp**
　　172.16.0.0/16 is variably subnetted, 8 subnets, 3 masks
D　　172.16.23.0/30 [**90/2681856**] via 172.16.34.1, 00:01:35, Serial0/0/0
　　　　　　　　　　[**90/2681856**] via 172.16.24.1, 00:01:35, Serial0/0/1
D　　172.16.3.0/24　[90/2172416] via 172.16.34.1, 00:01:35, Serial0/0/0
D　　172.31.0.0/16 [90/2681856] via 172.16.24.1, 00:01:36, Serial0/0/1

（2）查看路由协议配置和统计等信息

R2#**show ip protocols | begin Routing Protocol is "eigrp 1"**
Routing Protocol is "**eigrp 1**"　//EIGRP AS 号为 1
　Outgoing update filter list for all interfaces is not set
　Incoming update filter list for all interfaces is not set
　//以上两行表明出方向和入方向都没有配置分布列表（distribute-list）
　Default networks flagged in outgoing updates
　//允许出方向发送默认路由信息，通过路由模式下的 **default-information out** 命令配置
　Default networks accepted from incoming updates
　//允许入方向接收默认路由信息，通过路由模式下的 **default-information in** 命令配置
EIGRP-IPv4 Protocol for **AS(1)**　　//IPv4 EIGRP 进程 1 的信息
　　EIGRP **metric weight** K1=1, K2=0, K3=1, K4=0, K5=0　//计算度量值所用的 K 值
　　Soft SIA disabled　　//软 SIA 功能关闭
EIGRP **NSF-aware** route hold timer is 240s　//不间断转发的持续时间
Router-ID: 2.2.2.2　//EIGRP 路由器 ID
Topology : 0 (base)　　//拓扑 0 的信息
Active Timer: 3 min　//默认 SIA 计时器为 3 分钟
Distance: **internal** 90 **external** 170　//EIGRP 管理距离，内部为 90，外部为 170
　　Maximum path: 4　//默认支持负载均衡路径的条数，可以通过 **maximum-paths** *number-paths* 命令修改 EIGRP 支持等价路径的条数。EIGRP 只是将路由添加到本地路由表中，至于负载均衡，是路由器的交换硬件或者软件的功能，跟 EIGRP 本身没有必然的联系，本书实验环境 IOS 15.7 的最大值为 32
　　Maximum **hopcount** 100　//EIGRP 支持的最大跳数，默认为 100，最大为 255
　　　Maximum metric **variance** 1　//variance 值默认为 1，即默认时只支持等价路径的负载均衡
　　Automatic Summarization: enabled　//自动汇总已经开启，默认自动汇总是关闭的
　　　172.31.0.0/16 for Se0/0/1, Se0/1/0

```
        Summarizing 2 components with metric 2169856    //自动汇总 2 个子网及汇总后的度量值
    //以上两行表明自动汇总成 172.31.0.0/16 网络,并从接口 Se0/0/1 和 Se0/1/0 以初始度量值 2169856
发送出去
        172.16.0.0/16 for Se0/0/0
        Summarizing 5 components with metric 2169856    //自动汇总 5 个子网及汇总后的度量值
    //以上两行表明自动汇总成 172.16.0.0/16 网络,并从接口 Se0/0/0 以初始度量值 2169856 发送出去
    Maximum path: 4    //默认支持负载均衡路径的条数
    Routing for Networks:    //EIGRP 进程中命令 network 后的参数配置
        172.16.0.0
        172.31.0.0
    Routing Information Sources:    //路由信息源,即收到 EIGRP 路由更新信息的 EIGRP 邻居的接
                                        口地址
        Gateway            Distance            Last Update
        172.16.24.2          90                00:07:07
        172.16.23.2          90                00:07:07
        172.31.12.1          90                00:07:07
    Distance: internal 90 external 170    //EIGRP 管理距离
```

(3)查看 EIGRP 邻居信息

```
R2#show ip eigrp neighbors
IP-EIGRP neighbors for process 1    //EIGRP 进程 1 中关于 IPv4 的 EIGRP 的邻居信息
H   Address          Interface     Hold    Uptime      SRTT    RTO     Q      Seq
                                   (sec)               (ms)            Cnt    Num
2   172.16.24.2      Se0/1/0       12      01:45:34    15      1140    0      12
1   172.16.23.2      Se0/0/1       12      02:23:27    10      1140    0      19
0   172.31.12.1      Se0/0/0       12      02:30:39    10      1140    0      8
```

以上输出表明路由器 R2 有 3 个 EIGRP 邻居,各字段的含义如下。

① **H**: 用来跟踪邻居的编号。
② **Address**: 邻居路由器的接口地址。
③ **Interface**: 本路由器到邻居路由器的接口。
④ **Hold**: 认为邻居关系不存在所能等待的最长时间,单位为秒。

【提示】

① EIGRP 的 Hello 间隔可以在接口下通过命令 **ip hello-interval eigrp** *as-number seconds* 来修改。
② EIGRP 的 Hold 时间可以在接口下通过命令 **ip hold-time eigrp** *as-number seconds* 来修改。
③ Hello 间隔被修改后,Hold 时间并不会自动调整,需要手工配置。邻居之间 Hello 间隔和 Hold 时间的不同并不影响 EIGRP 邻居关系的建立,这点和 OSPF 是不同的。
④ Hold 时间到期或者更新重传 16 次都会使 EIGRP 重置邻居关系。
⑤ **Uptime**: 从邻居关系建立到目前的时间,以小时、分、秒计。
⑥ **SRTT**: 向邻居路由器发送一个数据包到本路由器收到确认数据包的时间,单位为 ms,用来确定重传间隔。
⑦ **RTO**: 路由器将重传队列中的数据包重传给邻居之前等待的时间,单位 ms。

⑧ **Q Cnt**：队列中等待发送的数据包数量，如果该值经常大于 0，则可能存在链路拥塞问题。
⑨ **Seq Num**：从邻居收到的最新的 EIGRP 数据包的序列号。

【技术要点】

运行 EIGRP 路由协议的路由器不能建立 EIGRP 邻居关系的可能原因：
- EIGRP 进程 ID 不同；
- 计算度量值的 K 值不同，可以通过命令 metric weights {*tos k1 k2 k3 k4 k5*} 修改 K 值，例如，R1(config-router)#metric weights 0 1 0 1 0 0；
- EIGRP 验证失败。

（4）查看 EIGRP 拓扑表信息

```
R2#show ip eigrp topology
IP-EIGRP Topology Table for AS(1)/ID(2.2.2.2)    //EIGRP 进程 1（ID 为 2.2.2.2）的拓扑表
Codes: P - Passive, A - Active, U - Update, Q - Query, R - Reply,
       r - reply Status, s - sia Status
P 172.16.34.0/30, 2 successors, FD is 2681856    //因为是等价路径，所以有 2 个后继路由器
        via 172.16.23.2 (2681856/2169856), Serial0/0/1    //FD 就是该路由的度量值
        via 172.16.24.2 (2681856/2169856), Serial0/1/0
P 172.16.24.0/30, 1 successors, FD is   2169856
        via Connected, Serial0/1/0
P 172.16.23.0/30, 1 successors, FD is   2169856
        via Connected, Serial0/0/1
P 172.31.1.0/24, 1 successors, FD is 2172416
        via 172.31.12.1 (2172416/28160), Serial0/0/0
P 172.31.0.0/16, 1 successors, FD is 2169856
        via Summary (2169856/0), Null0
P 172.16.4.0/24, 1 successors, FD is 2172416
        via 172.16.24.2 (2172416/28160), Serial0/1/0
P 172.16.0.0/16, 1 successors, FD is   2169856        //EIGRP 自动汇总路由
        via Summary (2169856/0), Null0
P 172.16.3.0/24, 1 successors, FD is 2172416
        via 172.16.23.2 (2172416/28160), Serial0/0/1
P 172.31.12.0/30, 1 successors, FD is 2169856
        via Connected, Serial0/0/0
```

以上输出表明 EIGRP 拓扑表中每条路由条目的信息包括路由条目的状态、后继路由器、可行距离、所有可行后继路由器及其通告距离等。以上输出信息的第二行和第三行状态代码的含义如下。

① P：代表 Passive，表示稳定状态，路由条目可用，可被加入到路由表中。
② A：代表 Active，当前路由条目不可用，正处于发送查询状态，不能加入到路由表中。
③ U：代表 Update，路由条目正在更新或者等待更新数据包确认状态。
④ Q：代表 Query，路由条目未被应答或者处于等待查询数据包确认的状态。
⑤ R：代表 Reply，路由器正在生成对该路由条目的应答或处于等待应答数据包确认的状态。
⑥ r：代表 Reply Status，发送查询并等待应答时设置的标记。

⑦ s：代表 sia Staus，默认经过 3 分钟，如果被查询的路由没有收到邻居的应答，该路由就被置为 **Stuck In Active** 状态，说明 EIGRP 网络的收敛发生了问题。路由模式下，可以通过命令 timers active-time [*time-limit* | disabled]修改 SIA 时间，其中 *time-limit* 的单位是分钟。

拓扑表中路由条目中 P 172.16.4.0/24, 1 successors, FD is 2172416 via 172.16.24.2 (2172416/28160), Serial0/1/0 的具体含义是：路由条目 172.16.4.0/24 处于被动状态，该条目有 1 个后继路由器，可行距离为 2172416，通告距离为 28160，下一跳地址为 172.16.24.2，本地出接口为 Se0/1/0。

（5）查看运行 EIGRP 路由协议的接口的信息

```
R2#show ip eigrp interfaces
IP-EIGRP interfaces for process 1        //参与运行 EIGRP 进程 1 的接口
                Xmit Queue    Mean    Pacing Time   Multicast    Pending
Interface  Peers Un/Reliable  SRTT    Un/Reliable   Flow Timer   Routes
Se0/0/0     1      0/0         10      5/190         230          0
Se0/0/1     1      0/0         10      5/190         234          0
Se0/1/0     1      0/0         15      5/190         242          0
```

以上输出表明路由器 R2 有 3 个接口运行 EIGRP，需要注意的是此命令的输出结果不包含配置的被动接口，各字段的含义如下。

① Interface：运行 EIGRP 协议的接口。
② Peers：接口的邻居的个数。
③ Xmit Queue Un/Reliable：在不可靠/可靠队列中存留的数据包的数量。
④ Mean SRTT：平均的往返时间，单位是毫秒。
⑤ Pacing Time Un/Reliable：用来确定不可靠/可靠队列中数据包被送出接口的时间间隔。
⑥ Multicast Flow Timer：组播数据包被发送前最长的等待时间，达到最长时间后，将从组播切换到单播。
⑦ Pending Routes：在传送队列中等待被发送的数据包携带的路由条目的数量。

（6）查看 EIGRP 发送和接收到的数据包的统计情况

```
R2#show ip eigrp traffic
IP-EIGRP Traffic Statistics for AS 1        // EIGRP 进程 1 的数据包统计
    Hellos sent/received: 6179/6148         //发送和接收的 Hello 数据包的数量
    Updates sent/received: 15/15            //发送和接收的 Update 数据包的数量
    Queries sent/received: 4/1              //发送和接收的 Query 数据包的数量
    Replies sent/received: 1/4              //发送和接收的 Reply 数据包的数量
    Acks sent/received: 10/10               //发送和接收的 ACK 数据包的数量
    SIA-Queries sent/received: 0/0          //发送和接收的 SIA 查询数据包的数量
    SIA-Replies sent/received: 0/0          //发送和接收的 SIA 应答数据包的数量
    Hello Process ID: 125                   //Hello 进程 ID
    PDM Process ID: 64                      //PDM（协议独立模块）进程 ID
    Socket Queue:  0/10000/2/0 (current/max/highest/drops)   // IP Socket 队列情况
    Input Queue:   0/10000/2/0 (current/max/highest/drops)   // IGRP 输入队列情况
```

【技术要点】

为了减少 EIGRP 邻接关系中断，进而提高网络的可靠性，EIGRP 数据包中新增 SIA-查询（SIA-Queries）和 SIA-应答（SIA-Replies）。它们是由主动过程改进（Active Process Enhancement）功能自动生成的，该功能让 EIGRP 路由器能够监控后继路由的查询进程，并确保邻居仍然是可达的。路由器在主动定时器的时间过去一半（默认为 1.5 分钟）后，使用 SIA-查询向邻居路由器查询路由状态，邻居路由器使用 SIA-应答数据包进行响应，收到 SIA-应答数据包后，路由器确定邻居路由器正常运行，即使到了 3 分钟没有收到查询响应，也不会终止和它的邻居关系。

（7）动态查看 EIGRP 邻居关系变化情况

在路由器 R2 上先执行 Se0/0/0 接口 **shutdown**，然后再执行 **no shutdown**，可以看到 EIGRP 邻居关系变化的过程。

```
R2#debug eigrp neighbors
    02:04:51:: %DUAL-5-NBRCHANGE: IP-EIGRP(0) 1: Neighbor 172.31.12.1 (Serial0/0/0) is down: interface down    //接口 down 导致 DUAL 邻居变化
    02:04:51: Going down: Peer 172.31.12.1 total=2 stub 0 template=2, iidb-stub=0 iid-all=0
    02:04:51:: EIGRP: Neighbor 172.31.12.1 went down on Serial0/0/0    //邻居 172.31.12.1 down
    02:04:51: EIGRP: New peer 172.31.12.1 total=3 stub 0 template=2 idbstub=0 iidball=1
    //显示增加新的邻居 172.31.12.1
    02:04:51:: %DUAL-5-NBRCHANGE: IP-EIGRP(0) 1: Neighbor 172.31.12.1 (Serial0/0/0) is up: new adjacency    //与 172.31.12.1 形成新的邻接关系
```

【注意】

EIGRP 默认显示邻居关系变化的日志信息，如果关闭该功能（路由模式下执行 **no eigrp log-neighbor-changes** 命令），则上面的输出中不会显示下面两条信息：

```
    02:04:51:: %DUAL-5-NBRCHANGE: IP-EIGRP(0) 1: Neighbor 172.31.12.1 (Serial0/0/0) is down: interface down
    02:04:51:: %DUAL-5-NBRCHANGE: IP-EIGRP(0) 1: Neighbor 172.31.12.1 (Serial0/0/0) is up: new adjacency
```

3.2.2 实验 2：配置高级 EIGRP

1. 实验目的

通过本实验可以掌握：
① EIGRP 等价负载均衡和非等价负载均衡的实现原理和配置。
② 修改 EIGRP 度量值的方法。
③ 可行距离（FD）、通告距离（RD）以及可行性条件（FC）的深层含义。
④ EIGRP 手工路由汇总配置。
⑤ EIGRP MD5 验证配置。
⑥ 向 EIGRP 网络中注入默认路由的方法。

2. 实验拓扑

配置高级 EIGRP 的实验拓扑如图 3-7 所示。

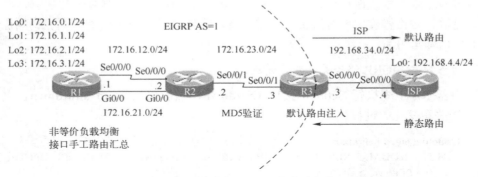

图 3-7 配置高级 EIGRP 的实验拓扑

3. 实验步骤

路由器 R1、R2 和 R3 之间运行 EIGRP，路由器 R3 和 ISP 之间配置静态路由，通过在边界路由器 R3 上重分布配置，使得路由器 R1 和 R2 学习到一条 EIGRP 的默认路由。同时在 R1 上对 4 条环回接口路由进行 EIGRP 路由手工汇总。为了提高链路利用率，在 R1 和 R2 之间配置 EIGRP 非等价负载均衡。为了增加安全性，在 R2 和 R3 间的串行链路上配置 EIGRP 的 MD5 验证。

（1）配置路由器 R1

```
R1(config)#router eigrp 1
R1(config-router)#eigrp router-id 1.1.1.1
R1(config-router)#variance 2          //配置 variance 的值，用于实现非等价负载均衡
R1(config-router)#network 172.16.0.0 0.0.3.255
//配置环回接口 0～3 运行 EIGRP
R1(config-router)#network 172.16.12.1 0.0.0.0
//通配符掩码全 0 表示精确激活运行 EIGRP 的接口
R1(config-router)#network 172.16.21.1 0.0.0.0
R1(config-router)#passive-interface loopback 0   //配置 EIGRP 被动接口
R1(config-router)#passive-interface loopback 1
R1(config-router)#passive-interface loopback 2
R1(config-router)#passive-interface loopback 3
R1(config)#interface GigabitEthernet0/0
R1(config-if)#ip summary-address eigrp 1 172.16.0.0 255.255.252.0
   //配置 EIGRP 手工路由汇总
R1(config-if)#ip hello-interval eigrp 1 10    //配置 EIGRP Hello 数据包发送间隔
R1(config-if)#ip hold-time eigrp 1 30          //配置 EIGRP Hold 时间
R1(config)#interface Serial0/0/0
R1(config-if)#ip summary-address eigrp 1 172.16.0.0 255.255.252.0
```

【技术要点】

① 配置 EIGRP 手工路由汇总时，仅当路由表中至少有一条该汇总路由的明细路由时，

汇总路由才能被通告出去。

② 只有当被汇总的明细路由全部 down 掉以后，汇总路由才会自动从路由表里被删除，从而可以有效避免路由抖动。

（2）配置路由器 R2

```
R2(config)#router eigrp 1
R2(config-router)#eigrp router-id 2.2.2.2
R2(config-router)#variance 128
R2(config-router)#network 172.16.12.2 0.0.0.0
R2(config-router)#network 172.16.21.2 0.0.0.0
R2(config-router)#network 172.16.23.2 0.0.0.0
R2(config)#key chain cisco    //配置密钥链，名称本地有效
R2(config-keychain)#key 1     //配置密钥 ID，范围为 1～2147483647
R2(config-keychain-key)#key-string cisco123    //配置密钥字符串
R2(config-keychain-key)#accept-lifetime 12:00:00 May 1 2019 12:00:00 May 31 2019    //密钥链中的密钥可以被有效接收的时间
R2(config-keychain-key)#send-lifetime 12:00:00 May 1 2019 12:00:00 May 31 2019
//密钥链中的密钥可以被有效发送的时间
R2(config-keychain-key)#exit
R2(config-keychain)#key 2
R2(config-keychain-key)#key-string cisco123
R2(config-keychain-key)#accept-lifetime 12:00:00 May 28 2019 infinite
R2(config-keychain-key)#send-lifetime 12:00:00 May 28 2019 infinite
```

【技术要点】

① EIGRP 支持使用密钥链来管理密钥，在密钥链中，为了提高安全性，可以定义多个密钥，每个密钥还可以指定特定的存活时间，存活时间包括接收时间和发送时间。

② 在配置 EIGRP 身份验证时，需要指定密钥 ID、密钥和密钥的存活期（可选，默认为 infinite），并使用第一个（基于密钥 ID）有效（根据寿命）的密钥进行验证。

③ 在非存活时间内，密钥将是不可用的。

④ 在密钥之间切换不会中断 EIGRP 进程。建议密钥的生存期彼此重叠，这样可以避免某一时刻没有存活的密钥。

⑤ 通过配置路由器时钟同步来确保所有路由器在同一时刻使用的密钥都是相同的，如 NTP（Network Time Protocol，网络时间协议）。

⑥ 用命令 show key chain 来查看密钥链以及每个密钥的字符串和存活时间等。

```
R1#show key chain cisco
Key-chain cisco:
    key 1 -- text "cisco123"    //key 1 的密钥字符串
        accept lifetime (12:00:00 UTC May 1 2019) - (12:00:00 UTC May 31 2019) [valid now]
        send lifetime (12:00:00 UTC May 1 2019) - (12:00:00 UTC May 31 2019) [valid now]
//以上 2 行说明 key 1 是有效的，即处于存活期
    key 2 -- text "cisco123"
        accept lifetime (12:00:00 UTC May 28 2019) - (infinite)
        send lifetime (12:00:00 UTC May 28 2019) - (infinite)
//以上 2 行说明 key 2 是无效的，即处于非存活期
```

```
R2(config)#interface Serial0/0/1
R2(config-if)#ip authentication mode eigrp 1 md5           //配置验证模式为 MD5
R2(config-if)#ip authentication key-chain eigrp 1 cisco    //在接口下调用密钥链
```

（3）配置路由器 R3

```
R3(config)#ip route 0.0.0.0 0.0.0.0 Serial0/0/0            //配置静态默认路由
R3(config)#router eigrp 1
R3(config-router)#eigrp router-id 3.3.3.3
R3(config-router)#redistribute static                     //重分布静态路由
R3(config-router)#network 172.16.23.3 0.0.0.0
R3(config)#key chain cisco
R3(config-keychain)#key 1
R3(config-keychain-key)#key-string cisco123
R3(config-keychain-key)#accept-lifetime 12:00:00 May 1 2019 12:00:00 May 31 2019
R3(config-keychain-key)#send-lifetime 12:00:00 May 1 2019 12:00:00 May 31 2019
R3(config-keychain-key)#exit
R3(config-keychain)#key 2
R3(config-keychain-key)#key-string cisco123
R3(config-keychain-key)#accept-lifetime 12:00:00 May 28 2019 infinite
R3(config-keychain-key)#send-lifetime 12:00:00 May 28 2019 infinite
R3(config-keychain-key)#exit
R3(config-keychain)#exit
R3(config)#interface Serial0/0/1
R3(config-if)#ip authentication mode eigrp 1 md5
R3(config-if)#ip authentication key-chain eigrp 1 cisco
```

（4）配置路由器 ISP

```
ISP(config)#ip route 172.16.0.0 255.255.0.0 Serial0/0/0    //配置静态汇总路由
```

4. 实验调试

（1）查看路由表

以下各命令的输出全部省略路由代码部分。

① R1#**show ip route eigrp**   //查看 EIGRP 路由表
```
Gateway of last resort is 172.16.21.2 to network 0.0.0.0   //默认路由的网关
  D*EX   0.0.0.0/0   [170/2684416] via 172.16.21.2, 00:22:06, GigabitEthernet0/0
                     [170/3193856] via 172.16.12.2, 00:22:06, Serial0/0/0
```
//路由器 R1 的默认路由有两条路径，而且度量值不相同，即 EIGRP 的非等价负载均衡，该路由由 R3 上通过重分布进入 EIGRP 网络，所以路由代码为 **D*EX**，**EX** 表示 external，*表示默认，管理距离为 170
```
       172.16.0.0/16 is variably subnetted, 14 subnets, 3 masks
  D        172.16.0.0/22 is a summary, 12:36:14, Null0
```
//接口执行手工路由汇总后，会在自己的路由表中产生一条以 Null0 为出接口的 EIGRP 汇总路由，主要是为了防止路由环路，该路由条目的管理距离默认为 5
```
  D        172.16.23.0/24 [90/2172416] via 172.16.21.2, 00:10:09, GigabitEthernet0/0
                          [90/2681856] via 172.16.12.2, 00:10:10, Serial0/0/0
```

//路由器 R1 到达 **172.16.23.0/24** 网络有两条路径,而且是在度量值不相同的同时被放入路由表中的两条路径

进一步查看 172.16.23.0/24 路由的详细信息:

R1#**show ip route 172.16.23.0 255.255.255.0**
Routing entry for **172.16.23.0/24**
Known via "eigrp 1", **distance** 90, **metric** 2172416, **type** internal
//EIGRP 进程、管理距离、度量值和路由类型
    Redistributing via **eigrp 1**   //通过 EIGRP 进程 1 进入路由表
    Last update from 172.16.12.2 on Serial0/0/0, 00:24:29 ago
//更新源地址、本地出接口以及离最近一次更新过去的时间
    Routing Descriptor Blocks: //路由描述区块
    * **172.16.21.2**, from 172.16.21.2, 00:24:35 ago, via **GigabitEthernet0/0**
        Route metric is 2172416, **traffic share count** is **120**
//路由条目度量值和流量分配数
        Total delay is 20100 microseconds, **minimum bandwidth** is 1544 Kbit
        Reliability 255/255, **minimum MTU** 1500 bytes
        Loading 1/255, Hops 1
//以上三行显示总延迟、最小带宽、可靠性、最小 MTU、负载和跳数,其中计算度量值时默认采用带宽和延迟
    172.16.12.2, from 172.16.12.2, 00:24:29 ago, via **Serial0/0/0**
        Route metric is 2681856, **traffic share count** is **97**
//路由条目度量值和流量分配数
        Total delay is 40000 microseconds, minimum bandwidth is 1544 Kbit
        Reliability 255/255, minimum MTU 1500 bytes
        Loading 1/255, Hops 1

以上输出信息表明两条路由流量分担的比例为 120:97。

② R2#**show ip route eigrp**
D*EX 0.0.0.0/0 [170/2681856] via 172.16.23.3, 00:00:21, Serial0/0/1
    172.16.0.0/16 is variably subnetted, 7 subnets, 3 masks
D   **172.16.0.0/22** [90/**156160**] via 172.16.21.1, 00:00:17, GigabitEthernet0/0
            [90/**2297856**] via 172.16.12.1, 00:00:11, Serial0/0/0
//在路由器 R2 上只收到 R1 上被汇总的路由条目 **172.16.0.0/22**,没有收到明细路由(R1 的环回接口 0~3 所在网络)条目。在进行手工路由汇总时,可以对学到的 EIGRP 路由进行汇总,这点和自动汇总是不同的。而汇总后路由的度量值以明细路由度量值中最小的值为汇总后的初始度量值。两条路由的度量值相差将近 15 倍,因此如果要两条路由同时出现在路由表中,则 variance 的值至少配置为 15
③ R3#**show ip route eigrp**
    172.16.0.0/16 is variably subnetted, 5 subnets, 3 masks
D   **172.16.0.0/22** [90/2300416] via 172.16.23.2, 00:05:10, Serial0/0/1
//路由器 R3 的路由表中也只收到 R1 手工汇总的路由条目
D   172.16.12.0/24 [90/2681856] via 172.16.23.2, 00:05:11, Serial0/0/1
D   172.16.21.0/24 [90/2172416] via 172.16.23.2, 00:05:11, Serial0/0/1

(2)查看 EIGRP 拓扑表的信息

R1#**show ip eigrp topology**
IP-EIGRP Topology Table for **AS 1/ID(1.1.1.1)**   //EIGRP 路由进程号和路由器 ID
Codes: P - Passive, A - Active, U - Update, Q - Query, R - Reply, r - Reply status

```
P 172.16.2.0/24, 1 successors, FD is 128256
        via Connected, Loopback2      //直连接口
P 172.16.0.0/22, 1 successors, FD is 128256
        via Summary (128256/0), Null0   //接口手工路由汇总产生，出接口为 Null0，RD 为 0
P 172.16.0.0/24, 1 successors, FD is 128256
        via Connected, Loopback0
P 172.16.21.0/24, 1 successors, FD is 28160
        via Connected, GigabitEthernet0/0
P 172.16.3.0/24, 1 successors, FD is 128256
        via Connected, Loopback3
P 0.0.0.0/0, 2 successors, FD is 2684416
        via 172.16.12.2 (3193856/2681856), Serial0/0/0
        via 172.16.21.2 (2684416/2681856), GigabitEthernet0/0
//以上三行表明默认路由通过重分布进入 EIGRP 网络，两个后继意味着两条路由都会进入路由表
P 172.16.12.0/24, 1 successors, FD is 2169856
        via Connected, Serial0/0/0
P 172.16.1.0/24, 1 successors, FD is 128256
        via Connected, Loopback1
P 172.16.23.0/24, 2 successors, FD is 2172416
        via 172.16.21.2 (2172416/2169856), GigabitEthernet0/0
        via 172.16.12.2 (2681856/2169856), Serial0/0/0
```

//从以上三行的输出中可以看到，第二条路径（选择 Se0/0/0 接口）的 RD 为 **2169856**，而最优路由（选择 **Gi0/0** 接口）的 FD 为 **2172416**，**RD<FD**，满足可行性条件，所以第二条路径（选择 Se0/0/0 接口）是最优路由（选择 Gi0/0 接口）的可行后继。因为 EIGRP 既支持等价负载均衡，也支持非等价负载均衡。EIGRP 的非等价负载均衡是通过 **variance** 命令来实现的。在前面路由器 R1 的配置中，配置了命令 **variance 2**，使得这两条路径在路由表中都可见和可用，上面 R1 的路由表已经验证了这一点

【技术要点】

EIGRP 非等价负载均衡是通过 **variance** 命令实现的，**variance** 默认值是 1（即等价路径的负载均衡），variance 值的范围为 1~128。这个参数代表了可以接受的非等价路径的度量值的倍数（即任何路径的度量值如果小于最优路径的度量值乘以 Variance 的值），在这个范围内的路由（满足 FC 条件）都将被接受并且被放入路由表中（前提是所有路由条目的数量≤EIGRP 支持负载均衡路径的默认最大条数）。

（3）调试 EIGRP 验证的输出信息

使用 **debug eigrp packets** 命令查看在路由器 R2 和 R3 之间串行链路的 EIGRP 验证的输出信息（下面的调试信息是在路由器 R3 上显示的）。

① 如果路由器 R2 的 Se0/0/1 接口的验证配置正确，而路由器 R3 的接口没有起用验证，则出现下面的提示信息：

```
    May   2 12:35:58.143: EIGRP: Serial0/0/1: ignored packet from 172.16.23.2, opcode = 5 (authentication off)
```
//验证关闭

② 如果路由器 R2 的 Se0/0/1 接口的验证配置正确，并且路由器 R3 的接口起用验证，但是没有调用钥匙链，则出现下面的提示信息：

May　2 12:37:30.387: EIGRP: Serial0/0/1: ignored packet from 172.16.23.2, opcode = 5 (**invalid authentication** or **key-chain missing**)　//无效的验证或者钥匙链丢失

③ 如果路由器 R2 和 R3 的 Se0/0/1 接口的验证配置正确，但是钥匙链的密匙不正确，则出现下面的提示信息：

　　May　2 12:39:48.095: EIGRP: **pkt key id = 1, authentication mismatch**　//key1 验证不匹配
　　May　2 12:39:48.095: EIGRP: Serial0/0/1: ignored packet from 172.16.23.2, opcode = 5 (**invalid authentication**)　//无效的验证

④ 如果路由器 R2 和 R3 的 Se0/0/1 接口的验证配置正确，但是路由器 R3 没有配置钥匙链，则出现下面的提示信息：

　　May　2 12:41:49.935: EIGRP: interface Serial0/0/1, **No live authentication keys**
　　//没有存活的验证 key
　　May　2 12:41:49.935: EIGRP: Serial0/0/1: ignored packet from 172.16.23.2, opcode = 5 (**invalid authentication**)　//无效的验证

⑤ 如果路由器 R2 和 R3 的 Se0/0/1 接口的验证配置正确，但是双方的 key id 配置不相同，比如 R2 用 key1，而 R3 用 key2，则出现下面的提示信息：

　　May　2 12:46:12.327: EIGRP: pkt **authentication key id = 1, key not defined**　//key 没有定义
　　May　2 12:46:12.327: EIGRP: Se0/0/1: ignored packet from 172.16.23.2, opcode = 5 (**invalid authentication**)　//无效的验证

（4）向 EIGRP 网络注入默认路由

通过 **ip default-network** 命令向 EIGRP 网络注入默认路由，路由器 R3 配置如下：

```
R3(config)#ip default-network 192.168.34.0
//由于网络地址后面没有子网掩码参数，所以此处必须为主类网络地址
R3(config)#router eigrp 1
R3(config-router)#no redistribute static
R3(config-router)#network 192.168.34.0
```

在路由器 R1、R2 和 R3 上查看路由表：

```
① R1#show ip route eigrp
     172.16.0.0/16 is variably subnetted, 14 subnets, 3 masks
D       172.16.0.0/22 is a summary, 01:08:41, Null0
D       172.16.23.0/24 [90/2172416] via 172.16.21.2, 01:23:10, GigabitEthernet0/0
                       [90/2681856] via 172.16.12.2, 01:23:10, Serial0/0/0
D*    192.168.34.0/24 [90/2684416] via 172.16.21.2, 00:00:07, GigabitEthernet0/0
                      [90/3193856] via 172.16.12.2, 00:00:07, Serial0/0/0
② R2#show ip route eigrp
     172.16.0.0/16 is variably subnetted, 7 subnets, 3 masks
D       172.16.0.0/22    [90/156160] via 172.16.21.1, 01:09:55, GigabitEthernet0/0
                         [90/2297856] via 172.16.12.1, 01:09:50, Serial0/0/0
D*    192.168.34.0/24 [90/2681856] via 172.16.23.3, 00:01:29, Serial0/0/1
```

以上①和②输出表明在 R3 上执行 **ip default-network** 命令确实向 EIGRP 网络中注入一条 D*的默认路由。

```
③ R3#show ip route | include 192.168.34.0
*     192.168.34.0/24 is variably subnetted, 2 subnets, 2 masks
C*        192.168.34.0/24 is directly connected, Serial0/0/0
//ip default-network 命令在本地路由器产生一条直连默认路由
```

以上①、②和③输出表明通过重分布静态默认路由可以向 EIGRP 网络注入默认路由，默认路由的表示方式是 **0.0.0.0**，路由代码是 **D *EX**，管理距离为 170；而通过 **ip default-network** 命令也可以向 EIGRP 网络注入默认路由，默认路由的表示方式是 **D*   192.168.34.0/24**，管理距离为 90，两种配置方法路由器产生的默认路由的表示方式是不一样的。

3.2.3 实验 3：配置 EIGRP stub

1. 实验目的

通过本实验可以掌握：
① EIGRP stub 的特征及应用场合。
② EIGRP stub 的配置和调试方法。
③ EIGRP stub 参数的区别。

2. 实验拓扑

配置 EIGRP stub 的实验拓扑如图 3-8 所示。

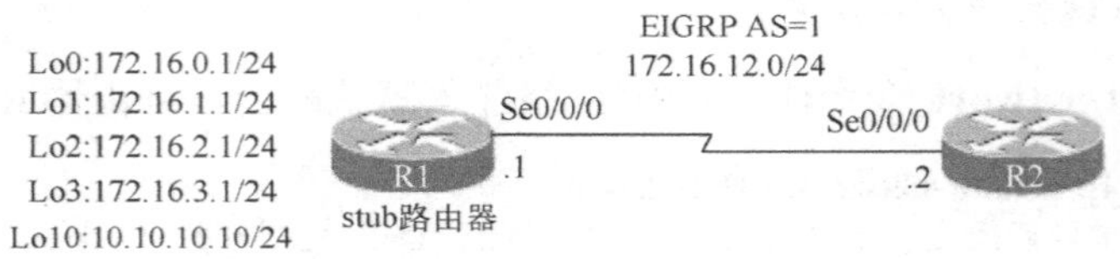

图 3-8 配置 EIGRP stub 的实验拓扑

3. 实验步骤

（1）配置路由器 R1

```
R1(config)#interface Serial0/0/0
R1(config-if)#ip summary-address eigrp 1 172.16.0.0 255.255.252.0
R1(config-if)#exit
R1(config)#router eigrp 1
R1(config-router)#network 172.16.12.1 0.0.0.0
R1(config-router)#network 172.16.0.0 0.0.3.255
R1(config-router)#redistribute connected
R1(config-router)#redistribute static
R1(config-router)#exit
R1(config)#ip route 100.100.100.0 255.255.255.0 Null0
```

（2）配置路由器 R2

```
R2(config)#router eigrp 1
R2(config-router)#network 172.16.12.2 0.0.0.0
```

4. 实验调试

为了验证 EIGRP stub 的各参数对路由的影响，在路由器 R1 上执行 **debug ip eigrp** 命令，各参数分别配置如下。

（1）配置 Connected 参数

该参数允许 EIGRP 末节路由器选择性地发送直连路由。如果直连路由不包括在 network 命令中，有必要使用 **redistribute connected** 命令来重分布直连路由，该选项是默认开启的，配置如下。

```
R1(config)#router eigrp 1
R1(config-router)#eigrp stub connected
04:29:28:IP-EIGRP(Default-IP-Routing-Table:1):Processing incoming UPDATE packet
04:29:28:IP-EIGRP(Default-IP-Routing-Table:1):172.16.12.0/24 - do advertise out Serial0/0/0
04:29:28:IP-EIGRP(Default-IP-Routing-Table:1):100.100.100.0/24 - denied by stub
04:29:28:IP-EIGRP(Default-IP-Routing-Table:1):10.10.10.0/24 - do advertise out Serial0/0/0
04:29:28:IP-EIGRP(Default-IP-Routing-Table:1):Ext 10.10.10.0/24 metric 128256 - 256 128000
04:29:28:IP-EIGRP(Default-IP-Routing-Table:1):172.16.0.0/24 - do advertise out Serial0/0/0
04:29:28:IP-EIGRP(Default-IP-Routing-Table:1):Int 172.16.0.0/24 metric 128256 - 256 128000
04:29:28:IP-EIGRP(Default-IP-Routing-Table:1):172.16.1.0/24 - do advertise out Serial0/0/0
04:29:28:IP-EIGRP(Default-IP-Routing-Table:1):Int 172.16.1.0/24 metric 128256 - 256 128000
04:29:28:IP-EIGRP(Default-IP-Routing-Table:1):172.16.2.0/24 - do advertise out Serial0/0/0
04:29:28:IP-EIGRP(Default-IP-Routing-Table:1):Int 172.16.2.0/24 metric 128256 - 256 128000
04:29:28:IP-EIGRP(Default-IP-Routing-Table:1):172.16.3.0/24 - do advertise out Serial0/0/0
04:29:28:IP-EIGRP(Default-IP-Routing-Table:1):Int 172.16.3.0/24 metric 128256 - 256 128000
04:29:28:IP-EIGRP(Default-IP-Routing-Table:1):172.16.0.0/22 - denied by stub
```

以上输出表明汇总路由条目 **172.16.0.0/22** 和静态路由条目 **100.100.100.0/24** 被拒绝，其他直连路由被通告，**Int** 和 **Ext** 分别表示内部路由和外部路由（重分布进 EIGRP 的路由）。在 R2 上查看路由表，显示如下。

```
R2#show ip route eigrp
      172.16.0.0/24 is subnetted, 5 subnets
D        172.16.0.0 [90/20640000] via 172.16.12.1, 00:00:44, Serial0/0/0
D        172.16.1.0 [90/20640000] via 172.16.12.1, 00:00:44, Serial0/0/0
D        172.16.2.0 [90/20640000] via 172.16.12.1, 00:00:44, Serial0/0/0
D        172.16.3.0 [90/20640000] via 172.16.12.1, 00:00:44, Serial0/0/0
      10.0.0.0/24 is subnetted, 1 subnets
D EX     10.10.10.0 [170/20640000] via 172.16.12.1, 00:00:44, Serial0/0/0
R2#show ip protocols
Routing Protocol is "eigrp 1"
  Outgoing update filter list for all interfaces is not set
  Incoming update filter list for all interfaces is not set
  Default networks flagged in outgoing updates
```

```
    Default networks accepted from incoming updates
    EIGRP metric weight K1=0, K2=0, K3=1, K4=0, K5=0
    EIGRP maximum hopcount 100
    EIGRP maximum metric variance 1
    EIGRP stub, connected    //本路由器是 stub 路由器，只发送直连路由，发送的直连路由信息也
包括重分布的直连路由
    Redistributing: connected, static, eigrp 1
（此处省略部分输出）
```

（2）配置 Static 参数

该参数允许 EIGRP 末节路由器选择性地发送静态路由。同样需要使用 **redistribute static** 命令来重分布静态路由，配置如下。

```
R1(config)#router eigrp 1
R1(config-router)#eigrp stub Static
04:31:45:IP-EIGRP(Default-IP-Routing-Table:1):Processing incoming UPDATE
04:31:45:IP-EIGRP(Default-IP-Routing-Table:1):172.16.12.0/24 - denied by stub
04:31:45:IP-EIGRP(Default-IP-Routing-Table:1):100.100.100.0/24 - do advertise out Serial0/0/0
04:31:45:IP-EIGRP(Default-IP-Routing-Table:1):Ext 100.100.100.0/24 metric 256 - 256 0
04:31:45:IP-EIGRP(Default-IP-Routing-Table:1):10.10.10.0/24 - denied by stub
04:31:45:IP-EIGRP(Default-IP-Routing-Table:1):172.16.0.0/24 - denied by stub
04:31:45:IP-EIGRP(Default-IP-Routing-Table:1):172.16.1.0/24 - denied by stub
04:31:45:IP-EIGRP(Default-IP-Routing-Table:1):172.16.2.0/24 - denied by stub
04:31:45:IP-EIGRP(Default-IP-Routing-Table:1):172.16.3.0/24 - denied by stub
04:31:45:IP-EIGRP(Default-IP-Routing-Table:1):172.16.0.0/22 - denied by stub
```

以上输出表明只有重分布进 EIGRP 的静态路由条目 **100.100.100.0/24** 被通告。在 R2 上查看路由表，显示如下。

```
R2#show ip route eigrp
    100.0.0.0/24 is subnetted, 1 subnets
D EX    100.100.100.0 [170/20512000] via 172.16.12.1, 00:01:32, Serial0/0/0
```

（3）配置 Summary 参数

该参数允许 EIGRP 末节路由器选择性地发送汇总路由条目，包括被手动或自动汇总的路由条目。这个选项默认是开启的，配置如下。

```
R1(config)#router eigrp 1
R1(config-router)#eigrp stub summary
04:34:04:IP-EIGRP(Default-IP-Routing-Table:1):Processing incoming UPDATE packet
04:34:04:IP-EIGRP(Default-IP-Routing-Table:1):172.16.12.0/24 - denied by stub
04:34:04:IP-EIGRP(Default-IP-Routing-Table:1):100.100.100.0/24 - denied by stub
04:34:04:IP-EIGRP(Default-IP-Routing-Table:1):10.10.10.0/24 - denied by stub
04:34:04:IP-EIGRP(Default-IP-Routing-Table:1):172.16.0.0/24 - denied by stub
04:34:04:IP-EIGRP(Default-IP-Routing-Table:1):172.16.1.0/24 - denied by stub
04:34:04:IP-EIGRP(Default-IP-Routing-Table:1):172.16.2.0/24 - denied by stub
04:34:04:IP-EIGRP(Default-IP-Routing-Table:1):172.16.3.0/24 - denied by stub
04:34:04:IP-EIGRP(Default-IP-Routing-Table:1):172.16.0.0/22 - do advertise out Serial0/0/0
04:34:04:IP-EIGRP(Default-IP-Routing-Table:1): Int 172.16.0.0/22 metric 128256 - 256 128000
```

以上输出表明只有汇总路由条目 **172.16.0.0/22** 被通告。在 R2 上查看路由表，显示如下。

```
R2#show ip route eigrp
        172.16.0.0/16 is variably subnetted, 2 subnets, 2 masks
D       172.16.0.0/22 [90/20640000] via 172.16.12.1, 00:00:35, Serial0/0/0
```

（4）配置 Redistributed 参数

该参数允许 EIGRP 末节路由器选择性地发送重分布进 EIGRP 的路由，配置如下。

```
R1(config)#router eigrp 1
R1(config-router)#eigrp stub redistributed
04:35:28:IP-EIGRP(Default-IP-Routing-Table:1):Processing incoming UPDATE packet
04:35:28:IP-EIGRP(Default-IP-Routing-Table:1):172.16.12.0/24 - denied by stub
04:35:28:IP-EIGRP(Default-IP-Routing-Table:1):100.100.100.0/24 - do advertise out Serial0/0/0
04:35:28:IP-EIGRP(Default-IP-Routing-Table:1):Ext 100.100.100.0/24 metric 256 - 256 0
04:35:28:IP-EIGRP(Default-IP-Routing-Table:1):10.10.10.0/24 - do advertise out Serial0/0/0
04:35:28:IP-EIGRP(Default-IP-Routing-Table:1):Ext 10.10.10.0/24 metric 128256 - 256 128000
04:35:28:IP-EIGRP(Default-IP-Routing-Table:1):172.16.0.0/24 - denied by stub
04:35:28:IP-EIGRP(Default-IP-Routing-Table:1):172.16.1.0/24 - denied by stub
04:35:28:IP-EIGRP(Default-IP-Routing-Table:1):172.16.2.0/24 - denied by stub
04:35:28:IP-EIGRP(Default-IP-Routing-Table:1):172.16.3.0/24 - denied by stub
04:35:28:IP-EIGRP(Default-IP-Routing-Table:1):172.16.0.0/22 - denied by stub
```

以上输出表明只有重分布路由条目 100.100.100.0/24 和 10.10.10.0/24 被通告。在 R2 上查看路由表，显示如下。

```
R2#show ip route eigrp
        100.0.0.0/24 is subnetted, 1 subnets
D EX    100.100.100.0 [170/20512000] via 172.16.12.1, 00:00:31, Serial0/0/0
        10.0.0.0/24 is subnetted, 1 subnets
D EX    10.10.10.0 [170/20640000] via 172.16.12.1, 00:00:31, Serial0/0/0
```

（5）配置 Receive-only 参数

该参数禁止路由器在 EIGRP 自治系统内和其他邻居分享它的路由。该参数使用后不能再使用其他参数，因为它阻止发送任何类型的路由，配置如下。

```
R1(config)#router eigrp 1
R1(config-router)#eigrp stub receive-only
04:36:27:IP-EIGRP(Default-IP-Routing-Table:1):Processing incoming UPDATE packet
04:36:27:IP-EIGRP(Default-IP-Routing-Table:1):172.16.12.0/24 - denied by stub
04:36:27:IP-EIGRP(Default-IP-Routing-Table:1):100.100.100.0/24 - denied by stub
04:36:27:IP-EIGRP(Default-IP-Routing-Table:1):10.10.10.0/24 - denied by stub
04:36:27:IP-EIGRP(Default-IP-Routing-Table:1):172.16.0.0/24 - denied by stub
04:36:27:IP-EIGRP(Default-IP-Routing-Table:1):172.16.1.0/24 - denied by stub
04:36:27:IP-EIGRP(Default-IP-Routing-Table:1):172.16.2.0/24 - denied by stub
04:36:27:IP-EIGRP(Default-IP-Routing-Table:1):172.16.3.0/24 - denied by stub
04:36:27:IP-EIGRP(Default-IP-Routing-Table:1):172.16.0.0/22 - denied by stub
```

以上输出表明所有路由条目都被拒绝。在 R2 上查看路由表，显示路由器 R2 的路由表中没有 EIGRP 路由条目。

（6）配置默认参数

配置命令后面不加任何参数，配置如下。

```
R1(config)#router eigrp 1
R1(config-router)#eigrp stub
04:39:36:IP-EIGRP(Default-IP-Routing-Table:1):Processing incoming UPDATE packet
04:39:36:IP-EIGRP(Default-IP-Routing-Table:1):172.16.12.0/24 - do advertise out Serial0/0/0
04:39:36:IP-EIGRP(Default-IP-Routing-Table:1):100.100.100.0/24 - denied by stub
04:39:36:IP-EIGRP(Default-IP-Routing-Table:1):10.10.10.0/24 - do advertise out Serial0/0/0
04:39:36:IP-EIGRP(Default-IP-Routing-Table:1):Ext 10.10.10.0/24 metric 128256 - 256 128000
04:39:36:IP-EIGRP(Default-IP-Routing-Table:1):172.16.0.0/24 - don't advertise out Serial0/0/0
04:39:36:IP-EIGRP(Default-IP-Routing-Table:1):172.16.1.0/24 - don't advertise out Serial0/0/0
04:39:36:IP-EIGRP(Default-IP-Routing-Table:1):172.16.2.0/24 - don't advertise out Serial0/0/0
04:39:36:IP-EIGRP(Default-IP-Routing-Table:1):172.16.3.0/24 - don't advertise out Serial0/0/0
04:39:36:IP-EIGRP(Default-IP-Routing-Table:1):172.16.0.0/22 - do advertise out Serial0/0/0
04:39:36:IP-EIGRP(Default-IP-Routing-Table:1):Int 172.16.0.0/22 metric 128256 - 256 128000
```

以上输出表明默认时直连路由和汇总路由条目被通告，如果不指定参数，则系统默认的参数是直连和汇总。在 R2 上查看路由表，显示如下。

```
R2#show ip route eigrp
        172.16.0.0/16 is variably subnetted, 2 subnets, 2 masks
D        172.16.0.0/22 [90/20640000] via 172.16.12.1, 00:00:44, Serial0/0/0
        10.0.0.0/24 is subnetted, 1 subnets
D EX    10.10.10.0 [170/20640000] via 172.16.12.1, 00:00:44, Serial0/0/0
```

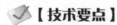【技术要点】

当 EIGRP 进程被关闭、执行 **no network** 命令、通过命令指定新的路由器 ID、修改 K 值或者配置被动接口等时，EIGRP 将通过 Hello 数据包广播一条 **Goodbye** 消息（所有 K 值设为 255），这样使得 EIGRP 不必等到保持计时器到期后才知道拓扑发生的变化，可以让 EIGRP 邻居保持高效同步并重新计算邻接关系。但是关闭接口或者重启路由器时不会发送 **Goodbye** 消息。比如在上面的实验中，在路由器 R1 上执行：

```
R1(config-router)#no network 172.16.12.1 0.0.0.0
```

R2 会显示如下信息：

```
    04:05:45: %DUAL-5-NBRCHANGE: IP-EIGRP(0) 1: Neighbor 172.16.12.1 (Serial0/0/0) is down:
Interface Goodbye received    //收到 R1 发送的 Goodbye 消息
```

3.2.4 实验 4：配置命名 EIGRP

1. 实验目的

通过本实验可以掌握：
① 命名 EIGRP 的特征。
② 命名 EIGRP 手工路由汇总配置。

③ 命名 EIGRP Hmac-sha-256 验证配置。

2. 实验拓扑

配置命名 EIGRP 的实验拓扑如图 3-9 所示。

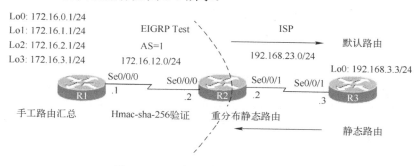

图 3-9 配置命名 EIGRP 的实验拓扑

3. 实验步骤

在路由器 R1 和 R2 之间运行命名 EIGRP，在路由器 R2 和 ISP（R3）之间配置静态路由，通过在边界路由器 R2 上进行重分布配置，使得路由器 R1 学习到一条 EIGRP 的默认路由。同时在 R1 上对 4 条环回接口路由进行 EIGRP 路由手工汇总。为了增加安全性，在 R1 和 R2 的串行链路上配置 EIGRP 的 Hmac-sha-256 验证。

（1）配置路由器 R1

```
R1(config)#router eigrp test   //进入命名 EIGRP 配置模式
R1(config-router)#address-family ipv4 unicast autonomous-system 1
//进入 IPv4 地址家族，配置 EIGRP 单播路由进程 ID 为 1
R1(config-router-af)#eigrp router-id 1.1.1.1   //配置 EIGRP 路由器 ID
R1(config-router-af)#network 172.16.12.1 0.0.0.0   //激活运行 EIGRP 的接口
R1(config-router-af)#af-interface Serial0/0/1   //进入地址家族接口配置模式
R1(config-router-af-interface)#summary-address 172.16.0.0 255.255.252.0
//配置在接口下进行 EIGRP 手工路由汇总
R1(config-router-af-interface)#exit-af-interface   //退出地址家族接口配置模式
R1(config-router-af)#af-interface Serial0/0/0
R1(config-router-af-interface)#authentication mode hmac-sha-256 cisco123
//配置验证模式为 Hmac-sha-256 并指定验证密钥，命名 EIGRP 支持 MD5 和 Hmac-sha-256 两种验
证方式，如果配置 MD5 验证模式，需要先定义钥匙链（key chain），然后通过 authentication mode md5 和
authentication key-chain key-chain 两条命令完成
R1(config-router-af-interface)#exit-af-interface//退出地址家族接口配置模式
R1(config-router-af)#exit-address-family   //退出地址家族配置模式
R1(config-router)#
```

【技术要点】

新版本的 IOS 支持命名 EIGRP 配置。传统方式配置 EIGRP 要求在接口和 EIGRP 路由模式下配置各种参数。为了配置 EIGRP IPv4 和 IPv6，必须要配置独立的 EIGRP 实例。传统 EIGRP

不支持虚拟路由和转发（VRF）在 IPv6 EIGRP 中实施。使用命名 EIGRP，一切都在 EIGRP 路由模式下完成配置。使用命名模式配置 EIGRP 仅需要创建单个实例，该实例可以用于所有地址家族类型。需要注意的是，传统 EIGRP 配置可以和命名 EIGRP 配置共存，即一台路由器采用传统方式配置 EIGRP，另一台路由器采用命名方式配置 EIGRP 是可以建立 EIGRP 邻居关系并且交换路由信息的。

（2）配置路由器 R2

```
R2(config)#ip route 0.0.0.0 0.0.0.0 serial 0/0/1
R2(config)#router eigrp test
R2(config-router)#address-family ipv4 unicast autonomous-system 1
R2(config-router-af)#eigrp router-id 2.2.2.2
R2(config-router-af)#network 172.16.12.2 0.0.0.0
R2(config-router-af)#topology base
//进入拓扑配置模式，可以是基本拓扑，也可以是 VRF 拓扑
R2(config-router-af-topology)#redistribute static      //重分布静态路由
R2(config-router-af-topology)#exit-af-topology          //退出拓扑配置模式
R2(config-router-af)#af-interface Serial0/0/0           //退出拓扑配置模式
R2(config-router-af-interface)#authentication mode hmac-sha-256 cisco123
R2(config-router-af-interface)#exit-af-interface
R2(config-router-af)#exit-address-family
R2(config-router)#
```

（3）配置路由器 R3

```
R3(config)#ip route 172.16.0.0 255.255.0.0 serial0/0/1
```

4. 实验调试

（1）查看 EIGRP IPv4 地址家族邻居信息

```
R2#show eigrp address-family ipv4 neighbors
//和命令 show ip eigrp neighbors 功能相同
EIGRP-IPv4 VR(test) Address-Family Neighbors for AS(1)
//EIGRP 虚拟实例（Virtual-Instance）的名字、地址家族以及进程 ID
H   Address        Interface   Hold Uptime   SRTT  RTO   Q    Seq
                               (sec)         (ms)        Cnt  Num
0   172.16.12.1    Se0/0/0     14   00:05:20  1    100   0    6
```

（2）查看 EIGRP 协议相关信息

```
R2#show eigrp protocols    //该命令输出是 show ip protocols 命令输出的一部分
EIGRP-IPv4 VR(test) Address-Family Protocol for AS(1)
  Metric weight K1=1, K2=0, K3=1, K4=0, K5=0 K6=0
  Metric rib-scale 128
  Metric version 64bit
  Soft SIA disabled
  NSF-aware route hold timer is 240
  Router-ID: 2.2.2.2
```

第 3 章　EIGRP

```
    Topology : 0 (base)          //基础拓扑 0 的信息
      Active Timer: 3 min
      Distance: internal 90 external 170
      Maximum path: 4
      Maximum hopcount 100
      Maximum metric variance 1
    Total Prefix Count: 6        //前缀的总数量
    Total Redist Count: 1        //重分布进入 EIGRP 进程的前缀数量
```

（3）查看 EIGRP 的 Hello 数据包的发送和接收情况

```
R2#debug eigrp packets hello
    (HELLO)
EIGRP Packet debugging is on
1d02h: EIGRP: received packet with HMAC-SHA-256 authentication
    //收到 HMAC-SHA-256 验证的 Hello 数据包
    1d02h: EIGRP: Received HELLO on Se0/0/0 - paklen 76 nbr 172.16.12.1     AS 1, Flags 0x0:(NULL),
Seq 0/0 interfaceQ 0/0 iidbQ un/rely 0/0 peerQ un/rely 0/0
```

以上两行输出信息中关键字段的含义如下。
① Se0/0/0：发送 Hello 数据包的出接口。
② paklen：Hello 数据包长度。
③ nbr：EIGRP 邻居地址。
④ AS：EIGRP 进程 ID。
⑤ Flags：是 EIGRP 数据包头中标记的状态，0x0 表示没有设置标记，0x1 表示设置了初始化位，指出附加路由条目是新邻居关系的开始。
⑥ Seq：表示一个数据包中的序列号 / 确认序列号。
⑦ iidbQ：表示接口下等待传送的不可靠组播数据包数量 / 可靠组播数据包数量。
⑧ peerQ：表示接口下等待传送的不可靠单播数据包数量 / 可靠单播数据包数量。

（4）查看 EIGRP 插件信息

```
R2#show eigrp plugins
EIGRP feature plugins:::
    eigrp-release          :    23.00.00 : Portable EIGRP Release
                           :    2.00.09 : Source Component Release(rel23)
    parser                 :    2.02.00 : EIGRP Parser Support
    igrp2                  :    2.00.00 : Reliable Transport/Dual Database
    manet                  :    3.00.00 : Mobile ad-hoc network (MANET)
    ipv6-af                :    2.01.01 : Routing Protocol Support
    ipv6-sf                :    2.01.00 : Service Distribution Support
    bfd                    :    2.00.00 : BFD Platform Support
    eigrp-pfr              :    1.00.01 : Performance Routing Support
    EVN/vNets              :    1.00.00 : Easy Virtual Network (EVN/vNets)
    ipv4-af                :    2.01.01 : Routing Protocol Support
    ipv4-sf                :    1.02.00 : Service Distribution Support
    vNets-parse            :    1.00.00 : EIGRP vNets Parse Support
    snmp-agent             :    2.00.00 : SNMP/SNMPv2 Agent Support
```

本 章 小 结

EIGRP 作为 Cisco 的私有路由协议，具有收敛快、支持非等价负载均衡等特点，可以应用于较大规模网络。本章介绍了 EIGRP 特征、DUAL 概念及有限状态机、EIGRP 数据包格式、SIA 及查询范围的限定等，并用实验演示和验证了 EIGRP 基本配置、负载均衡、路由汇总、验证、向 EIGRP 网络中注入默认路由、EIGRP stub 以及命名 EIGRP 配置。

第 4 章 OSPF

OSPF 路由协议是典型的链路状态路由协议，它克服了距离矢量路由协议依赖邻居做路由决策的缺点，应用广泛。1989 年，OSPFv1 规范在 RFC 1131 中发布，但是 OSPFv1 是一种实验性的路由协议，未获得实施。1991 年，OSPFv2 在 RFC 1247 中引入，到了 1998 年，OSPFv2 规范在 RFC 2328 中得以更新，也就是 OSPF 的现行 RFC 版本。1999 年，用于 IPv6 的 OSPFv3 在 RFC 2740 中发布。本章重点讨论 OSPFv2 路由协议，OSPFv3 在第 9 章讨论。

4.1 OSPF 概述

4.1.1 OSPF 特征

OSPF（Open Shortest Path First，开放最短链路优先）作为一种内部网关协议（Interior Gateway Protocol，IGP），用于在同一个自治系统（AS）中的路由器之间交换路由信息，运行 OSPF 的路由器彼此交换并保存整个网络的链路状态信息，从而掌握整个网络的拓扑结构，并独立计算路由。OSPF 的特征如下。

① 收敛速度快，适应规模较大的网络。
② 是无类别的路由协议，支持不连续子网、VLSM 和 CIDR 以及手工路由总结。
③ 采用组播方式（224.0.0.5 或 224.0.0.6）更新，支持等价负载均衡。
④ 支持区域划分，构成结构化的网络，提供路由分级管理，从而使得 SPF 的计算频率更低，链路状态数据库和路由表更小，链路状态更新的开销更小，同时可以将不稳定的网络限制在特定的区域。
⑤ 支持简单口令和 MD5 验证。
⑥ 采用触发更新，无路由环路，并且可以使用路由标记（Tag）对外部路由进行跟踪，便于监控和控制。
⑦ OSPF 路由协议的管理距离是 110，OSPF 路由协议采用开销（Cost）作为度量标准。
⑧ OSPF 维护邻居表（邻接数据库）、拓扑表（链路状态数据库）和路由表（转发数据库）。
⑨ 为了确保 LSDB（链路状态数据库）同步，OSPF 每隔 30 分钟对链路状态刷新一次。

4.1.2 OSPF 术语

① 链路（Link）：路由器上的一个接口。
② 链路状态（Link State）：有关各条链路状态的信息，用来描述路由器接口及其与邻居路由器的关系，这些信息包括接口的 IP 地址和子网掩码、网络类型、链路的开销以及链路上的所有相邻路由器。所有链路状态信息构成链路状态数据库。

③ 区域（Area）：共享链路状态信息的一组路由器。在同一个区域内的路由器有相同的 OSPF 链路状态数据库。

④ 自治系统（Autonomous System，AS）：采用同一种路由协议交换路由信息的路由器及其网络构成一个自治系统。

⑤ 链路状态通告（Link-State Advertisement，LSA）和链路状态更新（Link-State Update，LSU）：LSA 用来描述路由器和链路的状态，LSA 包括的信息有路由器接口的状态和所形成的邻接状态，不同类型的 LSA 的功能不同；LSU 可以包含一个或多个 LSA。

⑥ 最短路径优先（Shortest Path First，SPF）算法：OSPF 路由协议的基础和核心。SPF 算法也被称为 Dijkstra 算法，这是因为最短路径优先算法（SPF）是 Dijkstra 发明的。OSPF 路由器利用 SPF 算法独立地计算出到达目标网络的最佳路由。

⑦ 邻居（Neighbor）关系：如果两台路由器共享一条公共数据链路，并且能够协商 Hello 数据包中所指定的某些参数，它们就形成邻居关系。

⑧ 邻接（Adjacency）关系：相互交换 LSA 的 OSPF 邻居建立的关系，一般，在点到点、点到多点的网络上邻居路由器都能形成邻接关系，而在广播多路访问（Broadcast Multiaccess，BMA）和非广播多路访问（Non-Broadcast Multiaccess，NBMA）网络上，要选举 DR 和 BDR，DR 和 BDR 路由器与所有的邻居路由器形成邻接关系，但是 DRother 路由器（非 DR 路由器）之间不能形成邻接关系，只形成邻居关系。

⑨ 指定路由器（Designated Router，DR）和备份指定路由器（Backup Designated Router，BDR）：为了避免路由器之间建立完全邻接关系而引起的大量开销，OSPF 要求在多路访问的网络中选举一个 DR，每个路由器都与之建立邻接关系。选举 DR 的同时也选举出一个 BDR，当 DR 失效时，BDR 担负起 DR 的职责，而且在同一个 BMA 网络中所有其他路由器只与 DR 和 BDR 建立邻接关系。

⑩ OSPF 路由器 ID：运行 OSPF 路由器的唯一标识，长度为 32 比特，格式和 IP 地址相同。

4.1.3 OSPF 路由器类型

当一个 AS 划分成几个 OSPF 区域时，根据一个路由器在相应的区域的作用，可以将 OSPF 路由器进行分类，OSPF 路由器类型如图 4-1 所示。

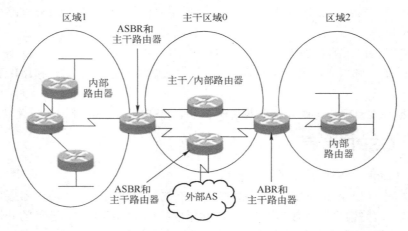

图 4-1　OSPF 路由器类型

① 内部路由器：OSPF 路由器上所有直连的链路都处于同一个区域。
② 主干路由器：具有连接区域 0 接口的路由器。
③ 区域边界路由器（Area Border Router，ABR）：路由器与多个区域相连，对于连接的每个区域，路由器都有一个独立的链路状态数据库。Cisco 建议每台路由器所属区域最多不要超过 3 个。
④ 自治系统边界路由器（Autonomous System Boundary Router，ASBR）：与 AS 外部的路由器相连并互相交换路由信息。同一台路由器可能属于多种类型 OSPF 路由器，比如可能既是 ABR，同时又是 ASBR。

4.1.4 OSPF 网络类型

OSPF 路由协议为了能够适应二层网络环境，根据路由器所连接的物理网络不同通常将网络划分为广播多路访问（Broadcast MultiAccess，BMA）、非广播多路访问（Non-Broadcast MultiAccess，NBMA）、点到点（Point-to-Point）和点到多点（Point-to-MultiPoint）4 种类型。在每种网络类型中，OSPF 的运行方式不同，包括是否需要 DR 选举等。OSPF 网络类型如表 4-1 所示，该表对 OSPF 不同网络类型进行了比较。

表 4-1 OSPF 网络类型

网络类型	物理网络举例	选举 DR	Hello 间隔	Dead 间隔	邻居
广播多路访问	以太网	是	10 秒	40 秒	自动发现
非广播多路访问	帧中继	是	30 秒	120 秒	管理员配置
点到点	PPP、HDLC	否	10 秒	40 秒	自动发现
点到多点	管理员配置	否	30 秒	120 秒	自动发现

4.1.5 OSPF 区域类型

OSPF 区域采用两级结构，一个区域所设置的特性控制着它所能接收到的链路状态信息的类型。区分不同 OSPF 区域类型的关键在于它们对区域外部路由的处理方式。OSPF 区域类型如下所述。

① 标准区域：可以接收链路更新信息、相同区域的路由信息、区域间路由信息以及外部 AS 的路由信息。标准区域通常与区域 0 连接。
② 主干区域：连接各个区域的中心实体，可以快速高效地传输 IP 数据包，其他的区域都要连接到该区域交换路由信息。主干区域也叫区域 0。
③ 末节区域（Stub Area）：不接受外部自治系统的路由信息。
④ 完全末节区域（Totally Stubby Area）：不接受外部自治系统的路由和自治系统内其他区域的路由汇总，完全末节区域是 Cisco 专有的特性。
⑤ 次末节区域（Not-So-Stubby Area，NSSA）：允许接收以 7 类 LSA 方式发送的外部路由信息，并且 ABR 要负责把类型 7 的 LSA 转换成类型 5 的 LSA。

4.1.6 OSPF LSA 类型

一台路由器中所有有效的 LSA 都被存放在它的链路状态数据库中，正确的 LSA 可以描述一个 OSPF 区域的网络拓扑结构。OSPFv2 中常见的 LSA 有 6 类，LSA 类型及相应的描述如表 4-2 所示。

表 4-2 LSA 类型及相应的描述

类型代码	名称及路由代码	描述
1	路由器 LSA（O）	所有的 OSPF 路由器都会产生这种 LSA，用于描述路由器上连接到某一个区域的链路或者某一接口的状态信息。该 LSA 只会在区域内扩散，而不会扩散至其他的区域。链路状态 ID 为本路由器 ID
2	网络 LSA（O）	由 DR 产生，用来描述一个多路访问网络和与之相连的所有路由器，只会在包含 DR 所属的多路访问网络的区域中扩散，不会扩散至其他的 OSPF 区域。链路状态 ID 为 DR 接口的 IP 地址
3	网络汇总 LSA（O IA）	由 ABR 产生，它将一个区域内的网络通告给 OSPF 自治系统中的其他区域。这些条目通过主干区域被扩散到其他的 ABR。链路状态 ID 为目标网络的地址
4	ASBR 汇总 LSA（O IA）	由 ABR 产生，描述到 ASBR 的可达性，由主干区域发送到其他 ABR。链路状态 ID 为 ASBR 路由器 ID
5	外部 LSA（O E1 或 E2）	由 ASBR 产生，含有关于自治系统外的链路信息。链路状态 ID 为外部网络的地址
7	NSSA 外部 LSA（O N1 或 N2）	由 ASBR 产生的关于 NSSA 的信息，可以在 NSSA 区域内扩散，ABR 可以将类型 7 的 LSA 转换为类型 5 的 LSA。链路状态 ID 为外部网络的地址

4.1.7 OSPF 数据包格式

每个 OSPF 数据包都具有 OSPF 数据包头部，长度为 24 字节。OSPF 数据包头和数据被封装到 IP 数据包中，在该 IP 数据包头中，协议字段被设为 89，TTL 值被设置为 1，目的地址则被设为以下两个组播地址之一：224.0.0.5 或 224.0.0.6。如果 OSPF 数据包被封装在以太网帧内，则目的 MAC 地址也是组播地址：01-00-5E-00-00-05 或 01-00-5E-00-00-06。

1. OSPF 数据包包头格式

OSPFv2 数据包包头格式如图 4-2 所示，各字段含义如下。

① 版本：OSPF 的版本号。
② 类型：OSPF 数据包类型。
③ 数据包长度：OSPF 数据包的长度，包括数据包头部的长度，单位为字节。
④ 路由器 ID：始发路由器的 ID。
⑤ 区域 ID：始发数据包的路由器所在区域 ID。
⑥ 校验和：对整个数据包的校验和。
⑦ 身份验证类型：验证类型有 3 种，其中 0 表示不验证，1 表示简单口令验证，2 表示 MD5 验证。
⑧ 身份验证：数据包验证的必要信息，如果验证类型为 0，将不检查该字段；如果验证类型为 1，则该字段包含的是一个最长为 64 位的口令；如果验证类型为 2，则该字段包含一个 key id、验证数据的长度和一个不会减小的加密序列号[用来防止重放（Replay）攻击]。这个摘要消息附加在 OSPF 报文的尾部，不作为 OSPF 数据包本身的一部分。

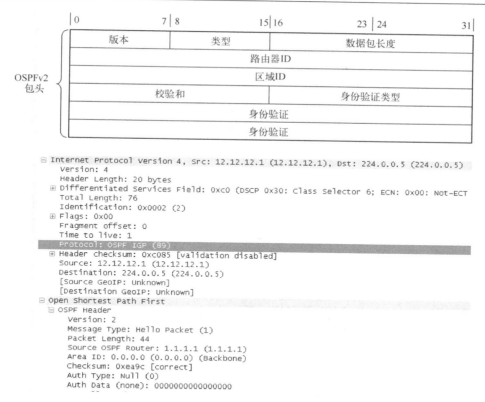

图 4-2 OSPFv2 数据包包头格式

2. OSPF 数据包类型

OSPF 数据包包括 5 种类型，每种数据包在 OSPF 路由过程中发挥各自的作用。

（1）**Hello** 数据包

用于建立和维持 OSPF 邻接关系。OSPF Hello 数据包格式如图 4-3 所示，除了 OSPF 包头，各字段的含义如下。

① 网络掩码：与发送方接口关联的子网掩码。

② Hello 间隔：连续两次发送 Hello 数据包之间的时间间隔，单位为秒。

③ 路由器优先级：8 比特，用于 DR/BDR 选举，范围为 0~255。

④ 路由器 Dead 间隔：宣告邻居路由器无效之前等待的最长时间。

⑤ 指定路由器（DR）：DR 路由器接口的 IP 地址，如果没有，该字段为 0.0.0.0。

⑥ 备用指定路由器（BDR）：BDR 路由器接口的 IP 地址，如果没有，该字段为 0.0.0.0。

⑦ 邻居列表：列出相邻路由器的 OSPF 路由器 ID。

（2）DBD 数据包

DBD（Database Description，数据库描述）：包含发送方路由器的链路状态数据库的简略列表，接收方路由器使用本数据包与其本地链路状态数据库对比。在同一区域内的所有链路状态路由器的 LSDB 必须保持一致，以构建准确的 SPF 树。OSPFv2 DBD 数据包格式如图 4-4 所示，除了包头，各字段含义如下。

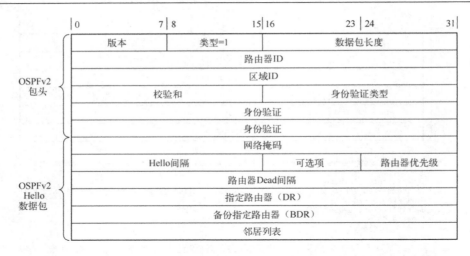

图 4-3　OSPFv2 Hello 数据包格式

图 4-4　OSPFv2 DBD 数据包格式

① 接口 MTU：在数据包不分段的情况下，路由器接口能发送的最大 IP 数据包的大小。
② I：初始位，发送第一个 DBD 数据包时 I 位置 1，后续 DBD 数据包的 I 位置 0。
③ M：后继位，最后一个 DBD 数据包的 M 位置 0，其他 M 位置 1。
④ MS：主从位，用于协商主/从路由器，置 1 表示主路由器，置 0 表示从路由器。
⑤ DD 序列号：在数据库同步过程中，用来确保路由器收到完整的 DBD 数据包，该序列号由主路由器在发送第一个 DBD 数据包时设置，后续数据包的序列号将依次增加。
⑥ LSA 头部：LSA 头部包含的信息可以唯一地标识一个 LSA，OSPFv2 LSA 头部格式如图 4-5 所示。

图 4-5　OSPFv2 LSA 头部格式

- 老化时间：发送 LSA 后经历的时间，单位为秒；
- 类型：LSA 的类型；
- 链路状态 ID：标识 LSA，LSA 类型不同标识方法也不同；
- 通告路由器：始发 LSA 通告的路由器 ID；
- 序列号：每当 LSA 被更新时都加 1，可以帮助识别最新的 LSA；
- 校验和：除老化时间之外的 LSA 全部信息的校验和；
- 长度：LSA 头部和 LSA 数据的总长度。

（3）LSR 数据包

LSR（Link-State Request，链路状态请求）：在 LSDB 同步过程中，路由器收到 DBD 数据包后，会查看自己的 LSDB 中不包括哪些 LSA，或者哪些 LSA 比自己的更新，然后把这些 LSA 记录在链路状态请求列表中，接着通过发送 LSR 数据包来请求 DBD 数据包中任何 LSA 条目的详细信息。OSPFv2 LSR 数据包格式如图 4-6 所示，除了包头，各字段的含义如下。
① 链路状态类型：LSA 的类型。
② 链路状态 ID：标识 LSA，LSA 类型不同标识方法也不同。
③ 通告路由器：始发 LSA 通告的路由器 ID。

图 4-6 OSPFv2 LSR 数据包格式

（4）LSU 数据包

LSU（Link-State Update，链路状态更新）：用于回复 LSR 或通告新的更新信息。OSPFv2 LSU 数据包格式如图 4-7 所示，除了包头，各字段的含义如下。

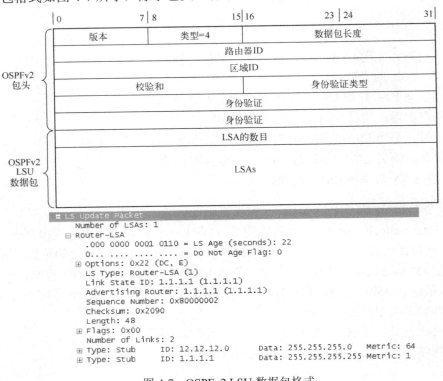

图 4-7 OSPFv2 LSU 数据包格式

① LSA 的数目：更新数据包中包含 LSA 的数量。
② LSAs：一个更新数据包中可以携带多个 LSA。

（5）LSAck 数据包

LSAck（Link-State Acknowledgement，链路状态确认）：路由器收到 LSU 数据包后，会发送一个 LSAck 数据包来确认接收到了 LSU 数据包。OSPFv2 LSAck 数据包格式如图 4-8 所示。多个 LSA 可以通过单个 LSAck 来确认。

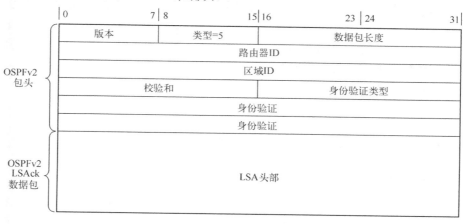

图 4-8　OSPFv2 LSAck 数据包格式

4.1.8　OSPF 邻居关系建立

在 OSPF 邻接关系建立的过程中，邻居关系的状态变化过程如下所述。

① Down（关闭）：路由器没有检测到 OSPF 邻居发送的 Hello 数据包。
② Init（初始）：路由器从运行 OSPF 协议的接口收到一个 Hello 数据包，但是邻居列表中没有自己的路由器 ID。

③ Two-way（双向）：路由器收到的 Hello 数据包中的邻居列表中包含自己的路由器 ID。如果所有其他需要的参数都匹配，则形成邻居关系。同时在多路访问的网络中将进行 DR 和 BDR 选举。

④ Exstart（准启动）：确定路由器主和从角色和 DBD 的序列号。路由器 ID 高的路由器成为主路由器。

⑤ Exchange（交换）：路由器间交换 DBD。

⑥ Loading（装载）：每个路由器将收到的 DBD 与自己的链路状态数据库进行比对，然后对缺少、丢失或者过期的 LSA 发出 LSR。每个路由器使用 LSU 对邻居的 LSR 进行应答。路由器收到 LSU 后，将进行确认。确认可以通过显示确认或者隐式确认完成。收到确认后，路由器将从重传列表中删除相应的 LSA 条目。

⑦ Full：链路状态数据库同步，建立了完全的邻接关系。

4.1.9 OSPF 运行步骤

OSPF 的运行过程分为如下 5 个步骤。

1. 建立邻居关系

所谓邻居关系是指 OSPF 路由器以交换路由信息为目的，在所选择的相邻路由器之间建立的一种关系。路由器首先发送拥有自身路由器 ID 信息的 Hello 数据包，与之相邻的路由器如果收到该 Hello 数据包，就将这个包内的路由器 ID 信息加入到自己的 Hello 数据包内的邻居列表中。如果路由器的某接口收到从其他路由器发送的含有自身路由器 ID 信息的 Hello 数据包，则它根据该接口所在网络类型确定是否可以建立邻接关系。在点对点网络中，路由器将直接和对端路由器建立邻居关系，并且该路由器将直接进入第三步操作。若为多路访问网络，该路由器将进入 DR 选举步骤。此过程完成后，路由器之间形成 Two-way 状态。

2. 选举 DR/BDR

多路访问网络通常有多个路由器，在这种状况下，OSPF 需要建立作为链路状态更新和 LSA 的中心节点，即 DR 和 BDR。DR 选举利用 Hello 数据包内的路由器 ID 和优先级（Priority）字段值来确定。优先级字段值最高的路由器成为 DR，优先级字段值次高的路由器成为 BDR。如果优先级相同，则路由器 ID 最高的路由器成为 DR。

3. 发现路由器

路由器与路由器之间首先利用 Hello 数据包中的路由器 ID 信息确认主从关系，然后主从路由器相互交换链路状态信息摘要。每个路由器对摘要信息进行分析比较，如果收到的信息有新的内容，路由器将要求对方发送完整的链路状态信息。这个过程完成后，路由器之间建立完全邻接（Full Adjacency）关系。

4. 选择适当的路由

当一个路由器拥有完整的链路状态数据库后，OSPF 路由器依据链路状态数据库的内容，独立地用 SPF 算法计算出到每一个目的网络的最优路径，并将路径存入路由表中。OSPF 利用量度（Cost）计算到目的网络的最优路径，Cost 最小者即为最优路径。

5. 维护路由信息

当链路状态发生变化时，OSPF 通过泛洪过程通告给网络上其他路由器。OSPF 路由器接收到包含有新信息的链路状态更新数据包后，将更新自己的链路状态数据库，然后用 SPF 算法重新计算路由表。在重新计算过程中，路由器继续使用旧路由表，直到 SPF 完成新的路由表计算。新的链路状态信息将被发送给其他路由器。值得注意的是，即使链路状态没有发生改变，OSPF 路由信息也会自动更新，默认时间为 30 分钟，称为链路状态刷新。

4.2 配置单区域 OSPF

4.2.1 实验 1：配置单区域 OSPF

1. 实验目的

通过本实验可以掌握：
① 启动 OSPF 路由进程的方法。
② 启用参与 OSPF 路由协议的接口的方法。
③ OSPF 度量值（Cost）的计算方法。
④ 配置 OSPF 计时器参数和计算度量值参考带宽的方法。
⑤ 点到点链路和广播多路访问链路上的 OSPF 的特征。
⑥ 修改 OSPF 接口优先级控制 DR 选举的方法。
⑦ 向 OSPF 网络注入默认路由的方法。
⑧ 查看和调试 OSPF 路由协议相关信息的方法。

2. 实验拓扑

配置单区域 OSPF 实验拓扑如图 4-9 所示。

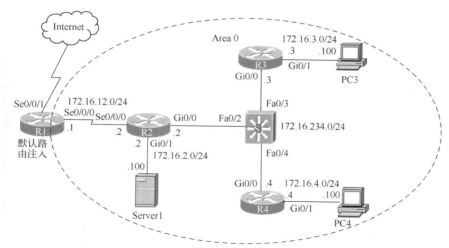

图 4-9 配置单区域 OSPF 实验拓扑

3. 实验步骤

（1）配置路由器 R1

```
R1(config)#ip route 0.0.0.0 0.0.0.0 Serial0/0/1    //配置到外网的静态默认路由
R1(config)#router ospf 1                            //启动 OSPF 进程
R1(config-router)#router-id 1.1.1.1                 //配置 OSPF 路由器 ID
R1(config-router)#auto-cost reference-bandwidth 1000
```
//修改 OSPF 计算度量值的参考带宽，单位为 Mbps，默认为 100，即 10^8。如果以太网接口的带宽为千兆，而采用默认的百兆参考带宽，计算出来的 Cost 值是 0.1，这显然是不合理的。修改参考带宽要在所有运行 OSPF 的路由器上配置，目的是确保计算度量值的参考标准一致。另外，当执行命令 **auto-cost reference-bandwidth** 时，系统也会提示如下信息：

```
       % OSPF: Reference bandwidth is changed.          //参考带宽改变
              Please ensure reference bandwidth is consistent across all routers.
              //请确保所有路由器的参考带宽一致
R1(config-router)#network 172.16.12.1 0.0.0.0 area 0
```
//配置参与 OSPF 的接口范围，该网络范围内的路由器的所有接口将激活 OSPF，通配符掩码越精确，激活接口的范围就越小，对于本实验而言，命令 **network 172.16.12.0 0.0.0.255 area 0** 的作用和命令 **network 172.16.12.1 0.0.0.0 area 0** 是一样的，只是前者的范围更大一些而已。在实际应用中，一般都用接口地址后跟通配符掩码 **0.0.0.0** 来精确匹配某一个接口地址

```
R1(config-router)#default-information originate    //向 OSPF 网络注入默认路由
```

【技术要点】

① OSPF 确定路由器 ID 遵循如下顺序。

- 最优先的是在 OSPF 进程中用命令 **router-id** 指定的路由器 ID。用 **clear ip ospf process** 命令可以使配置的新路由器 ID 生效。路由器 ID 并不是 IP 地址，只是格式和 IP 地址相同，通过该命令指定的路由器 ID 可以是任何 IP 地址格式的标识（路由器 ID 不能为 **0.0.0.0**），而该标识不一定要求接口下必须配置这样的 IP 地址。
- 如果没有在 OSPF 进程中指定路由器 ID，那么选择 IP 地址最大的环回接口的 IP 地址为路由器 ID。
- 如果没有配置环回接口，则选择最大的活动的物理接口的 IP 地址为路由器 ID。对于 ②和③，如果想要配置的新路由器 ID 生效，比如配置了更大的环回接口的 IP 地址，可行的方式是保存配置后重启路由器，或者删除 OSPF 配置，然后重新配置 OSPF。
- 建议用命令 **router-id** 来指定路由器 ID，这样可控性比较好。其次建议采用环回接口的 IP 地址作为路由器 ID，因为环回接口比较稳定。

② OSPF 路由进程 ID 的范围为 1~65535，而且只有本地含义，不同路由器的路由进程 ID 可以不同。

③ 区域 ID 为 0~4294967295 的十进制数，也可以是 IP 地址的格式。当网络区域 ID 为 0 或 0.0.0.0 时称为主干区域。

④ 在高版本的 IOS（如 IOS 15 版本以后）中使用 **network** 命令时，网络地址的后面可以跟通配符掩码，也可以跟网络掩码，IOS 系统会自动转换成通配符掩码。

⑤ 路由器上任何匹配 **network** 命令中的网络地址范围的接口都将启用 OSPF，可发送和接收 OSPF 数据包，在高版本的 IOS 中，可以在接口下通过命令 **ip ospf** *process-id* **area**

area-id 来激活参与 OSPF 的接口。

⑥ 向 OSPF 网络注入默认路由的命令为 **default-information originate [always] [cost** *metric***] [metric-type** *type***]**，各参数含义如下。

- **always**：无论路由表中是否存在默认路由，路由器都会向 OSPF 网络内注入一条默认路由。
- **cost**：指定初始度量值，默认为 20。
- **type**：指定注入 OSPF 默认路由的类型是 O E1 或 O E2，默认为 O E2。

```
R1(config)#interface serial0/0/0
R1(config-if)#ip ospf hello-interval 5    //修改 OSPF 接口 Hello 发送间隔
R1(config-if)#ip ospf dead-interval 20
//修改 OSPF 接口 Dead 时间，默认时是 Hello 数据发送间隔 4 倍。第一次修改 Hello 间隔后，Dead 时间、Wait 时间自动跟着变化，反之不可以。建立 OSPF 邻居关系路由器接口的这两个计时器值必须相同，所以 R2 的 Se0/0/0 接口也必须进行相应修改
```

（2）配置路由器 R2

```
R2(config)#router ospf 1
R2(config-router)#router-id 2.2.2.2
R2(config-router)#log-adjacency-changes
//当 OSPF 邻居关系状态变化时产生日志，是系统默认配置
R2(config-router)#auto-cost reference-bandwidth 1000
R2(config-router)#network 172.16.2.2 0.0.0.0 area 0
R2(config-router)#network 172.16.12.2 0.0.0.0 area 0
R2(config-router)#network 172.16.234.2 0.0.0.0 area 0
R2(config-router)#passive-interface gigabitEthernet0/1    //配置被动接口
R2(config)#interface serial0/0/0
R2(config-if)#ip ospf hello-interval 5
R2(config-if)#ip ospf dead-interval 20
R2(config)#interface gigabitEthernet0/0
R2(config-if)#ip ospf priority 20
//修改 OSPF 接口优先级，使得 R2 成为 DR，以太网接口 OSPF 优先级默认为 1
```

（3）配置路由器 R3

```
R3(config)#router ospf 1
R3(config-router)#router-id 3.3.3.3
R3(config-router)#auto-cost reference-bandwidth 1000
R3(config-router)#network 172.16.3.3 0.0.0.0 area 0
R3(config-router)#network 172.16.234.3 0.0.0.0 area 0
R3(config-router)#passive-interface gigabitEthernet0/1
R3(config)#interface gigabitEthernet0/0
R3(config-if)#ip ospf priority 10    //修改 OSPF 接口优先级，使得 R3 成为 BDR
```

（4）配置路由器 R4

```
R4(config)#router ospf 1
R4(config-router)#router-id 4.4.4.4
R4(config-router)#auto-cost reference-bandwidth 1000
```

```
R4(config-router)#network 172.16.4.4 0.0.0.0 area 0
R4(config-router)#network 172.16.234.4 0.0.0.0 area 0
R4(config-router)#passive-interface gigabitEthernet0/1
```

4. 实验调试

（1）查看 OSPF 邻居的基本信息

```
R2#show ip ospf neighbor
Neighbor ID     Pri   State            Dead Time   Address        Interface
3.3.3.3         10    FULL/BDR         00:00:36    172.16.234.3   GigabitEthernet0/0
4.4.4.4         1     FULL/DROTHER     00:00:37    172.16.234.4   GigabitEthernet0/0
1.1.1.1         0     FULL/ -          00:00:16    172.16.12.1    Serial0/0/0
```

以上输出表明路由器 R2 有三个 OSPF 邻居，它们的路由器 ID 分别为 1.1.1.1、3.3.3.3 和 4.4.4.4，其他参数解释如下。

① **Pri**：邻居路由器接口的 OSPF 优先级。
② **State**：当前邻居路由器的状态。
● **FULL**：表示建立了邻接关系；
● **BDR**：在 Gi0/0 接口所在的网络中，路由器 R3 是 BDR；
● **DROTHER**：在 Gi0/0 接口所在的网络中，路由器 R4 是 DROTHER，从而可以清楚地知道路由器 R2 在 Gi0/0 接口所在的网络中是 DR；
● **-**：表示点到点的链路上 OSPF 不进行 DR 和 BDR 选举。
③ **Dead Time**：重置 OSPF 邻居关系前等待的最长时间。
④ **Address**：邻居接口的 IP 地址。
⑤ **Interface**：路由器自己和邻居路由器相连的接口。

【技术要点 1】

在路由器 R1 上将 Se0/0/0 接口关闭（执行 **shutdown** 命令），然后再开启（执行 **no shutdown** 命令），通过 **debug** 命令查看 R1 和 R2 OSPF 邻接关系建立过程的详细信息如下：

```
R1#debug ip ospf adj
 *Apr 29 02:44:22.335: OSPF-1 ADJ    Se0/0/0: Route adjust notification: UP/UP
//路由调整通知
 *Apr 29 02:44:22.335: OSPF-1 ADJ    Se0/0/0: Interface going Up          //接口状态变为 Up
 *Apr 29 02:44:22.335: OSPF-1 ADJ    Se0/0/0: Interface state change to UP, new ospf state P2P  //接
口状态变为 UP，新的 OSPF 状态为点到点
 *Apr 29 02:44:22.335: OSPF-1 ADJ    Se0/0/0: 2 Way Communication to 2.2.2.2, state 2WAY
//邻居关系进入 2WAY 状态，即双向状态，表明 R1 从 R2 收到的 Hello 数据包的邻居列表中已经
看到自己路由器 ID
 *Apr 29 02:44:22.335: OSPF-1 ADJ    Se0/0/0: Nbr 2.2.2.2: Prepare dbase exchange
//和邻居路由器 R2 准备进行数据库交换
 *Apr 29 02:44:22.335: OSPF-1 ADJ    Se0/0/0: Send DBD to 2.2.2.2 seq 0x1953 opt 0x52 flag 0x7 len 32
 *Apr 29 02:44:22.335: OSPF-1 ADJ    Se0/0/0: Rcv DBD from 2.2.2.2 seq 0xF26 opt 0x52 flag 0x7 len 32  mtu 1500 state EXSTART
```

第 4 章　OSPF

```
        *Apr 29 02:44:22.335: OSPF-1 ADJ      Se0/0/0: NBR Negotiation Done. We are the SLAVE
```
//以上四行表示 R1 和 R2 通过发送和接收 First DBD 数据包确定了主从角色，本路由器为从路由器，R2 为主路由器，由于是 First DBD 数据包，所有 flag 的值为 7，表示 I、M、MS 位都为 1
```
        *Apr 29 02:44:22.335: OSPF-1 ADJ      Se0/0/0: Nbr 2.2.2.2: Summary list built, size 6
```
//构建 LSDB 摘要列表
```
        *Apr 29 02:44:22.335: OSPF-1 ADJ      Se0/0/0: Send DBD to 2.2.2.2 seq 0xF26 opt 0x52 flag 0x2 len 152
```
//以上两行表示 R1 发送 DBD 数据包，注意序列号是从主路由器 R2 继承来的 **0xF26**，而不是从路由器 R1 的序列号 **0x1953**
```
        *Apr 29 02:44:22.339: OSPF-1 ADJ      Se0/0/0: Rcv DBD from 2.2.2.2 seq 0xF27 opt 0x52 flag 0x1 len 72   mtu 1500 state EXCHANGE
```
//R1 收到主路由器发送的 DBD 数据包，主路由器 R2 负责将序列号加 1，即 **0xF27**
```
        *Apr 29 02:44:22.339: OSPF-1 ADJ      Se0/0/0: Send DBD to 2.2.2.2 seq 0xF27 opt 0x52 flag 0x0 len 32
```
//R1 向主路由器发送 DBD 数据包，从路由器 R1 保持原有从主路由器 R2 收到的序列号，即 **0xF27**，这个过程可以理解为 OSPF 的隐式确认，flag 为 0 表示没有后续的 DBD 数据包要发送
```
        *Apr 29 02:44:22.339: OSPF-1 ADJ      Se0/0/0: Exchange Done with 2.2.2.2
```
//R1 和 R2 的 LSDB 摘要信息交换过程完成
```
        *Apr 29 02:44:22.339: OSPF-1 ADJ      Se0/0/0: Send LS REQ to 2.2.2.2 length 36
```
//路由器 R1 向路由器 R2 发送 LSR，请求详细的 LSA 信息
```
        *Apr 29 02:44:22.339: OSPF-1 ADJ      Se0/0/0: Rcv LS UPD from Nbr ID 2.2.2.2 length 64 LSA count 1
```
//收到 R2 发送的 LSU 数据包，LSA 数量为 1
```
        *Apr 29 02:44:22.339: OSPF-1 ADJ      Se0/0/0: Synchronized with 2.2.2.2, state FULL
        *Apr 29 02:44:22.339: %OSPF-5-ADJCHG: Process 1, Nbr 2.2.2.2 on Serial0/0/0 from LOADING to FULL, Loading Done
```
//以上三行表明 R1 同 R2 的 LSDB 同步完成，达到 FULL 状态，形成邻接关系
```
        *Apr 29 02:44:22.339: OSPF-1 ADJ      Se0/0/0: Rcv LS REQ from 2.2.2.2 length 36 LSA count 1
        *Apr 29 02:44:42.339: OSPF-1 ADJ      Se0/0/0: Nbr 2.2.2.2: Clean-up dbase exchange
```
//以上两行表明从路由器 R2 收到 LSR，由于之前路由器 R2 已经有了 R1 的 LSA 信息（在 R2 上通过 show ip ospf database 命令查看），而且没有老化，所以清除该数据库交换信息

【技术要点2】

关于 OSPF DR 选举的细节如下。

① 在多路访问网络中，DROTHER 路由器只与 DR 和 BDR 建立邻接关系，DROTHER 路由器之间只建立邻居关系，通过命令 **show ip ospf interface** 查看运行 OSPF 接口的详细信息，如下所示。

```
R2#show ip ospf interface gigabitEthernet0/0
GigabitEthernet0/0 is up, line protocol is up      //接口状态为 up
  Internet Address 172.16.234.2/24, Area 0, Attached via Network Statement
```
//接口地址、掩码、所在区域以及接口激活 OSPF 的方式

接口激活 OSPF 的方式包括以下 2 种：
- 在路由模式下通过 network 命令激活该接口，显示为 Attached via Network Statement；
- 在接口模式下通过 ip ospf 1 area 0 命令激活该接口，显示为 Attached via Interface Enable。

```
  Process ID 1, Router ID 2.2.2.2, Network Type BROADCAST, Cost: 1
```
//OSPF 进程 ID、路由器 ID、网络类型和接口开销值

Topology-MTID	Cost	Disabled	Shutdown	Topology Name
0	10	no	no	Base

　　　　　　　　//以上两行显示了OSPF多拓扑路由的信息，包括多拓扑ID和名称等信息
　　Transmit Delay is 1 sec, **State** DR, **Priority** 20
//传输延时为1秒（可通过命令 **ip ospf transmit-delay** 命令修改）、接口状态为DR，接口优先级为20
　　　　Designated Router (ID) 2.2.2.2, Interface address 172.16.234.2
　　　　//DR的路由器ID以及接口地址
　　　　Backup Designated router (ID) 3.3.3.3, Interface address 172.16.234.3
　　　　//BDR的路由器ID以及接口地址
　　　　Timer intervals configured, **Hello** 10, **Dead** 40, **Wait** 40, **Retransmit** 5
　　　　//Hello 数据包发送周期、Dead 时间和 Wait 时间以及重传时间，其中 **Wait** 表示在选举 DR 和 BDR 之前等待邻居路由器 Hello 数据包的最长时间；**Retransmit** 表示在没有得到确认的情况下，重传 OSPF 数据包等待的时间，默认为5秒，可以通过 **ip ospf retransmit-interval** 命令来修改
　　　　　oob-resync timeout 40　　　　　　//oob（Out-of-band）同步超时时间
　　　　　Hello due in 00:00:03　　　　　　 //距离下一个Hello数据包到达的时间
　　　　（此处省略部分输出）
　　　　Neighbor Count is 2, **Adjacent** neighbor count is 2
　　　　//R2 是 DR，有2个邻居，并且与2个邻居全部形成邻接关系
　　　　　Adjacent with neighbor 3.3.3.3　(Backup Designated Router)　//与BDR形成邻接关系
　　　　　Adjacent with neighbor 4.4.4.4　　//与DROTHER形成邻接关系
　　　　Suppress hello for 0 neighbor(s)　　　　//接口没有抑制Hello数据包

② DR 和 BDR 有自己的组播地址 **224.0.0.6**。

③ DR 和 BDR 的选举是以独立的网络为基础的，也就是说 DR 和 BDR 选举是一个路由器的接口特性，而不是整个路由器的特性。例如，1台路由器可以是某个多路访问网络的 DR，也可以是另外多路访问网络的 BDR。

④ DR 选举的原则：
- 首要因素是时间，该时间就是启用 OSPF 路由协议接口下的 **wait** 时间。如果通过命令 **ip ospf hello-interval** 调整 Hello 间隔，或者通过命令 **ip ospf dead-interval** 调整 dead 时间，**wait** 时间都会跟着自动调整。这是等待进行 DR 选举的最长时间，过了该时间，将开始 DR 选举过程。
- 其次，如果在 **wait** 时间内网络内所有接口都加入 OSPF 进程，或者重新选举，则比较接口优先级（范围为 0~255），优先级最高的路由器被选举为 DR。默认情况下，多路访问网络的接口优先级为1，点到点网络接口优先级为0。修改接口优先级的命令是 **ip ospf priority** *priority*，如果接口的优先级被配置为0，那么该接口将不参与 DR 选举。
- 如果接口优先级相同，最后比较路由器 ID，路由器 ID 最高的路由器被选举为 DR。

⑤ DR 选举是非抢占的，但下列情况可以重新选举 DR：
- 路由器重新启动或者删除 OSPF 配置，然后再重新配置 OSPF 进程。
- 参与选举的路由器执行 **clear ip ospf process** 命令。
- DR 出现故障。
- 将 OSPF 接口的优先级设置为0。

⑥ 仅当 DR 出现故障，BDR 才会接管 DR 的任务，然后选举新的 BDR；如果 BDR 出现故障，将选举新的 BDR。所以大家经常说的 DR 选举实际上先选举的是 BDR，然后把 BDR 提升为 DR，接着再选出新的 BDR。所以在一个需要选举 DR 的网络中不可能出现有 BDR，而没有 DR 的情况；但是，一个网络有 DR，没有 BDR 是可能的。

【技术扩展】

与基于目的地址路由和基于策略路由方案不同，多拓扑路由（Multi-Topology Routing，MTR）根据流量类型动态选择路由，其最大的特点是能够将不同的流量（如语音流量）分开，使它们在不同的拓扑中根据网络结构独立进行选路和转发。多拓扑路由并不改变网络原有物理拓扑，而是在原有物理拓扑上划分多个逻辑子拓扑，这样可以给某些对链路质量敏感的业务流量划分出专门的网络通道，每个业务独立地进行路径选择和路由转发，防止因网络流量过大对某些业务造成影响。多拓扑路由知识已经超过本书的范围，更多的信息请读者在 Cisco 官网上查询，相关链接地址如下：

http://www.cisco.com/c/en/us/td/docs/ios/12_2sr/12_2srb/feature/guide/srmtrdoc.html#wp1054132
http://www.cisco.com/c/en/us/td/docs/ios/12_2sr/12_2srb/feature/guide/srmtrdoc.html
可以通过命令 show ip ospf 1 topology-info topology base 查看多拓扑路由具体信息。

【技术要点 3】

OSPF 邻居关系建立非常复杂，受到多种因素的限制，不能建立邻居关系的常见原因如下所述。

① OSPF 接口 Hello 间隔或 Dead 时间不同。同一链路上的 Hello 间隔和 Dead 间隔必须相同才能建立 OSPF 邻居关系。
② 建立 OSPF 邻居关系的两个接口所在区域 ID 不同。
③ OSPF 接口网络类型不同或者特殊区域（如 stub，nssa 等）的区域类型不匹配。
④ OSPF 身份验证类型或验证信息不一致。
⑤ 建立 OSPF 邻居关系的路由器 ID 相同。
⑥ 接口下应用了拒绝 OSPF 数据包的 ACL，如 **access-list 100 deny ospf any any**。
⑦ OSPF 链路上的 MTU 不匹配，可以通过命令 **ip ospf mtu-ignore** 忽略 MTU 检测。
⑧ 在 OSPF 多路访问网络中，接口的子网掩码不同或者网络类型不同。

（2）查看 IP 路由协议配置和统计信息

```
R2#show ip protocols | begin Routing Protocol is "ospf 1"
Routing Protocol is "ospf 1"          //当前路由器运行的 OSPF 进程 ID
  Outgoing update filter list for all interfaces is not set
  Incoming update filter list for all interfaces is not set
//以上两行表明入方向和出方向都没有配置分布列表
  Router ID 2.2.2.2                    //OSPF 路由器 ID
  Number of areas in this router is 1. 1 normal 0 stub 0 nssa
//本路由器接口所属的区域数量和区域类型
  Maximum path: 4   //默认支持等价路径数目，最大为 32 条（IOS 15.7，路由器为 2911）
  Routing for Networks:
  172.16.2.2 0.0.0.0 area 0
      172.16.12.1 0.0.0.0 area 0
      172.16.234.2 0.0.0.0 area 0
//以上四行表明路由模式下配置的通过 network 命令激活 OSPF 进程的接口匹配的范围及所在的区域
  Passive Interface(s):
```

```
        GigabitEthernet0/1
//以上两行表示 OSPF 进程中配置的被动接口
Routing Information Sources:
    Gateway         Distance        Last Update
    3.3.3.3         110             03:45:36
    4.4.4.4         110             03:45:11
    1.1.1.1         110             03:34:49
//以上五行表明 OSPF 路由信息源、管理距离和最后一次更新时间
Distance: (default is 110)   //OSPF 路由协议默认的管理距离为 110
```

（3）查看 OSPF 进程相关信息

```
R2#show ip ospf  //查看 OSPF 进程 ID、路由器 ID、OSPF 区域信息、OSPF 进程启动和持续时间
以及 SPF 算法执行次数等
Routing Process "ospf 1" with ID 2.2.2.2    //OSPF 路由进程 ID 和路由器 ID
  Start time: 02:08:33.472, Time elapsed: 04:26:14.164   //OSPF 进程启动时间和持续的时间
  Supports only single TOS(TOS0) routes    //只支持简单 TOS 路由
  Supports opaque LSA    //支持不透明 LSA
  Supports Link-local Signaling (LLS)    //支持链路本地信令
  Supports area transit capability    //支持区域传输能力
  Supports NSSA (compatible with RFC 3101)   //支持 NSSA（Not-So-Stubby Area）
  Supports Database Exchange Summary List Optimization (RFC 5243)
//支持 OSPF 数据库交换汇总列表优化
  Event-log enabled, Maximum number of events: 1000, Mode: cyclic
//启用事件日志功能，事件最大数量为 1000，模式为循环方式
  Router is not originating router-LSAs with maximum metric
  Initial SPF schedule delay 5000 msecs    //初始 SPF 运算计划延时
  Minimum hold time between two consecutive SPFs 10000 msecs
//防止路由器持续运行 SPF 算法的保留时间
  Maximum wait time between two consecutive SPFs 10000 msecs
//路由器运行完一次 SPF 算法后，等待 10 秒才再次运行该算法
（此处省略部分输出）
  Reference bandwidth unit is 1000 mbps   //计算 OSPF 接口开销的参考带宽
    Area BACKBONE(0)    //主干区域
        Number of interfaces in this area is 3   //区域 0 运行 OSPF 的接口的数量
        Area has no authentication   //区域没有启用验证
        SPF algorithm last executed 02:45:39.652 ago   //距离上次运行 SPF 算法的时间
        SPF algorithm executed 7 times   //SPF 算法运行的次数
        Area ranges are    //区域间路由汇总
        （此处省略部分输出）
```

（4）查看运行 OSPF 的接口信息摘要

```
R2#show ip ospf interface brief
Interface    PID    Area    IP Address/Mask      Cost    State    Nbrs F/C
Gi0/1        1      0       172.16.2.2/24        1       DR       0/0
Se0/0/0      1      0       172.16.12.2/24       647     P2P      1/1
Gi0/0        1      0       172.16.234.2/24      1       DR       2/2
```

以上输出显示运行 OSPF 接口的名字、进程 ID、接口所在区域、接口地址和掩码、接口 Cost

值、状态、邻居和邻接的数量。注意 Gi 0/1 所在接口没有 OSPF 邻居，自己成为 DR 的角色。

（5）查看 OSPF 链路状态数据库的信息

```
R2#show ip ospf database
        OSPF Router with ID (2.2.2.2) (Process ID 1)    //OSPF 路由器 ID 和进程 ID
              Router Link States (Area 0)    //类型 1 的 LSA
Link ID        ADV Router       Age        Seq#           Checksum       Link count
1.1.1.1        1.1.1.1          1710       0x8000000D     0x00CB95       2
2.2.2.2        2.2.2.2          1709       0x80000011     0x0005E8       4
3.3.3.3        3.3.3.3          1095       0x8000000E     0x003C9A       2
4.4.4.4        4.4.4.4          899        0x8000000F     0x0011BA       2
              Net Link States (Area 0)    //类型 2 的 LSA
Link ID        ADV Router       Age        Seq#           Checksum
172.16.234.2   2.2.2.2          1135       0x8000000A     0x008BCB
//172.16.234.2 是 OSPF DR 路由器接口的 IP 地址
              Type-5 AS External Link States   //类型 5 的 LSA
Link ID        ADV Router       Age        Seq#           Checksum       Tag
0.0.0.0        1.1.1.1          826        0x8000000A     0x000B9A       1
```

以上输出的是 R2 的区域 0 的链路状态数据库的信息，如果在 R1、R3 和 R4 上也查看 OSPF 链路状态数据库，会发现 R1～R4 的链路状态数据库是相同的。在以上输出中，标题行的含义解释如下。

① **Link ID**：标识每个 LSA。

② **ADV Router**：通告链路状态信息的路由器 ID。

③ **Age**：老化时间，范围为 0～60 分钟，老化时间达到 60 分钟的 LSA 条目将被从 LSDB 中删除。

④ **Seq#**：序列号，范围为 0x80000001～0x7fffffff，序列号越大，LSA 越新。为了确保 LSDB 同步，OSPF 每隔 30 分钟对链路状态刷新一次，序列号会自动加 1，刷新信息如下所示。

```
00:55:59: OSPF: Build router LSA for area 0, router ID 2.2.2.2, seq 0x80000007, process 1
01:29:33: OSPF: Build router LSA for area 0, router ID 2.2.2.2, seq 0x80000008, process 1
02:02:55: OSPF: Build router LSA for area 0, router ID 2.2.2.2, seq 0x80000009, process 1
```

⑤ **Checksum**：校验和，计算除 Age 字段以外的所有字段，LSA 存放在 LSDB 中，每 5 分钟进行一次校验，以确保该 LSA 没有损坏。

⑥ **Link count**：通告路由器在本区域内的链路数目。

⑦ **Tag**：外部路由的标识，默认为 1。

（6）查看路由表中 OSPF 路由信息

以下输出全部省略路由代码部分。

```
① R1#show ip route ospf
     172.16.0.0/16 is variably subnetted, 2 subnets, 2 masks
O       172.16.2.0/24 [110/649] via 172.16.12.2, 00:37:01, Serial0/0/0
O       172.16.3.0/24 [110/649] via 172.16.12.2, 00:37:01, Serial0/0/0
O       172.16.4.0/24 [110/649] via 172.16.12.2, 00:37:01, Serial0/0/0
O       172.16.234.0/24 [110/648] via 172.16.12.2, 00:37:01, Serial0/0/0
```

② R2#**show ip route ospf**
Gateway of last resort is **172.16.12.1** to network 0.0.0.0
O*E2   0.0.0.0/0 **[110/1]** via 172.16.12.1, 00:39:25, Serial0/0/0   //OSPF 默认路由
        172.16.0.0/16 is variably subnetted, 8 subnets, 2 masks
O   172.16.3.0/24 [110/2] via 172.16.234.3, 04:56:11, GigabitEthernet0/0
O   172.16.4.0/24 [110/2] via 172.16.234.4, 04:55:46, GigabitEthernet0/0
③ R3#**show ip route ospf**
Gateway of last resort is **172.16.234.2** to network 0.0.0.0
O*E2   0.0.0.0/0 **[110/1]** via 172.16.234.2, 00:40:27, GigabitEthernet0/0
        172.16.0.0/16 is variably subnetted, 7 subnets, 2 masks
O   172.16.2.0/24 [110/2]   via 172.16.234.2, 04:56:46, GigabitEthernet0/0
O   172.16.4.0/24   [110/2] via 172.16.234.4, 04:56:46, GigabitEthernet0/0
O   172.16.12.0/24 [110/648] via 172.16.234.2, 00:40:27, GigabitEthernet0/0
④ R4#**show ip route ospf**
Gateway of last resort is **172.16.234.2** to network 0.0.0.0
O*E2   0.0.0.0/0 **[110/1]** via 172.16.234.2, 00:42:29, GigabitEthernet0/0
        172.16.0.0/16 is variably subnetted, 7 subnets, 2 masks
O   172.16.2.0/24 [110/2]   via 172.16.234.2, 04:58:49, GigabitEthernet0/0
O   172.16.3.0/24   [110/2]   via 172.16.234.3, 04:58:49, GigabitEthernet0/0
O   172.16.12.0/24 [110/648] via 172.16.234.2, 00:42:29, GigabitEthernet0/0

以上①、②、③和④输出结果表明 OSPF 路由协议的管理距离是 **110**，同一个区域内通过 OSPF 路由协议学到的路由条目用代码 **O** 表示。路由器 R2、R3 和 R4 的路由表的输出表明，在 R1 上通过命令 **default-information originate** 确实可以向 OSPF 网络注入 1 条默认路由，路由类型为 **O E2**，默认度量值为 **1**。OSPF 接口 Cost 值计算公式为 10^9 / 接口带宽（bps），然后取整，而路由的度量值计算是路由传递方向的所有链路入口的 Cost 之和，环回接口的 Cost 值默认为 1。路由器 R4 的路由条目 **172.16.12.0/24** 的度量值为 **648**，计算过程如下：路由条目 **172.16.12.0/24** 到路由器 R4 经过的入接口包括路由器 R2 的 Se0/0/0 接口和路由器 R4 的 Gi0/0 接口，$10^9/1544000+10^9/10^9=648$。当然也可以直接通过命令 **ip ospf cost** *cost* 配置接口的 Cost 值，并且它是优先计算的 Cost 值。

【提示】

① 可以通过命令 **show interfaces** 查看接口的带宽。
② 如果网络中使用了环回接口，则其他路由器学到其所在网络的 OSPF 路由条目的掩码长度默认都是 32 位（不管接口的掩码长度实际是多少位），这是环回接口的特性，要使得路由条目的掩码长度和环回接口的掩码长度保持一致，解决的办法是在环回接口下修改网络类型为 **Point-to-Point**，接口下配置命令为 **ip ospf network point-to-point**。

4.2.2 实验 2：配置 OSPF 验证

1. 实验目的

通过本实验可以掌握：
① OSPF 验证的类型和意义。
② 配置基于区域的 OSPF 简单口令验证和 MD5 验证方法。

第 4 章 OSPF

③ 配置基于链路的 OSPF 简单口令验证和 MD5 验证方法。

2. 实验拓扑

配置 OSPF 验证实验拓扑如图 4-10 所示。

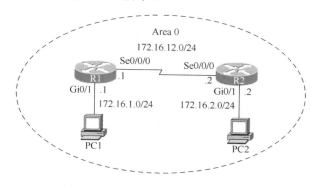

图 4-10　配置 OSPF 验证实验拓扑

3. 实验步骤

（1）配置路由器 R1

```
R1(config)#router ospf 1
R1(config-router)#router-id 1.1.1.1
R1(config-router)#auto-cost reference-bandwidth 1000
R1(config-router)#network 172.16.1.1 0.0.0.0 area 0
R1(config-router)#network 172.16.12.1 0.0.0.0 area 0
R1(config-router)#passive-interface gigabitEthernet0/1
R1(config-router)#area 0 authentication            //区域 0 启用简单口令验证
R1(config)#interface Serial0/0/0
R1(config-if)#ip ospf authentication-key cisco     //配置简单口令验证密码
```

（2）配置路由器 R2

```
R2(config)#router ospf 1
R2(config-router)#router-id 2.2.2.2
R2(config-router)#auto-cost reference-bandwidth 1000
R2(config-router)#network 172.16.2.2 0.0.0.0 area 0
R2(config-router)#network 172.16.12.2 0.0.0.0 area 0
R2(config-router)#passive-interface gigabitEthernet0/1
R2(config-router)#area 0 authentication
R2(config)#interface Serial0/0/0
R2(config-if)#ip ospf authentication-key cisco
```

4. 实验调试

（1）查看运行 OSPF 的接口信息

```
R2#show ip ospf interface Serial0/0/0
```

Internet Address 172.16.12.2/24, Area 0, Attached via Network Statement
　Process ID 1, Router ID 2.2.2.2, Network Type POINT_TO_POINT, Cost: 647
　（此处省略部分输出）
Simple password authentication enabled　　//接口启用了简单口令验证

（2）查看 OSPF 的进程信息

R1#**show ip ospf**
　Routing Process "ospf 1" with ID 1.1.1.1
（此处省略部分输出）
　　　Area BACKBONE(0)
Number of interfaces in this area is 2
　　　Area has simple password authentication　　//区域 0 启用简单口令验证
(此处省略部分输出)

（3）查看接口接收（IN）和发送（OUT）的 OSPF 数据包

R2#**debug ip ospf packet**
　　*Apr 29 **08:34:14**.377: OSPF-1 PAK: Se0/0/0: **OUT**: 172.16.12.2->**224.0.0.5**: **ver**:2 **type**:1 **len**:48 rid:2.2.2.2 area:0.0.0.0 chksum:E693 **auth**:1
　　*Apr 29 **08:34:16**.989: OSPF-1 PAK: Se0/0/0: **IN**: 172.16.12.1->**224.0.0.5**: **ver**:2 **type**:1 **len**:48 **rid**:1.1.1.1 **area**:0.0.0.0 **chksum**:E693 **auth**:1

以上输出表明运行 OSPF 的接口 Se0/0/0 接收和发送验证类型为 1 的 Hello 数据包，各部分含义如下。

① **ver**：OSPF 版本。
② **type**：OSPF 数据包类型，1 为 Hello 数据包，2 为 DBD 数据包，3 为 LSR 数据包，4 为 LSU 数据包，5 为 LSAck 数据包。
③ **len**：数据包长度，单位为字节。
④ **rid**：路由器 ID。
⑤ **area**：区域 ID。
⑥ **chksum**：校验和。
⑦ **auth**：验证类型，0 代表不进行验证，1 代表简单口令验证，2 代表 MD5 验证。

（4）启动验证

① 如果 R2 的区域 0 没有启动验证，而 R1 的区域 0 启动简单口令验证，则提示验证类型不匹配，在 R2 上执行命令 **debug ip ospf event**，提示信息如下：

　　05:37:00: OSPF: Rcv pkt from 172.16.12.1, Serial0/0/0 : **Mismatch Authentication type**. Input packet specified **type 1**, we use **type 0**

② 如果 R1 和 R2 的区域 0 都启动简单口令验证，但是 R2 的接口下没有配置密码或密码错误，则会提示验证 key 不匹配，提示信息如下：

　　05:38:57: OSPF: Rcv pkt from 172.16.12.1, Serial0/0/0 : **Mismatch Authentication Key - Clear Text**

③ 如果不想针对整个区域验证，而是针对某些关键的链路进行验证，则路由进程下不需要配置基于区域的验证，基于链路的简单口令验证配置步骤如下。

第 4 章 OSPF

A. 配置路由器 R1。

```
R1(config)#interface Serial0/0/0
R1(config-if)#ip ospf authentication              //接口启用简单口令验证
R1(config-if)#ip ospf authentication-key cisco    //配置简单口令验证密码
```

B. 配置路由器 R2。

```
R2(config)#interface Serial0/0/0
R2(config-if)#ip ospf authentication
R2(config-if)#ip ospf authentication-key cisco
```

（5）配置 OSPF MD5 验证

删除上述 OSPF 简单口令验证的配置，保留其他配置，然后配置 OSPF MD5 区域验证。

① 配置路由器 R1。

```
R1(config)#router ospf 1
R1(config-router)#area 0 authentication message-digest    //区域 0 启用 MD5 验证
R1(config)#interface Serial0/0/0
R1(config-if)#ip ospf message-digest-key 1 md5 cisco      //配置验证 key id 及密钥
```

② 配置路由器 R2。

```
R2(config)#router ospf 1
R2(config-router)#area 0 authentication message-digest
R2(config)#interface Serial0/0/0
R2(config-if)#ip ospf message-digest-key 1 md5 cisco
```

（6）验证 OSPF MD5 区域验证

① R2#**show ip ospf interface Serial0/0/0**
```
Serial0/0/0 is up, line protocol is up
  Internet Address 172.16.12.2/24, Area 0, Attached via Network Statement
  Process ID 1, Router ID 2.2.2.2, Network Type POINT_TO_POINT, Cost: 647
    （此处省略部分输出）
  Cryptographic authentication enabled
    Youngest key id is 1
```
//以上两行输出信息表明该接口启用了 MD5 验证，而且使用密钥 id 为 1 进行验证。OSPF 的 MD5 验证允许在接口下配置多个密钥，从而可以保证方便和安全地改变密钥。**Youngest key id** 和配置顺序有关，最后一次配置的就是 **Youngest key id**，和 id 数字本身大小没有关系

② R2#**show ip ospf**
（此处省略部分输出）
```
    Area BACKBONE(0)
        Number of interfaces in this area is 2
        Area has message digest authentication    //区域 0 采用 MD5 验证
        （此处省略部分输出）
```

③ R2#**debug ip ospf packet**
```
*Apr 29 08:59:17.733: OSPF-1 PAK  : Se0/0/0:  IN: 172.16.12.1->224.0.0.5: ver:2 type:1 len:48 rid:1.1.1.1 area:0.0.0.0 chksum:0 auth:2 keyid:1 seq:0x5723
*Apr 29 08:59:24.045: OSPF-1 PAK  : Se0/0/0:  OUT: 172.16.12.2->224.0.0.5: ver:2 type:1 len:48 rid:2.2.2.2 area:0.0.0.0 chksum:0 auth:2 keyid:1 seq:0x5723
```

以上输出表明运行 OSPF 的接口 Se0/0/0 收发验证类型为 2、keyid 为 1、序列号为 0x5723 的 Hello 数据包。

④ 如果 R1 的区域 0 启动 MD5 验证，而 R2 的区域 0 启动简单口令验证，则提示验证类型不匹配，在 R2 上执行命令 **debug ip ospf event**，提示信息如下：

```
05:51:30: OSPF: Rcv pkt from 172.16.12.1, Serial0/0/0 : Mismatch Authentication type. Input packet specified type 2, we use type 1
```

⑤ 如果 R1 和 R2 的区域 0 都启动 MD5 认证，但是 R2 的接口下没有配置 key id 和密码或密码错误，则提示验证 key 不匹配，提示信息如下：

```
05:54:04: OSPF: Rcv pkt from 172.16.12.1, Serial0/0/0 : Mismatch Authentication Key - Message Digest Key 1
```

（7）基于链路配置 MD5 验证

如果不想针对整个区域验证，而是针对某些关键的链路进行验证，则不需要在区域上启用 MD5 验证，基于链路的 MD5 验证配置步骤如下：

① 配置路由器 R1。

```
R1(config)#interface Serial0/0/0
R1(config-if)#ip ospf authentication message-digest    //接口启用 MD5 验证
R1(config-if)#ip ospf message-digest-key 1 md5 cisco   //配置 key id 及密钥
```

② 配置路由器 R2。

```
R2(config)#interface Serial0/0/0
R2(config-if)#ip ospf authentication message-digest
R2(config-if)#ip ospf message-digest-key 1 md5 cisco
```

【技术要点】

① OSPFv2 定义 3 种验证类型：0—表示不进行验证，是默认的类型；1—表示采用简单口令验证；2—表示采用 MD5 验证。

② 区域验证相当于开启了运行 OSPFv2 协议的所有接口的验证，而链路验证只是针对某个链路开启验证，OSPFv2 链路验证优于区域验证。

【知识扩展】

接下来的实验重点研究多个 key id 对 OSPF 的 MD5 验证的影响。

① 在上面的实验中，在路由器 R1 上配置 key id 2：

```
R1(config)#interface Serial0/0/0
R1(config-if)#ip ospf message-digest-key 2 md5 cisco2
```

在两台路由器上开启 debug ip ospf events，显示信息如下：

```
R1#
02:12:20: OSPF: Send with key 1
02:12:20: OSPF: Send hello to 224.0.0.5 area 0 on Serial0/0/0 from 172.16.12.1
```

第 4 章 OSPF

```
02:12:20: OSPF: Send with key 2
02:12:20: OSPF: Send hello to 224.0.0.5 area 0 on Serial0/0/0 from 172.16.12.1
R2#
02:13:44: OSPF: Send with youngest Key 1
02:13:44: OSPF: Send hello to 224.0.0.5 area 0 on Serial0/0/0 from 172.16.12.2
02:16:51: OSPF: Rcv pkt from 172.16.12.1, Serial0/0/0 : Mismatch Authentication Key - No message
digest key 2 on interface
```

以上输出信息表明当配置最新的 key id 时，R1 将新的和老的 key id 都发过去。由于 R2 上并没有配置 key id 2，所以仍旧发送 key id 1，并用老的 key id 1 进行匹配。此时虽然 R2 提示 **Mismatch Authentication Key**，但是不影响 OSPF 邻居关系的建立和路由信息的交换。此时在路由器 R1 上，可以看到的 OSFP 接口的信息如下：

```
R1#show ip ospf interface Serial0/0/0
Serial0/0/0 is up, line protocol is up
  Internet Address 172.16.12.1/24, Area 0
  （此处省略部分输出）
  Message digest authentication enabled   //启用 MD5 验证
     Youngest key id is 2    //最新的 key id 为 2
     Rollover in progress, 1 neighbor(s) using the old key(s):   //用老的 key 建立邻居关系
```

② 继续上面的实验，在路由器 R2 上配置 key id 2，但是密钥不同，如下：

```
R2(config)#interface serial 0/0/0
R2(config-if)#ip ospf message-digest-key 2 md5 cisco22
```

在两台路由器上开启 debug ip ospf events，显示信息如下：

```
R1#
02:44:02: OSPF: Rcv pkt from 172.16.12.2, Serial0/0/0 : Mismatch Authentication Key - Message
Digest Key 2
02:44:09: OSPF: Send with youngest Key 2
02:44:09: OSPF: Send hello to 224.0.0.5 area 0 on Serial0/0/0 from 172.16.12.1
R2#
02:43:54: OSPF: Rcv pkt from 172.16.12.1, Serial0/0/0 : Mismatch Authentication Key - Message
Digest Key 2
02:43:56: OSPF: Send with youngest Key 2
02:43:56: OSPF: Send hello to 224.0.0.5 area 0 on Serial0/0/0 from 172.16.12.2
```

以上输出信息表明当路由器 R2 配置最新的 key id 2 后，双方都检测到有 key id 2，就不再发 key id 1 了，由于 key id 2 的 key 不同，所以提示 **Mismatch Authentication Key - Message Digest Key 2**，此时不能建立邻居关系。

```
R1# show ip ospf interface Serial0/0/0
Serial0/0/0 is up, line protocol is up
（此处省略部分输出）
  Neighbor Count is 0, Adjacent neighbor count is 0    //没有形成邻居关系
  Suppress hello for 0 neighbor(s)
  Message digest authentication enabled
    Youngest key id is 2    //最新的 key id 为 2
```

③ 继续上面的实验，在路由器 R2 上删除 key id 2，如下：

```
R2(config)#interface serial 0/0/0
R2(config-if)#no ip ospf message-digest-key 2 md5 cisco22
```

在两台路由器上开启 **debug ip ospf events**,显示信息如下:

```
R1#
03:14:00: OSPF: Rcv hello from 2.2.2.2 area 0 from Serial0/0/0 172.16.12.2
03:14:00: OSPF: Send
immediate hello to nbr 2.2.2.2, src address 172.16.12.2, on Serial0/0/0
03:14:00: OSPF: Send with youngest Key 2
03:14:00: OSPF: Send hello to 224.0.0.5 area 0 on Serial0/0/0 from 172.16.12.1
R2#
03:14:41: OSPF: Send with youngest Key 1
03:14:41: OSPF: Rcv pkt from 172.16.12.1, Serial0/0/0 : Mismatch Authentication Key - No message digest key 2 on interface
```

以上输出信息表明,当路由器 R2 删除最新的 key id 2 后,R1 一直发送 key id 2。由于 R2 没有最新的 key id 了,此时发送次新的 key id 1 并且提示 **Mismatch Authentication Key**。

④ 清除上面实验接口下的验证信息,按照如表 4-3 所示的顺序重新配置 key id,该表给出了 key id 和 key。

表 4-3 key id 和 key

路由器 R1		路由器 R2	
key id 1	ccna	key id 2	ccnp
key id 2	ccnp	key id 1	ccna
key id 3	ccie	key id 3	ccie

路由器上配置如下:

```
R1(config)#interface Serial0/0/0
R1(config-if)#ip ospf message-digest-key 1 md5 ccna
R1(config-if)#ip ospf message-digest-key 2 md5 ccnp
R1(config-if)#ip ospf message-digest-key 3 md5 ccie

R2(config)#interface Serial0/0/0
R2(config-if)#ip ospf message-digest-key 2 md5 ccnp
R2(config-if)#ip ospf message-digest-key 1 md5 ccna
R2(config-if)#ip ospf message-digest-key 3 md5 ccie
```

配置好后,R1 和 R2 会发送最新的 key id 3。现在在路由器 R1 和 R2 上分别删除 key id 3,路由器 R1 和 R2 显示的信息如下:

```
R1#
03:35:21: OSPF: Send with youngest Key 2
03:35:21: OSPF: Send hello to 224.0.0.5 area 0 on Serial0/0/0 from 172.16.12.1
R2#
03:35:18: OSPF: Send with youngest Key 1
03:35:18: OSPF: Send hello to 224.0.0.5 area 0 on Serial0/0/0 from 172.16.12.2
```

以上输出信息表明虽然路由器 R1 和 R2 发送的 **youngest Key** 不同,但是邻接关系可以建立而且路由正常，这是因为彼此发送的 **youngest Key** 在对方都能找到相应 key id 及正确的密钥。

4.3 配置多区域 OSPF

4.3.1 实验 3：配置多区域 OSPF

1. 实验目的

通过本实验可以掌握：
① 在路由器上启动 OSPF 路由进程的方法。
② 启用参与 OSPF 路由协议接口的方法。
③ OSPF LSA 的类型和特征。
④ OSPF 不同类型的路由器功能。
⑤ OSPF 链路状态数据库的特征和含义。
⑥ E1 路由和 E2 路由的区别。
⑦ 查看和调试 OSPF 路由协议相关信息。

2. 实验拓扑

配置多区域 OSPF 实验拓扑如图 4-11 所示。

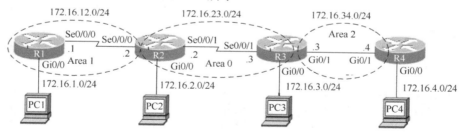

图 4-11 配置多区域 OSPF 实验拓扑

3. 实验步骤

路由器 R4 的 Gi0/0 接口不激活 OSPF，通过路由重分布进入 OSPF 网络。

（1）配置路由器 R1

```
R1(config)#router ospf 1
R1(config-router)#router-id 1.1.1.1
R1(config-router)#auto-cost reference-bandwidth 1000
R1(config-router)#network 172.16.1.1 0.0.0.0 area 1
R1(config-router)#network 172.16.12.1 0.0.0.0 area 1
R1(config-router)#passive-interface gigabitEthernet0/0
```

（2）配置路由器 R2

```
R2(config)#router ospf 1
```

```
R2(config-router)#router-id 2.2.2.2
R2(config-router)#auto-cost reference-bandwidth 1000
R2(config-router)#network 172.16.12.2 0.0.0.0 area 1
R2(config-router)#network 172.16.23.2 0.0.0.0 area 0
R2(config-router)#network 172.16.2.2 0.0.0.0 area 0
R2(config-router)#passive-interface gigabitEthernet0/0
```

（3）配置路由器 R3

```
R3(config)#router ospf 1
R3(config-router)#router-id 3.3.3.3
R3(config-router)#auto-cost reference-bandwidth 1000
R3(config-router)#network 172.16.23.3 0.0.0.0 area 0
R3(config-router)#network 172.16.3.3 0.0.0.0 area 0
R3(config-router)#network 172.16.34.3 0.0.0.0 area 2
R3(config-router)#passive-interface gigabitEthernet0/0
```

（4）配置路由器 R4

```
R4(config)#router ospf 1
R4(config-router)#router-id 4.4.4.4
R4(config-router)#auto-cost reference-bandwidth 1000
R4(config-router)#network 172.16.34.4 0.0.0.0 area 2
R4(config-router)#redistribute connected subnets    //将直连路由重分布进 OSPF
```

4. 实验调试

（1）查看路由表

① R1#show ip route ospf
```
     172.16.0.0/16 is variably subnetted, 9 subnets, 2 masks
O IA    172.16.2.0/24 [110/648] via 172.16.12.2, 00:56:46, Serial0/0/0
O IA    172.16.3.0/24 [110/1295] via 172.16.12.2, 00:55:29, Serial0/0/0
O E2    172.16.4.0/24 [110/20] via 172.16.12.2, 00:00:25, Serial0/0/0
O IA    172.16.23.0/24 [110/1294] via 172.16.12.2, 00:57:11, Serial0/0/0
O IA    172.16.34.0/24 [110/1295] via 172.16.12.2, 00:55:19, Serial0/0/0
```
② R2#show ip route ospf
```
     172.16.0.0/16 is variably subnetted, 10 subnets, 2 masks
O       172.16.1.0/24 [110/648] via 172.16.12.1, 00:59:09, Serial0/0/0
O       172.16.3.0/24 [110/648] via 172.16.23.3, 00:57:06, Serial0/0/1
O E2    172.16.4.0/24 [110/20] via 172.16.23.3, 00:02:03, Serial0/0/1
O IA    172.16.34.0/24 [110/648] via 172.16.23.3, 00:56:56, Serial0/0/10
```
③ R3#show ip route ospf
```
     172.16.0.0/16 is variably subnetted, 7 subnets, 2 masks
172.16.0.0/16 is variably subnetted, 10 subnets, 2 masks
O IA    172.16.1.0/24 [110/1295] via 172.16.23.2, 00:57:52, Serial0/0/1
O       172.16.2.0/24 [110/648] via 172.16.23.2, 00:57:52, Serial0/0/1
O E2    172.16.4.0/24 [110/20] via 172.16.34.4, 00:02:55, GigabitEthernet0/1
O IA    172.16.12.0/24 [110/1294] via 172.16.23.2, 00:57:52, Serial0/0/1
```
④ R4#show ip route ospf

```
172.16.0.0/16 is variably subnetted, 9 subnets, 2 masks
O IA      172.16.1.0/24 [110/1296] via 172.16.34.3, 00:03:36, GigabitEthernet0/1
O IA      172.16.2.0/24 [110/649] via 172.16.34.3, 00:03:36, GigabitEthernet0/1
O IA      172.16.3.0/24 [110/2] via 172.16.34.3, 00:03:36, GigabitEthernet0/1
O IA      172.16.12.0/24 [110/1295] via 172.16.34.3, 00:03:36, GigabitEthernet0/1
O IA      172.16.23.0/24 [110/648] via 172.16.34.3, 00:03:36, GigabitEthernet0/1
```

以上①、②、③和④输出表明，路由表中带有 **O** 的路由是区域内的路由，路由表中带有 **O IA** 的路由是区域间的路由，路由表中带有 **O E2** 的路由是外部自治系统网络被重分布到 OSPF 中的路由。这就是为什么在 R4 上要进行路由重分布，就是为了构造自治系统外部的路由。此外，在路由器 R1、R2 和 R3 上的 **O E2** 路由条目 **172.16.4.0/24** 的度量值都是 20，这是 **O E2** 路由的特征，当把外部自治系统的路由重分布到 OSPF 中时，如果不设置度量值和类型，默认度量值是 20，默认路由类型为 **O E2**。OSPF 的外部路由分为类型 1（在路由表中用代码 E1 表示）和类型 2（在路由表中用代码 E2 表示），它们计算路由度量值的方式不同，具体如下所述。

① 类型 1（E1）：外部路径成本＋数据包在 OSPF 网络所经过各链路成本。

② 类型 2（E2）：外部路径成本，即 ASBR 上的默认设置。

OSPF 的选路原则的优先顺序是：O ＞ O IA ＞ O E1 ＞ O E2。

在重分布时可以通过 metric-type 参数设置是类型 1 或 2，也可以通过 metric 参数设置外部路径成本，默认为 20。

（2）查看 OSPF 的链路状态数据库

```
show ip ospf database
① R1#show ip ospf database
       OSPF Router with ID (1.1.1.1) (Process ID 1)
          Router Link States (Area 1)    //区域1 类型1 的LSA
Link ID       ADV Router      Age       Seq#           Checksum    Link count
1.1.1.1       1.1.1.1         456       0x80000004     0x00A9E5    3
2.2.2.2       2.2.2.2         375       0x80000003     0x00B3A6    2
          Summary Net Link States (Area 1)   //区域1 类型3 的LSA
Link ID       ADV Router      Age       Seq#           Checksum
172.16.2.2    2.2.2.2         375       0x80000002     0x00A5CC
172.16.3.3    2.2.2.2         375       0x80000002     0x002E32
172.16.23.0   2.2.2.2         375       0x80000002     0x0065EA
172.16.34.0   2.2.2.2         375       0x80000002     0x0089AB
          Summary ASB Link States (Area 1)   //区域1 类型4 的LSA
Link ID       ADV Router      Age       Seq#           Checksum
4.4.4.4       2.2.2.2         130       0x80000002     0x00BF43
          Type-5 AS External Link States     //类型5 的LSA
Link ID       ADV Router      Age       Seq#           Checksum Tag
172.16.4.0    4.4.4.4         946       0x80000002     0x0065CB 0
② R2#show ip ospf database
       OSPF Router with ID (2.2.2.2) (Process ID 1)
          Router Link States (Area 0)    //区域0 类型1 的LSA
Link ID       ADV Router      Age       Seq#           Checksum Link count
2.2.2.2       2.2.2.2         412       0x80000003     0x006208    3
3.3.3.3       3.3.3.3         246       0x80000004 0x00EE73 3
```

Summary Net Link States (Area 0)　　//区域 0 类 3 的 LSA

Link ID	ADV Router	Age	Seq#	Checksum
172.16.1.1	2.2.2.2	412	0x80000002	0x00580C
172.16.12.0	2.2.2.2	412	0x80000002	0x00DE7C
172.16.34.0	3.3.3.3	246	0x80000002	0x00CD73

Summary ASB Link States (Area 0)　　//区域 0 类型 4 的 LSA

Link ID	ADV Router	Age	Seq#	Checksum
4.4.4.4	3.3.3.3	246	0x80000002	0x00040B

Router Link States (Area 1)　　//区域 1 类型 1 的 LSA

Link ID	ADV Router	Age	Seq#	Checksum	Link count
1.1.1.1	1.1.1.1	495	0x80000004	0x00A9E5	3
2.2.2.2	2.2.2.2	416	0x80000003	0x00B3A6	2

Summary Net Link States (Area 1)　　//区域 1 类型 3 的 LSA

Link ID	ADV Router	Age	Seq#	Checksum
172.16.2.2	2.2.2.2	416	0x80000002	0x00A5CC
172.16.3.3	2.2.2.2	416	0x80000002	0x002E32
172.16.23.0	2.2.2.2	416	0x80000002	0x0065EA
172.16.34.0	2.2.2.2	416	0x80000002	0x0089AB

Summary ASB Link States (Area 1)　　//区域 1 类型 4 的 LSA

Link ID	ADV Router	Age	Seq#	Checksum
4.4.4.4	2.2.2.2	170	0x80000002	0x00BF43

Type-5 AS External Link States　　//类型 5 的 LSA

Link ID	ADV Router	Age	Seq#	Checksum	Tag
172.16.4.0	4.4.4.4	986	0x80000002	0x0065CB	0

③ R3#**show ip ospf database**

OSPF Router with ID (3.3.3.3) (Process ID 1)

Router Link States (Area 0)　　//区域 0 类型 1 的 LSA

Link ID	ADV Router	Age	Seq#	Checksum	Link count
2.2.2.2	2.2.2.2	455	0x80000003	0x006208	3
3.3.3.3	3.3.3.3	287	0x80000004	0x00EE73	3

Summary Net Link States (Area 0)　　//区域 0 类型 3 的 LSA

Link ID	ADV Router	Age	Seq#	Checksum
172.16.1.1	2.2.2.2	455	0x80000002	0x00580C
172.16.12.0	2.2.2.2	455	0x80000002	0x00DE7C
172.16.34.0	3.3.3.3	287	0x80000002	0x00CD73

Summary ASB Link States (Area 0)　　//区域 0 类型 4 的 LSA

Link ID	ADV Router	Age	Seq#	Checksum
4.4.4.4	3.3.3.3	287	0x80000002	0x00040B

Router Link States (Area 2)　　//区域 2 类型 1 的 LSA

Link ID	ADV Router	Age	Seq#	Checksum	Link count
3.3.3.3	3.3.3.3	287	0x80000003	0x00CA4E	1
4.4.4.4	4.4.4.4	214	0x80000002	0x006FA4	1

Net Link States (Area 2)　　//区域 2 类型 2 的 LSA

Link ID	ADV Router	Age	Seq#	Checksum
172.16.34.3	3.3.3.3	526	0x80000001	0x0059D6

Summary Net Link States (Area 2)　　//区域 2 类型 3 的 LSA

Link ID	ADV Router	Age	Seq#	Checksum
172.16.1.1	3.3.3.3	291	0x80000002	0x00D778
172.16.2.2	3.3.3.3	291	0x80000002	0x002539
172.16.3.3	3.3.3.3	291	0x80000002	0x0072F9
172.16.12.0	3.3.3.3	291	0x80000002	0x005EE8

172.16.23.0	3.3.3.3	291	0x80000002	0x004705

      Type-5 AS External Link States //类型 5 的 LSA

Link ID	ADV Router	Age	Seq#	Checksum Tag
172.16.4.0	4.4.4.4	1027	0x80000002	0x0065CB 0

④ R4#**show ip ospf database**

      OSPF Router with ID (4.4.4.4) (Process ID 1)

      Router Link States (Area 2) //区域 2 类型 1 的 LSA

Link ID	ADV Router	Age	Seq#	Checksum Link count
3.3.3.3	3.3.3.3	347	0x80000003	0x00CA4E 2
4.4.4.4	4.4.4.4	268	0x80000002	0x006FA4 2

      Net Link States (Area 2) //区域 2 类型 2 的 LSA

Link ID	ADV Router	Age	Seq#	Checksum
172.16.34.3	3.3.3.3	526	0x80000001	0x0059D6

      Summary Net Link States (Area 2) //区域 2 类型 3 的 LSA

Link ID	ADV Router	Age	Seq#	Checksum
172.16.1.1	3.3.3.3	347	0x80000002	0x00D778
172.16.2.2	3.3.3.3	347	0x80000002	0x002539
172.16.3.3	3.3.3.3	347	0x80000002	0x0072F9
172.16.12.0	3.3.3.3	347	0x80000002	0x005EE8
172.16.23.0	3.3.3.3	347	0x80000002	0x004705

      Type-5 AS External Link States //类型 5 的 LSA

Link ID	ADV Router	Age	Seq#	Checksum Tag
172.16.4.0	4.4.4.4	1082	0x80000002	0x0065CB 0

  以上①、②、③和④输出结果包含了区域 1 的 LSA 类型 1、3、4 的链路状态信息，区域 0 的 LSA 类型 1、3、4 的链路状态信息，区域 2 的 LSA 类型 1、2、3 的链路状态信息以及 LSA 类型 5 的链路状态信息。同时可以看到路由器 R1 和 R2 的区域 1 的链路状态数据库完全相同，路由器 R2 和 R3 的区域 0 的链路状态数据库完全相同，路由器 R3 和 R4 的区域 2 的链路状态数据库完全相同。

【技术要点】

  ① 相同区域内的路由器具有相同的链路状态数据库，只是当为虚链路时略有不同。

  ② 命令 **show ip ospf database** 所显示的内容并不是数据库中存储的关于每条 LSA 的全部信息，而仅仅是 LSA 的头部信息。要查看 LSA 的全部信息，该命令后面还要跟详细的参数。显示 OSPF 链路状态数据库中 LSA 全部信息命令如表 4-4 所示。

表 4-4 显示 OSPF 链路状态数据库中 LSA 全部信息命令

命　　令	含　　义
show ip ospf database router	查看 OSPF LSDB 中类型 1 的 LSA 信息
show ip ospf database network	查看 OSPF LSDB 中类型 2 的 LSA 信息
show ip ospf database summary	查看 OSPF LSDB 中类型 3 的 LSA 信息
show ip ospf database asbr-summary	查看 OSPF LSDB 中类型 4 的 LSA 信息
show ip ospf database external	查看 OSPF LSDB 中类型 5 的 LSA 信息
show ip ospf database nssa-external	查看 OSPF LSDB 中类型 7 的 LSA 信息

③ 如果路由器所在区域过大，或者没有正确配置路由汇总，将收到大量的 LSA，这样会消耗大量的 CPU 和内存资源。可以使用命令 **max-lsa** *maximum-number* [*threshold-percentage*] [*warning-only*] [*ignore-time minutes*] [*ignore-count number*] [*reset-time minutes*]配置 OSPF LSDB 过载保护来防止这种问题发生，各参数的含义如下。

- *maximum-number*：非自身产生的 LSA 的最大数量；
- *threshold-percentage*：当超过最大数量的百分比之后，错误日志信息将被记录，默认为 75%；
- *warning-only*：当超过最大可接收的 LSA 数目之后，只是显示告警信息；
- *ignore-time*：当超过最大可接收的 LSA 数目之后，忽略所有 OSPF 邻居的时间，默认为 5 分钟；
- *ignore-count*：当超过最大可接收的 LSA 数目之后，OSPF 进程被连续置于忽略状态的次数，默认为 5 次；
- *reset-time*：定义了 *ignore-count* 被重置为 0 的时间，默认为 10 分钟。

当非自身产生的 LSA 的最大数量超过阈值时，产生的信息如下：

```
06:21:18: %OSPF-4-OSPF_MAX_LSA_THR: Threshold for maximum number of non self-generated LSA has been reached "ospf 1" - 0 LSAs
06:21:18: %OSPF-4-OSPF_MAX_LSA: Maximum number of non self-generated LSA has been exceeded "ospf 1" - 2 LSAs
```

（3）查看路由器 R4 的 OSPF 路由进程信息

```
R4#show ip ospf
Routing Process "ospf 1" with ID 4.4.4.4
（此处省略部分输出）
  It is an autonomous system boundary router    //路由器 R4 是一台 ASBR 路由器
  Redistributing External Routes from,
    connected with metric mapped to 20, includes subnets in redistribution
//以上两行表明该路由器将直连路由重分布到 OSPF 进程中，度量值为 20，重分布时携带子网信息
（此处省略部分输出）
```

（4）查看到达 ABR 和 ASBR 的 OSPF 路由表

```
R1#show ip ospf border-routers
OSPF Process 1 internal Routing Table    //OSPF 进程 1 内部路由表
Codes: i - Intra-area route, I - Inter-area route
//i 表示区域内的路由，I 表示区域间的路由
I 4.4.4.4 [1295] via 172.16.12.2, Serial0/0/0, ASBR, Area 1, SPF 3
//4.4.4.4 表示 ASBR 的路由器 ID，1295 表示到达 ASBR 的开销，Area 1 表示到达 ASBR 的路由是从区域 1 学到的，SPF 3 表示 SPF 的计算次数为 3
i 2.2.2.2 [647] via 172.16.12.2, Serial0/0/0, ABR, Area 1, SPF 3
```

4.3.2　实验 4：配置 OSPF 路由手工汇总

1. 实验目的

通过本实验可以掌握：

① 路由汇总的目的。
② OSPF 区域间路由汇总的配置和调试方法。
③ 外部自治系统路由汇总的配置和调试方法。

2. 实验拓扑

配置 OSPF 路由手工汇总实验拓扑如图 4-12 所示。

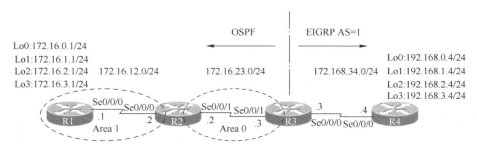

图 4-12 配置 OSPF 路由手工汇总实验拓扑

3. 实验步骤

路由器 R1、R2 和 R3 之间运行 OSPF，路由器 R3 和 R4 之间运行 EIGRP，路由器 R1 上的四个环回接口是为在路由器 R2 上进行区域间路由汇总准备的，路由器 R4 上的四个环回接口是为在路由器 R3 上进行外部路由汇总准备的。由于路由器 R3 同时运行 OSPF 和 EIGRP 两种路由协议，是边界路由器，所以要完成路由双向重分布，以便实现网络中路由信息的共享。

（1）配置路由器 R1

```
R1(config)#router ospf 1
R1(config-router)#router-id 1.1.1.1
R1(config-router)#auto-cost reference-bandwidth 1000
R1(config-router)#network 172.16.0.0 0.0.3.255 area 1
R1(config-router)#network 172.16.12.1 0.0.0.0 area 1
```

（2）配置路由器 R2

```
R2(config)#router ospf 1
R2(config-router)#router-id 2.2.2.2
R2(config-router)#auto-cost reference-bandwidth 1000
R2(config-router)#network 172.16.12.2 0.0.0.0 area 1
R2(config-router)#network 172.16.23.2 0.0.0.0 area 0
R2(config-router)#area 1 range 172.16.0.0 255.255.252.0
//配置 OSPF 区域间路由汇总，该路由条目初始度量值是被汇总路由条目度量值中最小的
```

（3）配置路由器 R3

```
R3(config)#router ospf 1
R3(config-router)#router-id 3.3.3.3
R3(config-router)#auto-cost reference-bandwidth 1000
```

```
R3(config-router)#network 172.16.23.3 0.0.0.0 area 0
R3(config-router)#summary-address 192.168.0.0 255.255.252.0
//配置 OSFP 外部自治系统路由汇总
R3(config-router)#redistribute eigrp 1 subnets    //将 EIGRP 路由重分布到 OSPF 中
R3(config)#router eigrp 1
R3(config-router)#network 192.168.34.3    0.0.0.0
R3(config-router)#redistribute ospf 1 metric 1000 100 255 1 1500
//将 OSPF 路由重分布到 EIGRP 中
```

(4)配置路由器 R4

```
R4(config)#router eigrp 1
R4(config-router)#network 192.168.34.4 0.0.0.0
R4(config-router)#network 192.168.0.0 0.0.3.255
```

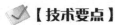【技术要点】

① OSPF 区域间路由汇总必须在 ABR 上完成。
② OSPF 外部路由汇总必须在 ASBR 上完成。

4. 实验调试

(1)在路由器 R2 上查看路由表

```
R2#show ip route ospf
     172.16.0.0/16 is variably subnetted, 8 subnets, 3 masks
O        172.16.1.1/32 [110/648] via 172.16.12.1, 00:05:53, Serial0/0/0
O        172.16.0.0/22 is a summary, 00:05:53, Null0
O        172.16.0.1/32 [110/648] via 172.16.12.1, 00:05:53, Serial0/0/0
O        172.16.3.1/32 [110/648] via 172.16.12.1, 00:05:53, Serial0/0/0
O        172.16.2.1/32 [110/648] via 172.16.12.1, 00:05:53, Serial0/0/0
O E2 192.168.34.0/24 [110/20] via 172.16.23.3, 00:03:26, Serial0/0/1
O E2 192.168.0.0/22 [110/20] via 172.16.23.3, 00:00:50, Serial0/0/1
```

以上输出表明路由器 R2 对 R1 的四条环回接口所在网络的路由汇总后,会产生一条指向 Null0 的路由,是为了避免路由环路,可以通过 **no discard-route** 命令阻止向路由表中添加这条路由;同时收到经路由器 R3 汇总的 EIGRP 的汇总路由,因为是重分布进来的外部自治系统的路由,所以路由代码为 **O E2**。

(2)在 R3 上查看路由表

```
R3#show ip route ospf
     172.16.0.0/16 is variably subnetted, 5 subnets, 3 masks
O IA     172.16.12.0/24 [110/1294] via 172.16.23.2, 00:05:12, Serial0/0/1
O IA     172.16.0.0/22 [110/1295] via 172.16.23.2, 00:05:12, Serial0/0/1
O        192.168.0.0/22 is a summary, 00:01:16, Null0
```

以上输出表明路由器 R3 对 R4 的四条环回接口所在网络的 EIGRP 路由汇总后,会产生一条指向 Null0 的路由,也是为了避免路由环路;同时收到经路由器 R2 汇总的路由,由于是区域间路由汇总,所以路由代码为 **O IA**。

4.3.3 实验 5：配置 OSPF 末节区域和完全末节区域

1. 实验目的

通过本实验可以掌握：
① 末节区域的条件。
② 末节区域的特征。
③ 完全末节区域的特征。
④ 末节区域的配置和调试方法。
⑤ 完全末节区域的配置和调试方法。

2. 实验拓扑

配置 OSPF 末节区域和完全末节区域实验拓扑如图 4-13 所示。

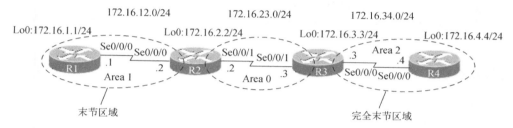

图 4-13 配置 OSPF 末节区域和完全末节区域实验拓扑

3. 实验步骤

本实验中，在路由器 R2 上将环回接口 0 重分布进入 OSPF 区域，用来构造 5 类的 LSA。所以路由器 R2 既是 ABR，又是 ASBR。将区域 1 配置成末节区域，将区域 2 配置成完全末节区域。

（1）配置路由器 R1

```
R1(config)#router ospf 1
R1(config-router)#router-id 1.1.1.1
R1(config-router)#auto-cost reference-bandwidth 1000
R1(config-router)#network 172.16.1.1 0.0.0.0 area 1
R1(config-router)#network 172.16.12.1 0.0.0.0 area 1
R1(config-router)#area 1 stub    //将区域 1 配置为末节区域
```

（2）配置路由器 R2

```
R2(config)#router ospf 1
R2(config-router)#router-id 2.2.2.2
R2(config-router)#auto-cost reference-bandwidth 1000
R2(config-router)#network 172.16.12.2 0.0.0.0 area 1
R2(config-router)#network 172.16.23.2 0.0.0.0 area 0
R2(config-router)#redistribute connected subnets    //重分布直连路由
R2(config-router)#area 1 stub
```

（3）配置路由器 R3

```
R3(config)#router ospf 1
R3(config-router)#router-id 3.3.3.3
R3(config-router)#auto-cost reference-bandwidth 1000
R3(config-router)#network 172.16.23.3 0.0.0.0 area 0
R3(config-router)#network 172.16.3.3 0.0.0.0 area 0
R3(config-router)#network 172.16.34.3 0.0.0.0 area 2
R3(config-router)#area 2 stub no-summary
//将区域 2 配置为完全末节区域，no-summary 参数阻止区域间的路由进入末节区域，所以叫完全末节区域。只需在 ABR 上启用本参数即可
```

（4）配置路由器 R4

```
R4(config)#router ospf 1
R4(config-router)#router-id 4.4.4.4
R4(config-router)#auto-cost reference-bandwidth 1000
R4(config-router)#network 172.16.4.4 0.0.0.0 area 2
R4(config-router)#network 172.16.34.4 0.0.0.0 area 2
R4(config-router)#area 2 stub
```

【技术要点】

末节和完全末节区域需要满足如下条件：
① 区域只有一个出口。
② 区域不需要作为虚链路的转接区域。
③ 区域内没有 ASBR。
④ 区域不是主干区域。
⑤ 对于末节区域，Hello 数据包中 Option 字段的 E 置位，建立邻居时该位必须匹配。

4. 实验调试

（1）在路由器 R1 上查看路由表

```
R1#show ip route ospf
     172.16.0.0/16 is variably subnetted, 6 subnets, 2 masks
O IA    172.16.34.0/24 [110/1941] via 172.16.12.2, 00:03:53, Serial0/0/0
O IA    172.16.23.0/24 [110/1294] via 172.16.12.2, 00:03:53, Serial0/0/0
O IA    172.16.4.4/32 [110/1942] via 172.16.12.2, 00:00:44, Serial0/0/0
O IA    172.16.3.3/32 [110/1295] via 172.16.12.2, 00:03:53, Serial0/0/0
O*IA 0.0.0.0/0 [110/648] via 172.16.12.2, 00:03:53, Serial0/0/0
```

以上的输出表明 R2 重分布进 OSPF 区域的环回接口所在网络的路由并没有在 R1 的路由表中出现，说明末节区域不接收类型 5 的 LSA；同时末节区域 1 的 ABR 路由器 R2 自动向该区域内传播度量值为 1 的默认路由；末节区域可以接收区域间路由。

（2）在路由器 R4 上查看路由表

```
R4#show ip route ospf
O*IA 0.0.0.0/0 [110/648] via 172.16.34.3, 00:05:41, Serial0/0/0
```

以上输出表明在完全末节区域 2 不接收外部路由和区域间路由，只有区域内的路由和一条由 ABR 路由器 R3 向该区域注入的默认路由。

（3）查看路由器 R1 的 OSPF 路由进程信息

```
R1#show ip ospf 1
（此处省略部分输出）
    Area 1
        Number of interfaces in this area is 2 (1 loopback)
        It is a stub area        //区域 1 是末节区域
        Area has no authentication
    （此处省略部分输出）
```

（4）查看路由器 R3 的 OSPF 路由进程信息

```
R3#show ip ospf 1
（此处省略部分输出）
    Area 2
        Number of interfaces in this area is 1
        It is a stub area, no summary LSA in this area
//区域 2 是末节区域，并且没有 summary LSA（即 3 类 LSA）
        generates stub default route with cost 1
//向该区域注入初始度量值为 1 的默认路由，可以在路由模式下通过命令 area area-id
default-cost cost 修改，如 area 2 default-cost cost 10
        Area has no authentication
    （此处省略部分输出）
```

4.3.4 实验 6：配置 OSPF NSSA 区域

1. 实验目的

通过本实验可以掌握：
① NSSA 区域的特征。
② NSSA 区域的配置和调试方法。
③ NSSA 产生默认路由的方法。

2. 实验拓扑

配置 OSPF NSSA 区域实验拓扑如图 4-14 所示。

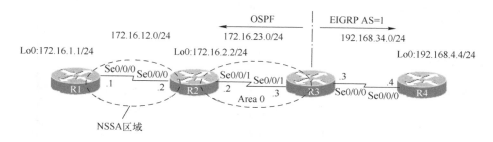

图 4-14　配置 OSPF NSSA 区域实验拓扑

3. 实验步骤

本实验中，在路由器 R1 上将环回接口 0 重分布进入 OSPF 区域，用来验证 5 类的 LSA 在 NSSA 区域的传递方式。在路由器 R2 上将环回接口 0 重分布进入 OSPF 区域。

（1）配置路由器 R1

```
R1(config)#router ospf 1
R1(config-router)#router-id 1.1.1.1
R1(config-router)#auto-cost reference-bandwidth 1000
R1(config-router)#network 172.16.12.1 0.0.0.0 area 1
R1(config-router)#redistribute connected subnets
R1(config-router)#area 1 nssa          //将区域 1 配置为 NSSA 区域
```

（2）配置路由器 R2

```
R2(config)#router ospf 1
R2(config-router)#router-id 2.2.2.2
R2(config-router)#auto-cost reference-bandwidth 1000
R2(config-router)#network 172.16.12.2 0.0.0.0 area 1
R2(config-router)#network 172.16.23.2 0.0.0.0 area 0
R2(config-router)#redistribute connected metric-type 1 subnets
R2(config-router)#area 1 nssa
```

（3）配置路由器 R3

```
R3(config)#router ospf 1
R3(config-router)#router-id 3.3.3.3
R3(config-router)#auto-cost reference-bandwidth 1000
R3(config-router)#network 172.16.23.3 0.0.0.0 area 0
R3(config-router)#redistribute eigrp 1 subnets
//将 EIGRP 路由重分布到 OSPF 区域
R3(config)#router eigrp 1
R3(config-router)#redistribute ospf 1 metric 1000 100 255 1 1500
//将 OSPF 路由重分布到 EIGRP 中
R3(config-router)#network 192.168.34.3 0.0.0.0
```

（4）配置路由器 R4

```
R4(config)#router eigrp 1
R4(config-router)#network 192.168.4.4 0.0.0.0
R4(config-router)#network 192.168.34.4 0.0.0.0
```

【技术要点】

① NSSA 区域是对 stub 区域的扩展，会阻止类型 4 和类型 5 的 LSA 进入。
② NSSA 区域打破了 stub 区域的规则，区域中可以存在 ASBR。
③ NSSA 区域的 Hello 数据包中的 Option 字段的 P 置位，告诉 ABR 将类型 7 的 LSA 转成类型 5 的 LSA。建立邻居时该位必须匹配。

④ NSSA 区域的 ABR 会将类型 7 的 LSA 转成类型 5 的 LSA，传播到其他区域，而 ABR 也同时成为了 ASBR。

4. 实验调试

（1）在 R1 上查看路由表

```
R1#show ip route ospf
      172.16.0.0/24 is subnetted, 4 subnets
O IA    172.16.23.0 [110/1294] via 172.16.12.2, 00:00:51, Serial0/0/0
O N1    172.16.2.0 [110/667] via 172.16.12.2, 00:00:51, Serial0/0/0
```

以上输出表明区域间的路由是可以进入 NSSA 区域的，但是在 R1 的路由表中并没有出现在 R3 上把 EIGRP 重分布进来的路由，因此说明 LSA 类型 5 的外部路由不能在 NSSA 区域中传播，ABR 也没有能力把类型 5 的 LSA 转成类型 7 的 LSA。但是在 ABR 路由器 R2 上重分布的路由可以进入 NSSA 区域，路由代码为 **O N1**。

【技术要点】

① 如果不想在 NSSA 区域中出现区域间的路由，只需要在 ABR 路由器上配置 NSSA 区域时加上 **no-summary** 参数即可。这时 ABR 也会自动向 NSSA 区域注入一条 **O* IA** 的默认路由，配置如下：

```
R2(config-router)#area 1 nssa no-summary
```

查看 R1 的路由表如下：

```
R1#show ip route ospf
      172.16.0.0/24 is subnetted, 3 subnets
O N1    172.16.2.0 [110/667] via 172.16.12.2, 00:02:30, Serial0/0/0
O*IA 0.0.0.0/0 [110/648] via 172.16.12.2, 00:00:02, Serial0/0/0
```

② 本实验中，如果在路由器 R2 配置 NSSA 时没有加 **no-summary** 参数，那么对路由器 R1 来讲，EIGRP 部分的路由是不可达的，为了解决此问题，在路由器 R2 上配置 NSSA 区域时加上 **default-information-originate** 参数即可，此时 ABR 路由器 R2 会向 NSSA 区域注入一条 **O*N2** 的默认路由，配置如下：

```
R2(config-router)#area 1 nssa default-information-originate
R1#show ip route ospf
      172.16.0.0/24 is subnetted, 4 subnets
O IA    172.16.23.0 [110/1294] via 172.16.12.2, 00:00:13, Serial0/0/0
O N1    172.16.2.0 [110/667] via 172.16.12.2, 00:03:43, Serial0/0/0
O*N2 0.0.0.0/0 [110/1] via 172.16.12.2, 00:00:08, Serial0/0/0
```

以上输出表明 **default-information-originate** 参数会向 NSSA 区域注入一条 **O*N2** 的默认路由，但是并没有像 **no-summary** 参数那样阻止区域间的路由进入 NSSA 区域。

③ 如果在 R2 上同时配置 **no-summary** 参数和 **default-information-originate** 参数，如下所示：

```
R2(config-router)#area 1 nssa default-information-originate   no-summary
```

则 R1 的路由表如下：

```
R1#show ip route ospf
    172.16.0.0/24 is subnetted, 3 subnets
O N1       172.16.2.0 [110/667] via 172.16.12.2, 00:05:15, Serial0/0/0
O*IA 0.0.0.0/0 [110/648] via 172.16.12.2, 00:00:02, Serial0/0/0
```

以上输出表明 **O IA** 的路由优于 **O N2** 的路由进入路由表。

④ 如果在 R2 路由器上不想让重分布的路由进入 NSSA 区域，但是可以进入区域 0，则需要配置 **no-redistribution** 参数，如下所示：

```
R2(config-router)#area 1 nssa   no-summary no-redistribution
```

则 R1 的路由表如下：

```
R1#show ip route ospf
O*IA 0.0.0.0/0 [110/782] via 172.16.12.2, 00:01:26, Serial0/0/0
```

以上输出表明 **no-redistribution** 参数可以阻止重分布路由进入 NSSA 区域，但是不会阻止进入区域 0，在路由器 R3 上查看路由表，如下所示：

```
R3#show ip route ospf
    172.16.0.0/24 is subnetted, 4 subnets
O IA       172.16.12.0 [110/1562] via 172.16.23.2, 00:11:54, Serial0/0/1
O E2       172.16.1.0 [110/20] via 172.16.23.2, 00:11:38, Serial0/0/1
O E1       172.16.2.0 [110/801] via 172.16.23.2, 00:12:18, Serial0/0/1
```

（2）在 R2 上查看路由表

```
R2#show ip route ospf
    172.16.0.0/24 is subnetted, 4 subnets
O N2       172.16.1.0 [110/20] via 172.16.12.1, 00:07:43, Serial0/0/0
O E2 192.168.4.0/24 [110/20] via 172.16.23.3, 00:07:43, Serial0/0/1
O E2 192.168.34.0/24 [110/20] via 172.16.23.3, 00:07:43, Serial0/0/1
```

以上输出表明 NSSA 区域的路由代码为 **O N2** 或 **O N1**。

（3）在 R3 上查看路由表

```
R3#show ip route ospf
    172.16.0.0/24 is subnetted, 4 subnets
O IA       172.16.12.0 [110/1294] via 172.16.23.2, 00:17:00, Serial0/0/1
O E2       172.16.1.0 [110/20] via 172.16.23.2, 00:16:44, Serial0/0/1
O E1       172.16.2.0 [110/667] via 172.16.23.2, 00:17:24, Serial0/0/1
```

以上输出表明 ABR 路由器 R2 将类型 7 的 LSA 转换成类型 5 的 LSA，并且扩散到路由器 R3。

（4）在 R2 上查看 LSDB

```
R2#show ip ospf database
```

```
            OSPF Router with ID (2.2.2.2) (Process ID 1)
                Router Link States (Area 0)
Link ID         ADV Router      Age         Seq#            Checksum        Link count
2.2.2.2         2.2.2.2         798         0x8000000C      0x0063CD        2
3.3.3.3         3.3.3.3         1731        0x80000002      0x001423        2
               Summary Net Link States (Area 0)
Link ID         ADV Router      Age         Seq#            Checksum
172.16.12.0     2.2.2.2         794         0x80000001      0x00E07B
                Router Link States (Area 1)
Link ID         ADV Router      Age         Seq#            Checksum        Link count
1.1.1.1         1.1.1.1         799         0x8000000D      0x00A8A5        2
2.2.2.2         2.2.2.2         794         0x8000000B      0x004FFA        2
               Summary Net Link States (Area 1)
Link ID         ADV Router      Age         Seq#            Checksum
0.0.0.0         2.2.2.2         270         0x80000002      0x00FA32
            Type-7 AS External Link States (Area 1)
Link ID         ADV Router      Age         Seq#            Checksum        Tag
0.0.0.0         2.2.2.2         489         0x80000001      0x00D0D8        0
172.16.1.0      1.1.1.1         4           0x80000002      0x007990        0
                Type-5 AS External Link States
Link ID         ADV Router      Age         Seq#            Checksum        Tag
172.16.1.0      2.2.2.2         779         0x80000001      0x00F11F        0
172.16.2.0      2.2.2.2         820         0x80000001      0x008CCD        0
192.168.4.0     3.3.3.3         1585        0x80000001      0x00AF77        0
192.168.34.0    3.3.3.3         1681        0x80000003      0x0060A6        0
```

以上输出结果表明,路由器 R2 的区域 1 的 LSDB 中类型 7 的 LSA 和类型 5 的 LSA 并存。同时存在两条 **0.0.0.0** 的 LSA,一条为类型 3,另一条为类型 7,但是路由器 R1 会把类型 3 的 LSA 放到路由表中。

(5)查看路由器 R2 的 OSPF 路由进程信息

```
R2#show ip ospf 1
    (此处省略部分输出)
Area 1
        Number of interfaces in this area is 1
        It is a NSSA area, no redistribution into this area
        //区域 1 是 NSSA 区域,在 R2 上重分布的路由没有进入该区域
        Perform type-7/type-5 LSA translation
        //完成 LSA 类型 7 到 LSA 类型 5 的转换
        generates NSSA default route with cost 1
        //产生初始度量值为 1 的 NSSA 默认路由
        Area has no authentication
    (此处省略部分输出)
```

4.3.5　实验 7:配置虚链路

在实际 OSPF 网络中,可能会存在主干区域不连续或者某一个区域与主干区域物理不直接相连的情况,在这两种情况下,可以通过虚链路技术来解决。

1. 实验目的

通过本实验可以掌握：
① 虚链路的特征。
② 虚链路的配置和调试方法。

2. 实验拓扑

虚链路配置实验拓扑如图 4-15 所示。

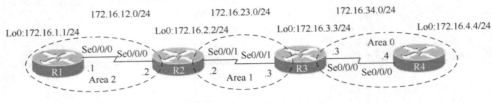

图 4-15　虚链路配置实验拓扑

3. 实验步骤

（1）配置路由器 R1

```
R1(config)#router ospf 1
R1(config-router)#router-id 1.1.1.1
R1(config-router)#auto-cost reference-bandwidth 1000
R1(config-router)#network 172.16.1.1 0.0.0.0 area 2
R1(config-router)#network 172.16.12.1 0.0.0.0 area 2
```

（2）配置路由器 R2

```
R2(config)#router ospf 1
R2(config-router)#router-id 2.2.2.2
R2(config-router)#auto-cost reference-bandwidth 1000
R2(config-router)#area 0 authentication message-digest
R2(config-router)#area 1 virtual-link 3.3.3.3 message-digest-key 1 md5 cisco
//配置虚链路及验证方式
R2(config-router)#network 172.16.2.2 0.0.0.0 area 1
R2(config-router)#network 172.16.23.2 0.0.0.0 area 1
R2(config-router)#network 172.16.12.2 0.0.0.0 area 2
```

【技术要点】

① 一条虚链路不能穿越多个区域，也不能穿越末节区域和完全末节区域。
② 虚链路一般用于建立临时的连接或者作为备份连接。
③ 虚链路必须配置在两台 ABR 之间，在配置虚链路时，命令 virtual-link 后一定要互指对方的路由器 ID。虚链路属于区域 0，所以在进行区域 0 验证时，不要忘记在配置虚链路的两台 ABR 路由器上也要配置验证方式。

④ LSDB 中通过虚链路学到的 LSA 不会老化。

（3）配置路由器 R3

```
R3(config)#interface Serial0/0/0
R3(config-if)#ip ospf message-digest-key 1 md5 cisco
R3(config)#router ospf 1
R3(config-router)#router-id 3.3.3.3
R3(config-router)#auto-cost reference-bandwidth 1000
R3(config-router)#area 0 authentication message-digest
R3(config-router)#area 1 virtual-link 2.2.2.2 message-digest-key 1 md5 cisco
R3(config-router)#network 172.16.3.3 0.0.0.0 area 0
R3(config-router)#network 172.16.34.3 0.0.0.0 area 0
R3(config-router)#network 172.16.23.3 0.0.0.0 area 1
```

（4）配置路由器 R4

```
R4(config)#interface Serial0/0/0
R4(config-if)#ip ospf message-digest-key 1 md5 cisco
R4(config)#router ospf 1
R4(config-router)#router-id 4.4.4.4
R4(config-router)#auto-cost reference-bandwidth 1000
R4(config-router)#area 0 authentication message-digest
R4(config-router)#network 172.16.4.4 0.0.0.0 area 0
R4(config-router)#network 172.16.34.4 0.0.0.0 area 0
```

4. 实验调试

（1）查看路由器 R4 的路由表

```
R4#show ip route ospf
     172.16.0.0/16 is variably subnetted, 7 subnets, 2 masks
O IA    172.16.23.0/24 [110/1294] via 172.16.34.3, 00:12:47, Serial0/0/0
O IA    172.16.12.0/24 [110/1941] via 172.16.34.3, 00:11:48, Serial0/0/0
O       172.16.3.3/32 [110/648] via 172.16.34.3, 00:12:47, Serial0/0/0
O IA    172.16.2.2/32 [110/1295] via 172.16.34.3, 00:11:53, Serial0/0/0
O IA    172.16.1.1/32 [110/1942] via 172.16.34.3, 00:11:18, Serial0/0/0
```

从 R4 路由表中可以看出，区域 2 的路由能通过转接区域到达区域 0。

（2）查看路由器 R2 的链路状态数据库

```
R2#show ip ospf database
            OSPF Router with ID (2.2.2.2) (Process ID 1)
                Router Link States (Area 0)
Link ID         ADV Router      Age           Seq#          Checksum     Link count
2.2.2.2         2.2.2.2         793           0x80000004    0x0055D6     1
3.3.3.3         3.3.3.3         1     (DNA)   0x80000004    0x007DC6     4
4.4.4.4         4.4.4.4         45    (DNA)   0x80000003    0x00A19D     3
                Summary Net Link States (Area 0)
Link ID         ADV Router      Age           Seq#          Checksum
172.16.1.1      2.2.2.2         758           0x80000001    0x005A0B
```

Link ID	ADV Router	Age	Seq#	Checksum	
172.16.2.2	2.2.2.2	813		0x80000001	0x00A7CB
172.16.2.2	3.3.3.3	48	(DNA)	0x80000001	0x002738
172.16.12.0	2.2.2.2	788		0x80000001	0x00E07B
172.16.23.0	2.2.2.2	813		0x80000001	0x0067E9
172.16.23.0	3.3.3.3	48	(DNA)	0x80000001	0x004904

Router Link States (Area 1)

Link ID	ADV Router	Age	Seq#	Checksum	Link count
2.2.2.2	2.2.2.2	807	0x80000005	0x006AF9	3
3.3.3.3	3.3.3.3	811	0x80000002	0x001D17	2

Summary Net Link States (Area 1)

Link ID	ADV Router	Age	Seq#	Checksum
172.16.1.1	2.2.2.2	760	0x80000001	0x005A0B
172.16.3.3	3.3.3.3	858	0x80000001	0x0074F8
172.16.4.4	3.3.3.3	848	0x80000001	0x00FC5E
172.16.12.0	2.2.2.2	791	0x80000001	0x00E07B
172.16.34.0	3.3.3.3	848	0x80000001	0x00CF72

Router Link States (Area 2)

Link ID	ADV Router	Age	Seq#	Checksum	Link count
1.1.1.1	1.1.1.1	766	0x80000002	0x00ADE3	3
2.2.2.2	2.2.2.2	765	0x80000002	0x00B5A5	2

Summary Net Link States (Area 2)

Link ID	ADV Router	Age	Seq#	Checksum
172.16.2.2	2.2.2.2	791	0x80000001	0x00A7CB
172.16.3.3	2.2.2.2	792	0x80000001	0x003031
172.16.4.4	2.2.2.2	792	0x80000001	0x00B896
172.16.23.0	2.2.2.2	793	0x80000001	0x0067E9
172.16.34.0	2.2.2.2	793	0x80000001	0x008BAA

以上输出表明虚链路的路由被拉进区域 0，并带有标记（**DNA**），表示不老化。通过 OSPF Virtual Link 发送的 LSA 都会带有标记（DNA）。

（3）查看 OSPF 虚链路的信息

```
R2#show ip ospf virtual-links
Virtual Link OSPF_VL0 to router 3.3.3.3 is up    //虚链路工作正常
  Run as demand circuit
  DoNotAge LSA allowed.    //LSA 不老化
  Transit area 1, via interface Serial0/0/1, Cost of using 647
//转接区域、建立虚链路的接口和接口 Cost 值
  Transmit Delay is 1 sec, State POINT_TO_POINT,
  Timer intervals configured, Hello 10, Dead 40, Wait 40, Retransmit 5
    Hello due in 00:00:00
    Adjacency State FULL (Hello suppressed)
//虚链路只在建邻居时发送 Hello 数据包，当邻居关系建立后，不再发送 Hello 数据包
    Index 1/2, retransmission queue length 0, number of retransmission 0
    First 0x0(0)/0x0(0) Next 0x0(0)/0x0(0)
    Last retransmission scan length is 0, maximum is 0
    Last retransmission scan time is 0 msec, maximum is 0 msec
  Message digest authentication enabled
    Youngest key id is 1
```

（4）查看 OSPF 邻居的信息

```
R2#show ip ospf neighbor
Neighbor ID     Pri    State       Dead Time    Address        Interface
3.3.3.3         0      FULL/  -    -            172.16.23.3    OSPF_VL0
3.3.3.3         0      FULL/  -    00:00:37     172.16.23.3    Serial0/0/1
1.1.1.1         0      FULL/  -    00:00:38     172.16.12.1    Serial0/0/0
```

以上输出表明路由器 R2 通过虚链路与 R3 建立了邻接关系，接口为 **OSPF_VL0**。

本 章 小 结

OSPF 是目前应用最为广泛的链路状态路由协议，通过区域划分很好地实现了路由的分级管理，在大规模网络的情况下，OSPF 可以通过划分区域来规划和限制网络规模。本章介绍了 OSPF 特征、数据包格式、路由器类型、LSA 类型、区域类型、LSDB、邻居关系建立和运行等，并用实验演示和验证了单区域和多区域 OSPF 基本配置、路由手工汇总、简单口令和 MD5 验证、末节区域和完全末节区域、NSSA 区域和虚链路配置。

第 5 章 IS-IS

近几年来，随着 IS-IS 在 ISP 中的广泛应用，IS-IS 路由协议已经变得很普及。IS-IS 最初是由 DECnet 公司开发的，1985 年被 ISO 采纳并更名为 IS-IS，是工作在 OSI 无连接网络服务（CLNS）环境中的链路状态路由协议。1991 年 Cisco 公司的 IOS 开始支持 IS-IS。IS-IS 仅支持 CLNS 路由选择，而集成的 IS-IS 支持 IP 和 CLNS 路由选择。本章重点讨论集成的 IS-IS 路由协议。

5.1 IS-IS 概述

5.1.1 IS-IS 特征

IS-IS（Intermediate System-to-Intermediate System，中间系统到中间系统）是一个非常灵活的路由协议，具有很好的可扩展性，而且已经整合了诸如 MPLS（多协议标记交换）之类的特性，其主要特征如下：

① 维护一个链路状态数据库，并使用 SPF 算法来计算最佳路径。
② 用 Hello 数据包建立和维护邻居关系。
③ 使用区域来构造两级层次化的拓扑结构。
④ 在区域之间可以使用路由汇总来减少路由器的负担。
⑤ 支持 VLSM 和 CIDR，支持明文和 MD5 验证。
⑥ 在广播多路访问网络中，通过选举指定 IS（DIS）来管理和控制网络上的泛洪扩散。
⑦ IS-IS 管理距离为 115，采用 Cost（开销）作为度量值。
⑧ 快速收敛，适合大型网络。

虽然 IS-IS 和 OSPF 都是链路状态路由协议，但是二者之间是有区别的。OSPF 和集成 IS-IS 的区别如表 5-1 所示。

表 5-1　OSPF 和集成 IS-IS 的区别

OSPF	集成 IS-IS
区域边界在 ABR 上	区域边界在链路上
每条链路只属于一个区域	每台路由器只属于一个区域
扩展主干时复杂	扩展主干时简单
运行在 IP 上	运行在 CLNS 上
需要 IP 地址	需要 IP 地址和 CLNS 地址
默认度量值与接口带宽成反比	所有接口 Cisco 默认度量值为 10
难以扩展	容易使用 TLV 支持新的协议，如 IPv6

5.1.2 IS-IS 术语

1) CLNS（Connectionless Network Service，无连接网络服务）：提供数据的无连接传送，

在数据传输之前不需要建立连接,它描述提供给传输层的服务。

2) CLNP (Connectionless Network Protocol, 无连接网络协议): 是 OSI 参考模型中网络层的一种无连接的网络协议,和 IP 有相同的特质。

3) ES (End System, 端系统): 没有路由能力的网络节点。

4) IS (Intermediate System, 中间系统): 有数据包转发能力的网络节点,即路由器。

5) LSP (Link State Packet, 链路状态数据包): 在 IS-IS 协议中 LSP 在区域中交换链路状态信息,以建立链路状态数据库。

6) SAP (Network Service Access Point, 网络服务访问点): 是 CLNS 的地址,类似于 IP 包头中的 IP 地址,与 IP 地址不同,CLNS 的地址不是代表接口而是节点,IS-IS 的 LSP 通过 NSAP 地址来标识路由器并建立拓扑表和底层的 IS-IS 路由选择树,因此即使纯粹的 IP 环境也必须有 NSAP 地址。NSAP 地址长度范围为 8~20 字节,NSAP 地址结构如图 5-1 所示,各部分含义如下。

图 5-1 NSAP 地址结构

① AFI 和 IDI 构成 NSAP 地址的初始域部分,其中 AFI 是机构格式标识符,如 39 代表 ISO 数据国别编码,45 代表 E.164,49 表示本地管理,相当于 RFC 1918 的私有地址。IDI 是 AFI 的子域。

② 高位 DSP、系统 ID 和 NSEL 构成 NSAP 地址的特定域部分,其中,高位 DSP 用来将域划分为不同的区域;系统 ID 用来标识 OSI 设备;NSEL 标识设备中的进程。在 IGP 中运行 IS-IS 时,Cisco 使用最简单的 NSAP 地址格式,即区域地址、系统 ID 和 NSEL 三个部分,如 49.0001.2222.2222.2222.00。

● 区域地址: 至少 1 字节,由 AFI 和区域标识符组成。上例中,AFI 的值为 49,区域标识符为 0001。
● 系统 ID: 6 字节长的标识符。上例中,系统 ID 为 2222.2222.2222。
● NSEL (网络选择标识): 对于路由器,NSEL 总是为 0。

7) NET (Network Entity Titles, 网络实体标题): 是当 NSAP 地址格式中 NSEL 为 0 时的 NSAP 地址。

8) SNPA (Subnetwork Point of Attachment, 子网连接点): 是和三层地址对应的二层地址,它通常被定义为 LAN 环境中的 MAC 地址,在 HDLC 接口中 SNPA 被设置为"HDLC"。由于 NSAP 和 NET 相当于一个设备或节点,那么 SNPA 就相当于用来区分该设备上的不同接口。

9) SNP (Sequence Number PDU, 序列号 PDU): 确保 IS-IS 的链路状态数据库同步并使用最新的 LSP 计算路由。

10) PSNP (Partial SNP, 部分 SNP): 确认和请求丢失的链路状态信息,是链路状态数据库中的完整 LSP 的一个子集。

11) CSNP (Complete SNP, 完整 SNP): 描述链路状态数据库中的完整 LSP 列表。

12) DIS (Designated Intermediate System, 指定中间系统): 在 IS-IS 中广播链路本身被

视为一个伪节点,需要选举一个路由器作为 DIS 来代表该伪节点。

13）Level（级别）：IS-IS 规范定义了 4 种类型路由级别,如图 5-2 所示。

① Level 0：根据 ES 和 IS 进行路由。

② Level 1：在 IS-IS 区域内根据区域内的系统 ID 进行路由。

③ Level 2：在 IS-IS 区域之间根据区域 ID 进行路由。

④ Level 3：在 IS-IS 域之间进行路由,类似 IP 中的 BGP。Cisco 没有实现 Level 3 路由,而是通过 ICMP 路由器发现协议（ICMP Router Discovery Protocol, IRDP）来完成这种功能的。

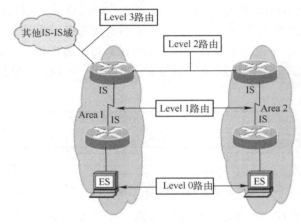

图 5-2　IS-IS 路由级别

5.1.3　IS-IS 路由器类型

IS-IS 路由器的类型,如图 5-3 所示。

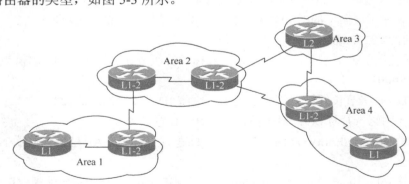

图 5-3　IS-IS 路由器的类型

① L1 路由器：通过 LSP 获得所在区域内的路由信息,类似 OSPF 的内部路由器。

② L1/2 路由器：通过 LSP 获得所在区域内和区域间的路由信息,类似 OSPF 的 ABR。

③ L2 路由器：通过 LSP 获得区域间的路由信息,类似 OSPF 的主干路由器。连接到 L2 和 L1/2 路由器的路径被称为主干,所有主干都必须是连续的。

5.1.4　IS-IS 数据包格式

IS-IS 路由协议使用三大类数据包：Hello 数据包、LSP 数据包和 SNP 数据包,共计 9 种具体的数据包。每种数据包都有一个特定的类型号,在 IS-IS 的数据包包头中,有一个数据包类型字段,此字段中所包含的信息就是 IS-IS 数据包的类型号,路由器就是通过类型号来识别所收到的数据包的类型的。

IS-IS 数据包包头字段是相同的,长度为 8 字节。IS-IS 数据包头部如图 5-4 所示,各字段含义如下。

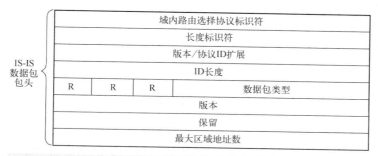

图 5-4 IS-IS 数据包头部

① 域内路由选择协议标识符:这是 ISO 9577 分配给 IS-IS 的一个固定的值,用于标识网络层协议数据单元的类型,对于 IS-IS 数据包,该字段的值永远都为 0x83。

② 长度标识符:标识该固定头部字段的长度。

③ 版本 / 协议 ID 扩展:当前始终为 1。

④ ID 长度:表示系统 ID 的长度,值为 0 表示长度为 6 字节;值为 255 表示长度为 0,即为空;值为 1~8 的整数,表示系统 ID 具有相同长度的字节数。在 Cisco 的路由器上,该字段始终为 0。

⑤ R:保留位,没有使用的比特,始终为 0。

⑥ 数据包类型:是一个 5 比特的字段,标识 IS-IS 数据包的类型,IS-IS 数据包类型和对应的类型号如表 5-2 所示。

表 5-2 IS-IS 数据包类型和对应的类型号

IS-IS 数据包类别	IS-IS 数据包类型	IS-IS 数据包类型号
Hello 数据包	Level 1 LAN Hello 数据包	15
	Level 2 LAN Hello 数据包	16
	点到点 Hello 数据包	17
LSP 数据包	Level 1 LSP 数据包	18
	Level 2 LSP 数据包	20
SNP 数据包	Level 1 CSNP 数据包	24
	Level 2 CSNP 数据包	25
	Level 1 PSNP 数据包	26
	Level 2 PSNP 数据包	27

⑦ 版本：当前为 1。
⑧ 保留：当前设置为全 0。
⑨ 最大区域地址数：IS 区域所允许的最大区域地址数量。值为 1～254 的整数，值为 0 表示最多支持 3 个区域地址数，Cisco 默认情况下的值为 0。

1. IS-IS LAN Hello 数据包格式

Level 1 和 Level 2 LAN 的 Hello 数据包格式是相同的，只是头部的数据包类型字段的值不同。IS-IS LAN Hello 数据包格式如图 5-5 所示，相关字段含义如下。

图 5-5　IS-IS LAN Hello 数据包格式

① 电路类型：2 比特，01 表示 L1 路由器，10 表示 L2 路由器，11 表示 L1/2 路由器，如果为 00，该数据包被忽略。

② 源 ID：发送该数据包的路由器的系统 ID。

③ 抑制时间：用来通知它的邻居路由器在认为这台路由器失效之前应该等待的时间。如果在抑制时间内收到邻居发送的 Hello 数据包，将认为邻居依然处于存活状态。在 IS-IS 中，默认情况下抑制时间是发送 Hello 数据包间隔的 3 倍。

④ 数据包长度：整个数据包的长度，包括包头和 TLV 字段。

⑤ 优先级：接口的 DIS 优先级，用来在 BMA 或者 NBMA 网络中选举 DIS。优先级数

值越高，路由器成为 DIS 的可能性越大。L1 和 L2 的 DIS 是分别选举的。

⑥ LAN ID：由 DIS 的系统 ID 和 1 字节的伪节点 ID 组成，LAN ID 用来区分同一台 DIS 上的不同 LAN。

2. IS-IS 点到点 Hello 数据包格式

IS-IS 点到点 Hello 数据包格式如图 5-6 所示，从 IS-IS 点到点 Hello 数据包的格式可以看出，大部分字段与 LAN Hello 数据包的格式相同。但是在点到点 Hello 数据包中没有优先级字段，因为在点到点链路上不需要选举 DIS。而且使用本地电路 ID 字段代替了 LAN 数据包中的 LAN ID 字段。本地电路 ID 是由发送 Hello 数据包的路由器分配给这条电路的标识，并且在路由器的接口下是唯一的。在点到点链路的另一端，Hello 数据包中的本地电路 ID 的值可能相同，也可能不同。

图 5-6　IS-IS 点到点 Hello 数据包格式

3. IS-IS LSP 数据包格式

LSP 数据包分为 Level 1 LSP 和 Level 2 LSP，它们各自承载了 IS-IS 不同层次的路由选择信息，但是它们有着相同的数据包格式。IS-IS LSP 数据包格式如图 5-7 所示，相关字段含义如下。

```
⊟ ISO 10589 ISIS Link State Protocol Data Unit
    PDU length: 86
    Remaining lifetime: 1199
    LSP-ID: 2222.2222.2222.00-00
    Sequence number: 0x00000004
  ⊞ Checksum: 0x1f73 [correct]
  ⊟ Type block(0x03): Partition Repair:0, Attached bits:0, Overload bit:0, IS type:3
      0... .... = Partition Repair: Not supported
    ⊞ .000 0... = Attachment: 0
      .... .0.. = Overload bit: Not set
      .... ..11 = Type of Intermediate System: Level 2 (3)
  ⊟ Area address(es) (4)
      Area address (3): 49.0001
  ⊟ Protocols supported (1)
      NLPID(s): IP (0xcc)
  ⊟ Hostname (2)
      Hostname: R2
  ⊟ IP Interface address(es) (4)
      IPv4 interface address: 172.16.2.2 (172.16.2.2)
  ⊟ IP Internal reachability (24)
    ⊞ IPv4 prefix: 172.16.12.0/24
    ⊞ IPv4 prefix: 172.16.2.0/24
  ⊟ IS Reachability (12)
      IsNotVirtual
    ⊞ IS Neighbor:   2222.2222.2222.01
```

图 5-7　IS-IS LSP 数据包格式

① 数据包长度：整个数据包的长度。

② 剩余生存时间：LSP 到期前的生存时间。当生存时间为 0 时，LSP 将从链路状态数据库中被清除。

③ LSP ID：用来标识不同的 LSP 和生成 LSP 的源路由器。LSP ID 包括三个部分：系统 ID、伪节点标识符和 LSP 编号。

④ 序列号：32 比特的无符号数，主要作用是让路由器能够识别一个 LSP 的新旧版本。

⑤ 校验和：主要用于检查被破坏的 LSP 或者还没有从网络中清除的过期 LSP。

⑥ P：分区，表示区域划分或者分段区域的修复位。当 P 被设置为 1 时，表明始发路由器支持自动修复区域的分段情况。对于 Cisco 路由器来说，该位始终为 0。

⑦ ATT：区域关联，L1/L2 路由器在其生成的 L1 LSP 中设置该字段以通知同一区域中的 L1 路由器自己与其他区域相连。当 L1 区域中的路由器收到 L1/2 路由器发送的 ATT 被置位的 L1 LSP 后，它将创建一条指向 L1/2 路由器的默认路由，以便数据可以被路由到其他区域。虽然 ATT 同时在 L1 LSP 和 L2 LSP 中进行了定义，但是它只会在 L1 LSP 中被置位，并且只有 L1/2 路由器会设置这个字段。

⑧ OL：超载，表示路由器的资源状态。如果该位被置位，就表示路由器发生了超载。被设置了超载位的 LSP 不会在网络中进行泛洪，并且当其他路由器收到设置了超载位的 LSP 后，在计算路径信息时不会考虑此 LSP，因此最终计算出来的到达目的地的路径将绕过超载的路由器。

⑨ IS 类型：表示 LSP 是来自 L1 路由器还是 L2 路由器，也表示了收到此 LSP 的路由器将把这个 LSP 放到 L1 链路状态数据库中还是 L2 链路状态数据库中。01 表示 L1，11 表示 L2，00 与 10 未使用。

4. IS-IS SNP 数据包格式

SNP 数据包分为 CSNP 和 PSNP。CSNP 与 PSNP 都包含了路由器本地链路状态数据库中 LSP 的摘要信息。其中 CSNP 包含的是所有 LSP 的摘要信息，PSNP 包含的是部分 LSP 的摘要信息。IS-IS CSNP 数据包格式如图 5-8 所示，其中起始 LSP ID 表示 TLV 字段中描述的 LSP 范围的第一个 LSP ID，结束 LSP ID 表示 TLV 字段中描述的 LSP 范围的最后一个 LSP ID。

```
□ ISO 10589 ISIS Complete Sequence Numbers Protocol Data Unit
    PDU length: 67
    Source-ID:    2222.2222.2222.00
    Start LSP-ID: 0000.0000.0000.00-00
    End LSP-ID:   ffff.ffff.ffff.ff-ff
  □ LSP entries (32)
    □ LSP-ID: 1111.1111.1111.00-00, Sequence: 0x00000007, Lifetime: 1198s, Checksum: 0x1fa7
        LSP-ID:               : 1111.1111.1111.00-00
        LSP Sequence Number   : 0x00000007
        Remaining Lifetime    : 1198s
        LSP checksum          : 0x1fa7
    □ LSP-ID: 2222.2222.2222.00-00, Sequence: 0x00000002, Lifetime: 1199s, Checksum: 0x84c6
        LSP-ID:               : 2222.2222.2222.00-00
        LSP Sequence Number   : 0x00000002
        Remaining Lifetime    : 1199s
        LSP checksum          : 0x84c6
```

图 5-8 IS-IS CSNP 数据包格式

IS-IS PSNP 数据包格式如图 5-9 所示，通过对比可以看出，PSNP 数据包的格式与 CSNP 相似，只不过没有起始 LSP ID 和结束 LSP ID 两个字段。由于 PSNP 携带的只是部分 LSP 的摘要信息，所以不需要起始和结束 LSP ID 字段。

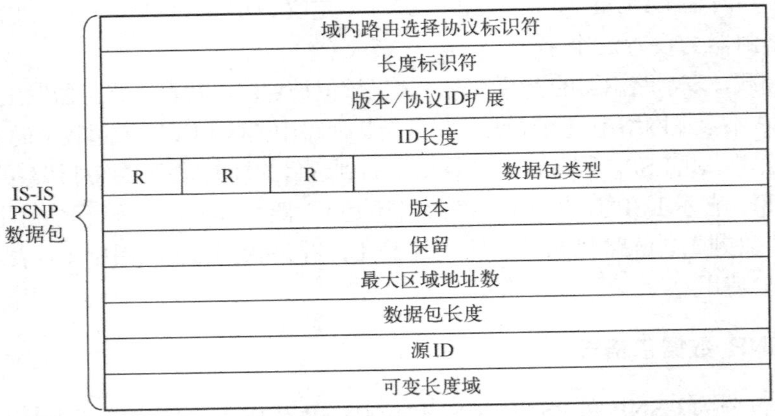

```
□ ISO 10589 ISIS Partial Sequence Numbers Protocol Data Unit
    PDU length: 35
    Source-ID: 1111.1111.1111.00
□ LSP entries (16)
  □ LSP-ID: 2222.2222.2222.00-00, Sequence: 0x00000002, Lifetime: 1198s, Checksum: 0x62e8
      LSP-ID:              : 2222.2222.2222.00-00
      LSP Sequence Number  : 0x00000002
      Remaining Lifetime   : 1198s
      LSP checksum         : 0x62e8
```

图 5-9 IS-IS PSNP 数据包格式

5.2 配置集成 IS-IS

5.2.1 实验 1：配置单区域集成 IS-IS

1. 实验目的

通过本实验可以掌握：
① 在路由器上启动 IS-IS 路由进程的方法。
② 启用参与路由协议接口的方法。
③ IS-IS 度量值的计算方法。
④ NET 地址配置。
⑤ DIS 选举的原则及选举控制方法。
⑥ 查看和调试 IS-IS 路由协议相关信息的方法。

2. 实验拓扑

配置单区域集成 IS-IS 实验拓扑如图 5-10 所示。

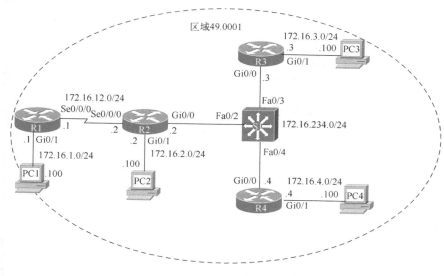

图 5-10 配置单区域集成 IS-IS 实验拓扑

3. 实验步骤

（1）配置路由器 R1

```
R1(config)#router isis cisco                        //启动 IS-IS 路由进程，进程名称为 cisco
R1(config-router)#net 49.0001.1111.1111.1111.00     //配置 NET 地址
R1(config)#interface gigabitEthernet0/1
R1(config-if)#ip router isis cisco                  //接口下启用 IS-IS
R1(config)#interface Serial0/0/0
R1(config-if)#ip router isis cisco
```

（2）配置路由器 R2

```
R2(config)#router isis cisco
R2(config-router)#net 49.0001.2222.2222.2222.00
R2(config)#interface gigabitEthernet0/1
R2(config-if)#ip router isis cisco
R2(config)#interface gigabitEthernet0/0
R2(config-if)#ip router isis cisco
R2(config)#interface Serial0/0/0
R2(config-if)#ip router isis cisco
```

（3）配置路由器 R3

```
R3(config)#router isis cisco
R3(config-router)#net 49.0001.3333.3333.3333.00
R3(config)#interface gigabitEthernet0/1
R3(config-if)#ip router isis cisco
R3(config)#interface gigabitEthernet0/0
R3(config-if)#ip router isis cisco
```

（4）配置路由器 R4

```
R4(config)#router isis cisco
R4(config-router)#net 49.0001.4444.4444.4444.00
R4(config)#interface gigabitEthernet0/1
R4(config-if)#ip router isis cisco
R4(config)#interface gigabitEthernet0/0
R4(config-if)#ip router isis cisco
```

4. 实验调试

（1）查看 IS-IS 的邻居

```
R2#show clns neighbors
System Id    Interface    SNPA              State    Holdtime    Type Protocol
R1           Se0/0/0      *HDLC*            Up       27          L1L2 IS-IS
R3           Gi0/0        f872.ea69.18b8    Up       24          L1L2 IS-IS
R4           Gi0/0        f872.eac8.4f98    Up       9           L1L2 IS-IS
```

从以上输出可以看出，路由器 R2 有 3 个邻居，而且都是 **L1L2** 类型的，这也是运行 IS-IS 的路由器的默认类型。由于 R1 和 R2 是串行连接的，所以 SNPA 为 ***HDLC***，而 R2 与 R3 和 R4 是通过以太网连接的，所以 SNPA 分别是 R3 和 R4 以太网接口 **Gi0/0** 的 MAC 地址。

【技术要点】

① IS-IS 进程的名字只有本地含义，一台路由器可以启动多个 IS-IS 进程。
② Cisco 路由器支持动态主机名字映射，可以通过命令 **show isis hostname** 查看:

```
R2#show isis hostname
Level    System ID            Dynamic Hostname    (cisco)
 2       4444.4444.4444       R4
 2       3333.3333.3333       R3
 2       1111.1111.1111       R1
 *       2222.2222.2222       R2
```

上面的输出清楚地显示了系统 ID 和动态主机名的映射关系，其中 * 表示本地路由器 R2。
③ 默认情况下，IS-IS 发送 Hello 数据包周期为 10 秒，Hold 时间为 30 秒，即 3 倍的关系。可以在接口下通过 **isis hello-interval** 命令修改 Hello 数据包发送的周期，同时通过 **isis hello-multiplier** 命令定义 Hold 时间是 Hello 周期的倍数。

（2）查看和 CLNS 路由协议相关的信息

```
R2#show clns protocol
IS-IS Router: cisco       //IS-IS 路由进程名字，如果不指定，则显示为 <Null Tag>
  System Id: 2222.2222.2222.00    IS-Type: level-1-2   //系统 ID 以及 IS-IS 路由器类型
  Manual area address(es):
        49.0001
  Routing for area address(es):
        49.0001
```

```
                Interfaces supported by IS-IS:
                    Serial0/0/0 - IP
                    GigabitEthernet0/1 - IP
                    GigabitEthernet0/0 – IP
//以上四行表示运行 IS-IS 路由协议的接口
                Redistribute:
                    static (on by default)
                Distance for L2 CLNS routes: 110    //L2 CLNS 路由的管理距离
                RRR level: none
                Generate narrow metrics: level-1-2
                Accept narrow metrics:   level-1-2
//以上两行表示使用和接受"窄"度量
                Generate wide metrics:    none
                Accept wide metrics:      none
```

（3）查看 CLNS 接口状态的基本信息

```
R2#show clns interface serial0/0/0
Serial0/0/0 is up, line protocol is up
    Checksums enabled, MTU 1500, Encapsulation HDLC
    ERPDUs enabled, min. interval 10 msec.
    CLNS fast switching enabled         //CLNS 快速交换启动
    CLNS SSE switching disabled         //CLNS SSE 交换关闭
    DEC compatibility mode OFF for this interface
    Next ESH/ISH in 47 seconds
    Routing Protocol: IS-IS
        Circuit Type: level-1-2       //电路类型
        Interface number 0x1, local circuit ID 0x100   //接口号和本地电路 ID
        Neighbor System-ID: R1    //IS-IS 邻居路由器系统 ID
        Level-1 Metric: 10, Priority: 64, Circuit ID: R2.00
//接口 Level-1 的度量值、接口优先级以及电路 ID
        Level-1 IPv6 Metric: 10
        Number of active level-1 adjacencies: 1    //该接口活动 L1 邻居的个数
        Level-2 Metric: 10, Priority: 64, Circuit ID: R2.00
//接口 Level-2 的度量值、接口优先级以及电路 ID，接口度量值默认为 10
        Level-2 IPv6 Metric: 10
        Number of active level-2 adjacencies: 1    //该接口活动 L2 邻居的个数
        Next IS-IS Hello in 7 seconds    //距离发送下一个 Hello 数据包的时间
        if state UP    //接口状态
```

（4）查看 CLNS Level 2 路由信息

```
R2#show clns route
Codes: C - connected, S - static, d - DecnetIV
       I - ISO-IGRP, i - IS-IS, e - ES-IS
       B - BGP,    b - eBGP-neighbor
```

因为这条命令用于 OSI 路由选择，所以以上输出没有太多的信息。

（5）查看 IS-IS 的拓扑结构信息

```
R2#show isis topology
IS-IS paths to level-1 routers
System Id           Metric          Next-Hop        Interface   SNPA
R1                  10              R1              Se0/0/0     *HDLC*
R2                  --
R3                  10              R3              Gi0/0       f872.ea69.18b8
R4                  10              R4              Gi0/0       f872.eac8.4f98

IS-IS paths to level-2 routers
System Id           Metric          Next-Hop        Interface   SNPA
R1                  10              R1              Se0/0/0     *HDLC*
R2                  --
R3                  10              R3              Gi0/0       f872.ea69.18b8
R4                  10              R4              Gi0/0       f872.eac8.4f98
```

以上输出表明 IS-IS 为 L1 路由器和 L2 路由器分别存放拓扑结构数据库，其中 **Metric** 是到达目的网络的开销之和。

（6）查看 IS-IS 链路状态数据库

```
R2#show isis database
Tag cisco:
IS-IS Level-1 Link State Database:     // IS-IS Level-1 链路状态数据库
LSPID              LSP Seq Num       LSP Checksum    LSP Holdtime/Rcvd    ATT/P/OL
R1.00-00           * 0x00000005      0x98DA          1152       / *       0/0/0
R2.00-00           0x00000007        0x720D          424        /1199     0/0/0
R2.01-00           0x00000001        0xB487          0 (436)    /0        0/0/0
R3.00-00           0x00000005        0x7785          1111       /1198     0/0/0
R4.00-00           0x00000005        0xCCC5          1191       /1198     0/0/0
R4.01-00           0x00000001        0xD0C5          425        /1198     0/0/0
IS-IS Level-2 Link State Database:    // IS-IS Level-2 链路状态数据库
LSPID              LSP Seq Num       LSP Checksum    LSP Holdtime/Rcvd    ATT/P/OL
R1.00-00           * 0x00000008      0xFBF2          1176       /*        0/0/0
R2.00-00           0x0000000A        0x1544          430        /1199     0/0/0
R2.01-00           0x00000001        0xB487          0 (437)    /0        0/0/0
R3.00-00           0x00000007        0x7CE7          430        /1198     0/0/0
R4.00-00           0x00000005        0x2FCD          432        /1198     0/0/0
R4.01-00           0x00000003        0x7E9E          1119       /1198     0/0/0
```

以上输出表明：IS-IS 为 L1 路由和 L2 路由分别维护独立的链路状态数据库。由于 IS-IS 是链路状态路由协议，而且四台路由器具有相同区域，所以它们的链路状态数据库是相同的；IS-IS 的 LSP 老化时间为 20 分钟，采用倒计时，路由器每隔 15 分钟刷新链路状态一次，序列号会加 1；路由器 R4 是 DIS，LSPID（链路状态协议数据单元 ID）由以下三个部分构成。

① 系统 ID：长度为 6 字节。

② 伪节点 ID：长度为 1 字节，它代表了 1 个 LAN，当这个值非 0 时，表示该路由器为 DIS，系统 ID 和伪节点就构成了电路 ID（Circuit ID），如 R4.01。

③ LSP 分段号：长度为 1 字节，如果是 00 表示所有数据都在单个的 LSP 中。

【技术要点】

IS-IS 中 DIS 的选举原则如下：
① 只有形成邻接关系的路由器才有资格参与选举。
② 接口优先级最高的路由器成为 DIS。
③ 如果接口优先级相同，则接口具有最高 MAC 地址的路由器成为 DIS。
④ DIS 选举是抢占的。
⑤ 接口优先级为 0 的路由器也有可能成为 DIS，这点和 OSPF DR 选举不同。

修改接口优先级的命令是 **isis priority** *priority*，默认是 64，取值范围为 0～127。可以针对 L1 和 L2 分别指定接口优先级。在本例中可以将 R2 的以太网接口的接口优先级改为 100，则 R2 被选为 DIS，显示如下：

```
R2#show isis database | include R2
R2.00-00            * 0x0000000B      0xEF58         1065/*        0/0/0
R2.01-00            * 0x00000003      0x66FA         1066/*        0/0/0
R2.00-00            * 0x00000010      0x46D9         1070/*        0/0/0
R2.01-00            * 0x00000002      0x1AD0         1060/*        0/0/0

R2#show clns interface gigabitEthernet 0/0 | include DR
    DR ID: R2.01    //Level 1 的 DIS
    DR ID: R2.01    //Level 2 的 DIS
```

（7）查看 CLNS Level 1 的路由信息

```
R2#show isis route
IS-IS not running in OSI mode (*) (only calculating IP routes)
(*) Use "show isis topology" command to display paths to all routers
```

由于该命令是针对 OSI 路由选择协议的，所以没有具体的输出。

（8）查看 IS-IS 路由协议相关信息

```
R2#show ip protocols   | begin Routing Protocol is "isis cisco"
Routing Protocol is "isis cisco"
  Invalid after 0 seconds, hold down 0, flushed after 0
//更新计时器全部为 0，表示 IS-IS 路由协议采用触发方式更新
  Outgoing update filter list for all interfaces is not set
  Incoming update filter list for all interfaces is not set
//以上两行表明入方向和出方向都没有配置分布列表
  Redistributing: isis cisco
  Address Summarization:    //地址汇总信息
    None
  Maximum path: 4    //默认支持等价路径数目
  Routing for Networks:
    GigabitEthernet0/0
    GigabitEthernet0/1
    Serial0/0/0
//以上四行表示运行 IS-IS 路由协议的接口
```

```
Routing Information Sources:
    Gateway          Distance        Last Update
    172.16.4.4       115             00:02:29
    172.16.3.3       115             00:02:29
    172.16.1.1       115             00:02:29
//以上五行表示路由信息源、管理距离和距离最近一次更新的时间
    Distance: (default is 115)    //IS-IS 默认管理距离
```

（9）查看路由表中 IS-IS 路由

以下输出全部省略路由代码部分。

```
① R1#show ip route isis
         172.16.0.0/16 is variably subnetted, 8 subnets, 2 masks
i L1     172.16.2.0/24      [115/20] via 172.16.12.2, 00:42:36, Serial0/0/0
i L1     172.16.3.0/24      [115/30] via 172.16.12.2, 00:42:00, Serial0/0/0
i L1     172.16.4.0/24      [115/30] via 172.16.12.2, 00:41:28, Serial0/0/0
i L1     172.16.234.0/24    [115/20] via 172.16.12.2, 00:42:36, Serial0/0/0
② R2#show ip route isis
         172.16.0.0/24 is subnetted, 6 subnets
i L1     172.16.1.0/24 [115/20] via 172.16.12.1, 00:09:41, Serial0/0/0
i L1     172.16.3.0/24 [115/20] via 172.16.234.3, 00:09:41, GigabitEthernet0/0
i L1     172.16.4.0/24 [115/20] via 172.16.234.4, 00:09:41, GigabitEthernet0/0
③ R3#show ip route isis
         172.16.0.0/16 is variably subnetted, 8 subnets, 2 masks
i L1     172.16.1.0/24    [115/30] via 172.16.234.2, 00:42:46, GigabitEthernet0/0
i L1     172.16.2.0/24    [115/20] via 172.16.234.2, 00:42:56, GigabitEthernet0/0
i L1     172.16.4.0/24    [115/20] via 172.16.234.4, 00:42:23, GigabitEthernet0/0
i L1     172.16.12.0/24   [115/20] via 172.16.234.2, 00:42:56, GigabitEthernet0/0
④ R4#show ip route isis
         172.16.0.0/16 is variably subnetted, 8 subnets, 2 masks
i L1     172.16.1.0/2     [115/30] via 172.16.234.2, 00:14:11, GigabitEthernet0/0
i L1     172.16.2.0/24    [115/20] via 172.16.234.2, 00:43:32, GigabitEthernet0/0
i L1     172.16.3.0/24    [115/20] via 172.16.234.3, 00:14:11, GigabitEthernet0/0
i L1     172.16.12.0/24   [115/20] via 172.16.234.2, 00:43:32, GigabitEthernet0/0
```

以上①、②、③和④输出表明区域内的路由代码为 **i L1**，即 Level-1 路由。默认情况下，IS-IS 使用窄度量计算度量值，所有链路都使用 10 作为度量值。

5.2.2 实验 2：配置多区域集成 IS-IS

1. 实验目的

通过本实验可以掌握：
① 在路由器上启动 IS-IS 路由进程的方法。
② 启用参与路由协议接口的方法。
③ L1 和 L2 路由的区别。
④ 配置 L1 或 L2 路由器的方法。
⑤ 配置 IS-IS 电路类型的方法。

⑥ 配置 IS-IS 区域间路由汇总的方法。
⑦ 向 IS-IS 网络注入默认路由的方法。
⑧ 配置 IS-IS 验证的方法。
⑨ 查看和调试多区域集成 IS-IS 路由协议相关信息的方法。

2. 实验拓扑

配置多区域集成 IS-IS 实验拓扑如图 5-11 所示。

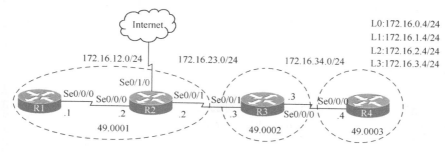

图 5-11 配置多区域集成 IS-IS 实验拓扑

3. 实验步骤

IS-IS 区域的划分是基于路由器的，也就是说一个路由器只能属于一个区域，而 OSPF 区域的划分是基于链路的。

（1）配置路由器 R1

```
R1(config)#router isis cisco
R1(config-router)#net 49.0001.1111.1111.1111.00
R1(config-router)#is-type level-1              //将 R1 配置成 IS-IS L1 路由器
R1(config-router)#area-password ccna           //启用 IS-IS 区域验证
R1(config)#interface Serial0/0/0
R1(config-if)#ip router isis cisco
R1(config-if)#isis password ccnp level-1       //启用 IS-IS Level-1 邻居验证
```

（2）配置路由器 R2

```
R2(config)#router isis cisco
R2(config-router)#net 49.0001.2222.2222.2222.00
R2(config-router)#default-information originate //向 IS-IS 区域注入默认路由
R2(config-router)#area-password ccna
R2(config-router)#domain-password ccie          //启用 IS-IS 域验证
R2(config)#interface Serial0/0/0
R2(config-if)#ip router isis cisco
R2(config-if)#isis password ccnp level-1
R2(config)#interface Serial0/0/1
R2(config-if)#ip router isis cisco
```

（3）配置路由器 R3

```
R3(config)#router isis cisco
R3(config-router)#net 49.0002.3333.3333.3333.00
R3(config-router)#is-type level-2-only        //将 R3 配置成 IS-IS L2 路由器
R3(config-router)#domain-password ccie
R3(config)#interface Serial0/0/0
R3(config-if)#ip router isis cisco
R3(config-if)#isis circuit-type level-2-only  //配置 IS-IS 接口电路类型
R3(config-if)#isis password ccsp level-2      //启用 IS-IS Level 2 邻居验证
R3(config)#interface Serial0/0/1
R3(config-if)#ip router isis cisco
```

（4）配置路由器 R4

```
R4(config)#router isis cisco
R4(config-router)#net 49.0003.4444.4444.4444.00
R4(config-router)#summary-address 172.16.0.0 255.255.252.0
//配置 IS-IS 区域间路由汇总
R4(config-router)#is-type level-2-only
R4(config-router)#domain-password ccie
R4(config-router)#passive-interface loopback 0
R4(config-router)#passive-interface loopback 1
R4(config-router)#passive-interface loopback 2
R4(config-router)#passive-interface loopback 3
R4(config)#interface range loopback 0 -3
R4(config-if-range)#ip router isis cisco
R4(config)#interface Serial0/0/0
R4(config-if)#ip router isis cisco
R4(config-if)#isis circuit-type level-2-only
R4(config-if)#isis password ccsp level-2
```

【技术要点】

本节介绍的 IS-IS 的验证被认为是旧命令，对于不支持基于钥匙链验证的低版本的 IOS，采用此方式验证。而对于较新的 IOS 版本，建议采用基于钥匙链的验证（在后面 5.2.3 实验 3 中介绍）。Cisco 的 IOS 支持 3 个级别的验证，可以单独使用，也可以同时使用。下面是对旧命令的三类验证规则的解释。

① 邻居验证：相互连接的路由器接口必须配置相同的密码，同时必须为 L1 和 L2 类型的邻居关系配置各自的验证，L1 邻居验证的密码和 L2 邻居验证的密码可以不同。邻居验证通过命令 **isis password** 配置。本实验中 R1 和 R2 之间的串行链路启用 Level-1 的邻居验证，而 R3 和 R4 之间的串行链路启用 Level-2 的邻居验证。

② 区域验证：区域内的每台路由器必须执行验证，并且必须使用相同的密码。区域验证通过命令 **area-password** 配置。本实验中区域 49.0001 启用区域验证。

③ 域验证：域内的每一台 L2 和 L1/L2 类型的路由器必须执行验证，并且必须使用相同的密码。域验证通过命令 **domain-password** 配置。本实验中 R2、R3 和 R4 都配置域验证，因为路由器 R1 是 L1 路由器，所以不用配置域验证。

第 5 章　IS-IS

4. 实验调试

（1）查看 IS-IS 链路状态数据库

① R1#**show isis database**
Tag cisco:
IS-IS **Level-1** Link State Database:

LSPID	LSP Seq Num	LSP Checksum	LSP Holdtime/Rcvd	ATT/P/OL
R1.00-00	* 0x00000005	0x47DC	875/*	0/0/0
R2.00-00	0x00000006	0xEDB2	977/1199	1/0/0

//ATT 被置位，表明当 L1 区域中的路由器 R1 收到 L1/2 路由器 R2 发送的 ATT 被置位的 L1 LSP 后，它将创建一条指向 L1/2 路由器的默认路由，由于 R1 是 L1 路由器，所以只有 L1 的链路状态数据库

② R2#**show isis database**
Tag cisco:
IS-IS Level-1 Link State Database:

LSPID	LSP Seq Num	LSP Checksum	LSP Holdtime/Rcvd	ATT/P/OL
R1.00-00	0x00000005	0x47DC	884/1199	0/0/0
R2.00-00	* 0x00000006	0xEDB2	987/*	1/0/0

IS-IS Level-2 Link State Database:

LSPID	LSP Seq Num	LSP Checksum	LSP Holdtime/Rcvd	ATT/P/OL
R2.00-00	* 0x00000005	0x18C5	981/*	0/0/0
R3.00-00	0x00000006	0xA074	1067/1199	0/0/0
R4.00-00	0x00000005	0x5EC7	1068/1198	0/0/0

③ R3#**show isis database**
Tag cisco:
IS-IS Level-2 Link State Database:

LSPID	LSP Seq Num	LSP Checksum	LSP Holdtime/Rcvd	ATT/P/OL
R2.00-00	0x00000005	0x18C5	841/1199	0/0/0
R3.00-00	* 0x00000006	0xA074	927/*	0/0/0
R4.00-00	0x00000005	0x5EC7	927/1199	0/0/0

④ R4#**show isis database**
Tag cisco:
IS-IS Level-2 Link State Database:

LSPID	LSP Seq Num	LSP Checksum	LSP Holdtime/Rcvd	ATT/P/OL
R2.00-00	0x00000005	0x18C5	841/1199	0/0/0
R3.00-00	* 0x00000006	0xA074	927/*	0/0/0
R4.00-00	0x00000005	0x5EC7	927/1199	0/0/0

以上①、②、③和④输出表明：R1 路由器为 L1 路由器，只维护 Level-1 的链路状态数据库；R2 路由器为 L1/L2 路由器，同时为 Level-1 和 Level-2 维护单独的链路状态数据库，也表明所在区域有另一台路由器 R1；R3 和 R4 路由器为 L2 路由器，只维护 Level-2 的链路状态数据库。

（2）查看路由表中 IS-IS 路由

以下输出全部省略路由代码部分。

① R1#**show ip route isis**
　　172.16.0.0/24 is subnetted, 2 subnets

```
i L1      172.16.23.0 [115/20] via 172.16.12.2, Serial0/0/0
i*L1      0.0.0.0/0   [115/10] via 172.16.12.2, Serial0/0/0
//该默认路由是由 L1/L2 路由器 R2 注入的
② R2#show ip route isis
     172.16.0.0/16 is variably subnetted, 4 subnets, 2 masks
i L2      172.16.34.0/24 [115/20] via 172.16.23.3, Serial0/0/1
i L2      172.16.0.0/22 [115/30] via 172.16.23.3, Serial0/0/1
③ R3#show ip route isis
     172.16.0.0/16 is variably subnetted, 4 subnets, 2 masks
i L2      172.16.12.0/24 [115/20] via 172.16.23.2, Serial0/0/1
i L2      172.16.0.0/22 [115/20] via 172.16.34.4, Serial0/0/0
i*L2      0.0.0.0/0 [115/10] via 172.16.23.2, Serial0/0/1
④ R4#show ip route isis
     172.16.0.0/16 is variably subnetted, 8 subnets, 2 masks
i L2      172.16.23.0/24 [115/20] via 172.16.34.3, Serial0/0/0
i L2      172.16.12.0/24 [115/30] via 172.16.34.3, Serial0/0/0
i su      172.16.0.0/22 [115/10] via 0.0.0.0, Null0
i*L2      0.0.0.0/0 [115/20] via 172.16.34.3, Serial0/0/0
```

以上①、②、③和④输出表明：由于 R1 为 L1 路由器，所以只有 i L1 的路由和一条到最近的 L1/L2 路由器的默认路由 i*L1；由于 R1 和 R2 在同一个区域，所以 R2 应该既有 i L1 路由，又有 i L2 的路由；R3 和 R4 都是 L2 路由器，所以只有 i L2 路由；R3 和 R4 都收到一条由 R2 注入的默认路由 i*L2；R2 和 R3 都收到路由器 R4 四个环回接口汇总的路由条目，同时 R4 的路由表自动生成一条 i su 的汇总路由条目，主要是为了避免路由环路。

（3）查看 IS-IS 接口的电路类型

```
R3#show clns interface s0/0/0
Serial0/0/0 is up, line protocol is up
  （此处省略部分输出）    Routing Protocol: IS-IS
    Circuit Type: level-2
    （此处省略部分输出）
```

从以上输出可以看到接口的电路类型为 level-2。

（4）查看 IS-IS 路由协议相关信息

```
R4#show ip protocols | begin Routing Protocol is "isis cisco"
Routing Protocol is "isis cisco"
  Outgoing update filter list for all interfaces is not set
  Incoming update filter list for all interfaces is not set
  Redistributing: isis
  Address Summarization:
    172.16.0.0/255.255.252.0 into level-2
//以上两行表示地址汇总信息，该路由类型为 level-2
  Maximum path: 4
  Routing for Networks:
    Serial0/0/0
Passive Interface(s):
```

```
        Loopback0
        Loopback1
        Loopback2
        Loopback3
    Routing Information Sources:
        Gateway            Distance      Last Update
        (this router)      115           14:02:24
        172.16.23.3        115           00:01:30
        172.16.23.2        115           00:01:30
    Distance: (default is 115)
```

（5）查看 IS-IS 路由器类型

```
① R1#show clns protocol
IS-IS Router: cisco
    System Id: 1111.1111.1111.00    IS-Type: level-1
（此处省略部分输出）
② R2#show clns protocol
IS-IS Router: cisco
    System Id: 2222.2222.2222.00    IS-Type: level-1-2
（此处省略部分输出）
③ R3#show clns protocol
IS-IS Router: cisco
    System Id: 3333.3333.3333.00    IS-Type: level-2
（此处省略部分输出）
④ R4#show clns protocol
IS-IS Router: cisco
    System Id: 4444.4444.4444.00    IS-Type: level-2
（此处省略部分输出）
```

以上①、②、③和④输出表明路由器 R1 的 IS-IS 路由器类型为 L1，路由器 R2 的 IS-IS 路由器类型为 L1/2，路由器 R3 的 IS-IS 路由器类型为 L2，路由器 R4 的 IS-IS 路由器类型为 L4。

5.2.3 实验 3：配置集成 IS-IS 验证

1. 实验目的

通过本实验可以掌握：
① IS-IS 验证的类型和意义。
② IS-IS 明文验证的配置和调试方法。
③ IS-IS MD5 验证的配置和调试方法。

2. 实验拓扑

配置集成的 IS-IS 验证实验拓扑如图 5-12 所示。

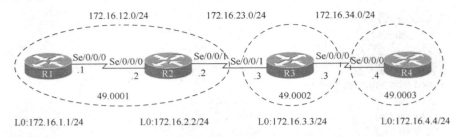

图 5-12 配置集成的 IS-IS 验证实验拓扑

3. 实验步骤

新版本的 IOS 支持 IS-IS 基于钥匙链的明文和 HMAC MD5 两种方式验证。本实验中路由器 R1 和 R2 的邻居采用 Level-1 明文验证，路由器 R3 和 R4 的邻居采用 Level-2 MD5 验证，区域 49.0001 采用 MD5 验证，域采用 MD5 验证。

（1）配置路由器 R1

```
R1(config)#key chain L1_neighbor_auth
R1(config-keychain)#key 1
R1(config-keychain-key)#key-string cisco
R1(config)#key chain Area_auth
R1(config-keychain)#key 1
R1(config-keychain-key)#key-string ccna
R1(config)#interface Loopback0
R1(config-if)#ip router isis cisco
R1(config)#interface Serial0/0/0
R1(config-if)#ip router isis cisco
R1(config-if)#isis authentication mode text level-1
//配置 IS-IS 邻居验证模式，启用 Level-1 邻居验证
R1(config-if)#isis authentication key-chain L1_neighbor_auth level-1
//调用 IS-IS 邻居验证钥匙链
R1(config)#router isis cisco
R1(config-router)#net 49.0001.1111.1111.1111.00
R1(config-router)#is-type level-1
R1(config-router)#authentication mode md5 level-1
//配置 IS-IS 区域验证模式，启用 Level-1 区域验证
R1(config-router)#authentication key-chain Area_auth level-1
//调用 IS-IS 区域验证钥匙链
```

（2）配置路由器 R2

```
R2(config)#key chain L1_neighbor_auth
R2(config-keychain)#key 1
R2(config-keychain-key)#key-string cisco
R2(config)#key chain Area_auth
R2(config-keychain)#key 1
R2(config-keychain-key)#key-string ccna
R2(config)#key chain Domain_auth
R2(config-keychain)#key 1
```

第 5 章 IS-IS

```
R2(config-keychain-key)#key-string ccie
R2(config)#interface Loopback0
R2(config-if)#ip router isis cisco
R2(config)#interface Serial0/0/0
R2(config-if)#ip router isis cisco
R2(config-if)#isis authentication mode text level-1
R2(config-if)#isis authentication key-chain L1_neighbor_auth level-1
R2(config)#interface Serial0/0/1
R2(config-if)#ip router isis cisco
R2(config)#router isis cisco
R2(config-router)#net 49.0001.2222.2222.2222.00
R2(config-router)#authentication mode md5
R2(config-router)#authentication key-chain Area_auth level-1
R2(config-router)#authentication key-chain Domain_auth level-2
//配置 IS-IS 域验证模式，启用 Level-2 域验证
```

（3）配置路由器 R3

```
R3(config)#key chain Domain_auth
R3(config-keychain)#key 1
R3(config-keychain-key)#key-string ccie
R3(config)#key chain L2_neighbor_auth
R3(config-keychain)#key 1
R3(config-keychain-key)#key-string ccnp
R3(config)#interface Loopback0
R3(config-if)#ip router isis cisco
R3(config)#interface Serial0/0/0
R3(config-if)#ip router isis cisco
R3(config-if)#isis circuit-type level-2-only
R3(config-if)#isis authentication mode md5 level-2
//配置邻居验证模式，启用 Level-2 邻居验证
R3(config-if)#isis authentication key-chain L2_neighbor_auth level-2
R3(config)#interface Serial0/0/1
R3(config-if)#ip router isis cisco
R3(config)#router isis cisco
R3(config-router)#net 49.0002.3333.3333.3333.00
R3(config-router)#is-type level-2-only
R3(config-router)#authentication mode md5 level-2
R3(config-router)#authentication key-chain Domain_auth level-2
```

（4）配置路由器 R4

```
R4(config)#key chain Domain_auth
R4(config-keychain)#key 1
R4(config-keychain-key)#key-string ccie
R4(config)#key chain L2_neighbor_auth
R4(config-keychain)#key 1
R4(config-keychain-key)#key-string ccnp
R4(config)#interface Loopback0
R4(config-if)#ip router isis cisco
R4(config)#interface Serial0/0/0
```

```
R4(config-if)#ip router isis cisco
R4(config-if)#isis circuit-type level-2-only
R4(config-if)#isis authentication mode md5 level-2
R4(config-if)#isis authentication key-chain L2_neighbor_auth level-2
R4(config)#router isis cisco
R4(config-router)#net 49.0003.4444.4444.4444.00
R4(config-router)#is-type level-2-only
R4(config-router)#authentication mode md5 level-2
R4(config-router)#authentication key-chain Domain_auth level-2
```

【技术要点】

① 邻居验证、区域验证和域验证可以根据需要使用明文或 MD5 验证方式。

② 在邻居之间进行验证时，可以对 L1 和 L2 类型的邻居分别验证，可以使用不同的钥匙链。

③ 在进行区域验证时，区域内每台路由器都必须使用相同的验证模式。

④ 在进行域验证时，IS-IS 域内的每台 L2 和 L1/L2 路由器都必须使用相同的验证模式。

⑤ 当使用 **isis authentication mode** 和 **isis authentication key-chain** 命令时，如果没有指定关键字 **level-1** 或 **level-2**，默认是 **level-1** 和 **level-2**。

⑥ 邻居验证只用来验证邻居关系的建立，而区域验证可以验证 L1 类型的链路状态数据库信息的交换。例如，区域 49.0001 验证没有配置正确，邻居验证配置正确，路由器 R1 和 R2 仍然可以形成邻居关系，但是不会进行 L1 的 LSP 交换。

⑦ 域验证可以用来验证 L2 路由信息的交换，但是不会去验证 L2 类型的邻居关系。

4. 实验调试

以下输出全部省略路由代码部分。

（1）查看路由器 R1 的 IS-IS 路由

```
R1#show ip route isis
     172.16.0.0/24 is subnetted, 4 subnets
i L1    172.16.23.0 [115/20] via 172.16.12.2, Serial0/0/0
i L1    172.16.2.0 [115/20] via 172.16.12.2, Serial0/0/0
i*L1    0.0.0.0/0 [115/10] via 172.16.12.2, Serial0/0/0
```

（2）查看路由器 R2 的 IS-IS 路由

```
R2#show ip route isis
     172.16.0.0/24 is subnetted, 7 subnets
i L2    172.16.34.0 [115/20] via 172.16.23.3, Serial0/0/1
i L2    172.16.4.0 [115/30] via 172.16.23.3, Serial0/0/1
i L1    172.16.1.0 [115/20] via 172.16.12.1, Serial0/0/0
i L2    172.16.3.0 [115/20] via 172.16.23.3, Serial0/0/1
```

（3）查看路由器 R3 的 IS-IS 路由

```
R3#show ip route isis
```

```
           172.16.0.0/24 is subnetted, 7 subnets
    i L2   172.16.12.0 [115/20] via 172.16.23.2, Serial0/0/1
    i L2   172.16.4.0 [115/20] via 172.16.34.4, Serial0/0/0
    i L2   172.16.1.0 [115/30] via 172.16.23.2, Serial0/0/1
    i L2   172.16.2.0 [115/20] via 172.16.23.2, Serial0/0/1
```

(4) 查看路由器 R4 的 IS-IS 路由

```
R4#show ip route isis
           172.16.0.0/24 is subnetted, 7 subnets
    i L2   172.16.23.0 [115/20] via 172.16.34.3, Serial0/0/0
    i L2   172.16.12.0 [115/30] via 172.16.34.3, Serial0/0/0
    i L2   172.16.1.0 [115/40] via 172.16.34.3, Serial0/0/0
    i L2   172.16.2.0 [115/30] via 172.16.34.3, Serial0/0/0
    i L2   172.16.3.0 [115/20] via 172.16.34.3, Serial0/0/0
```

以上输出表明邻居验证、区域验证和域验证通过，路由信息正确。

本 章 小 结

作为链路状态路由协议，IS-IS 基于 ISO CLNS，设计初是为了实现 ISO CLNP 路由，后来增加了对 IP 路由的支持，具有良好的灵活性、可扩展性和稳定性。从选择方向看，IS-IS 更适合运营商级的网络，而 OSPF 非常适合企业级网络。本章介绍了 IS-IS 特征、数据包格式、路由器类型等，并用实验演示和验证了单区域集成 IS-IS 的基本配置、多区域集成 IS-IS 配置和 IS-IS 验证配置。

第 6 章　路由重分布与路径控制

当网络中运行多种路由协议时，必须在这些不同的路由选择协议之间共享路由信息，才能保证网络连通性。在路由选择协议之间交换路由信息的过程被称为路由重分布（Route Redistribution）。为了保证网络的伸缩性、稳定性、安全性和快速收敛，有必要对路由信息的更新进行控制和优化。

6.1　路由重分布概述

6.1.1　路由重分布种子度量值

路由重分布为在同一个互联网络中高效地支持多种路由协议提供了可能，执行路由重分布的路由器被称为边界路由器，因为它们位于两个或多个自治系统的边界上。

路由重分布时度量标准和管理距离是必须要考虑的。每一种路由协议都有自己的度量标准，所以在进行路由重分布时必须转换度量标准，使得它们兼容。种子度量值（Seed Metric）是在路由重分布时定义的，它是一条通过外部重分布进来的路由的初始度量值。路由协议默认的种子度量值如表 6-1 所示。

表 6-1　路由协议默认的种子度量值

路由重分布进来的路由协议	默认种子度量值
RIP	无限大
EIGRP	无限大
OSPF	BGP 为 1，其他为 20
IS-IS	0
BGP	IGP 的度量值

6.1.2　路由重分布存在的问题

进行路由重分布时应该考虑如下一些问题。

① 路由反馈：路由器有可能将从一个自治系统学到的路由信息发送回该自治系统，特别是在进行双向重分布时，一定要注意。

② 次优路由：每一种路由协议的度量标准不同，所以路由器通过重分布所选择的路径可能并非是最佳路径。一般可通过路由过滤或者修改管理距离等方式来解决路由重分布的次优路由问题。

③ 收敛时间不一致：因为不同的路由协议收敛的时间不同，如 EIGRP 的收敛速度要比 RIP 快得多。

6.2 路径控制概述

为了保证网络的高效运行以及在进行路由重分布时避免次优路由或者路由环路，有必要对路由更新进行控制，常用的方法有被动接口、默认路由、静态路由、路由映射表、分布列表、前缀列表、偏移列表、Cisco IOS IP 服务等级协议（SLA）和策略路由。在对路径进行控制时，可能是多种方法的组合。

6.2.1 路由映射表（Route Map）

可以将路由映射表比作复杂的访问控制列表，主要用于路由重分布的路由控制和策略路由等。定义路由映射表的命令为 **route-map** *map-tag* **[permit | deny]** *[sequence-number]*。通常每个 Route Map 陈述中都包含 **match** 和 **set**。

match 用来匹配条件，常用的匹配条件包括 IP 地址、接口、度量值、tag、路由类型以及数据包长度等。当有多个匹配条件时，逻辑关系必须搞清楚。如格式为 **match ip address a b c**，表示逻辑或，只要有一个条件匹配即可。如果格式为

match ip address a
match ip address b
match ip address c

则表示逻辑与，必须匹配所有的条件。

set 定义符合匹配条件时所采取的行为，通常，set 的行为如表 6-2 所示。

表 6-2 set 的行为

set 行为	描述
set ip next hop	设定数据包的下一跳地址
set interface	设定数据包出接口
set ip default next hop	设定默认的下一跳地址，用于当路由表里没有到达目的地址路由条目的时候
set default interface	设定默认的出接口
set ip tos	设定 IP 数据包的 ToS 值
set ip precedence	设定 IP 数据包的优先级
set metric	设定路由的度量值
set tag	设定路由的标记值

定义路由映射表时，要注意以下几点。

① 一个 Route Map 的末尾默认行为是 deny any。这个 deny 的使用结果依赖于这个 Route Map 是怎样使用的。比如，在执行策略路由时，如果一个数据包对于 Route Map 没有匹配项，它会正常转发数据包；而在路由重分布时，对于路由条目，如果 Route Map 没有匹配项，则被过滤掉。

② 一个 Route Map 可以包含多个 Route Map 陈述，它们的执行顺序像 ACL 一样，按照从上到下的顺序执行。如果 Route Map 陈述没有 match，则意味着匹配所有。

③ 序号指定了 Route Map 陈述条件检查的顺序，编号默认为 10，序号不会自动递增。
④ 在 Route Map 陈述中不写 deny 或 permit 则默认为 permit。
⑤ 在删除 Route Map 语句时，没写编号则删除整个 Route Map。

6.2.2 分布列表、前缀列表和偏移列表

（1）分布列表（Distribute List）

ACL 通常用于过滤用户数据流，而不是路由协议产生的流量，同时 ACL 对自身产生的流量不能进行过滤，而分布列表提供控制路由更新的另一种方法，通常与 ACL、路由映射表或者前缀列表结合使用。可以对单一路由协议的路由进行过滤，也可在路由协议之间做重分布的时候进行路由过滤，防止路由反馈和路由环路等。

使用分布列表对允许哪些路由更新，拒绝哪些路由更新等具有很大的灵活性，但是值得注意的是对于同一个区域的链路状态路由协议（如 OSPF），由于要保持链路状态数据库的同步，所以不能使用分布列表对出站路由进行过滤。

在配置分布列表时，不同的路由协议，可供选择的参数可能不同。

（2）前缀列表（Prefix List）

前缀列表的作用类似于 ACL，但比 ACL 更为灵活，且更易于理解。前缀列表的特点如下所述。

① 编辑的方便性：在配置前缀列表时，可以指定序列号，只要序列号不是连续的，以后就可以方便地插入条目，或者删除针对某个序号的条目，而不是整个前缀列表。

② 执行的高效性：在大型列表的加载和路由查找方面性能比 ACL 有显著的改进。

③ 灵活性：可以在前缀列表中指定掩码的长度，也可以指明掩码长度的范围。

（3）偏移列表（Offset List）

偏移列表用于在出站或者入站方向增加 EIGRP 或者 RIP 路由条目的度量值。偏移列表只适用于距离矢量路由协议。可以通过 ACL 来限制偏移列表的作用范围。

6.2.3 IP SLA

企业的边缘路由器可以检测出连接到 ISP 的链路是否出现故障，对于存在备份链路的情况，如果去往一个 ISP 的链路出现故障，边缘路由器将使用备份 ISP 链路转发数据包。但是，如果 ISP 的基础设施出现故障，而边缘路由器连接到 ISP 的链路仍然能够正常工作，这时仍然会选择有故障的 ISP，这显然是不合理的。Cisco IOS IP SLA 通过向网络发送模拟数据流，以连续、可靠和可预测的方式来主动监控网络的性能和可达性。收集的信息包括响应时间、单向延迟、抖动、丢包率、语音品质、网络资源可用性、应用程序性能以及服务器响应时间等。通过使用 Cisco IOS IP SLA 收集的信息，可以核实服务保证、检查网络性能、改善网络的可靠性和预先发现网络问题，进而调整和配置网络。因此，Cisco IOS IP SLA 在性能测量、性能监控以及建立网络基线数据方面非常有用。Cisco IOS IP SLA 可以测量 Cisco 设备之间以及 Cisco 设备和主机之间的性能，并向 IP 应用程序和服务提供有关服务等级的信息。所有的 Cisco IOS IP SLA 探测操作都是在 IP SLA 源上配置的，由源向目标发送探测数据包。

Cisco IOS IP SLA 有两种操作：一种是目标设备运行了 IP SLA 响应器，如 Cisco 路由器；另一种是目标设备没有运行 IP SLA 响应器组件，如 Web 服务器或 IP 主机。实施 IP SLA 必须执行以下任务。

① 启用 IP SLA 响应方（如果需要）。
② 配置所需要的 IP SLA 操作类型。
③ 配置操作类型的选项。
④ 配置阈值上限或下限的条件（如果需要）。
⑤ 调度 IP SLA 的操作。
⑥ 收集信息。

6.2.4 策略路由（PBR）

策略路由提供了根据网络管理者制定的策略来进行数据包转发的一种机制。基于策略的路由比传统路由能力更强，使用更灵活，它使网络管理者不仅能够根据目的地址而且能够根据协议类型、报文大小、应用或 IP 源地址来选择转发路径。策略路由的策略由路由映射图表来定义。

6.2.5 VRF

Cisco 的虚拟化技术可以把通过公共物理网络传输的数据流进行分离。VRF（VPN Routing and Forwarding）是 Cisco 的一种虚拟化技术，允许在逻辑上把共享的网络设备的路由功能分开。每一个 VRF 可以看作一台虚拟的路由器，包括独立的路由表/转发表、一组归属于该 VRF 的接口集合和一组只用于本 VRF 的路由协议。VRF 经常与 MPLS VPN 一起使用，所以首先介绍以下几个术语。

① CE 路由器：客户端路由器，与 PE 路由器相连。
② PE 路由器：服务供应商控制的路由器。
③ RD（Route Distinguisher，路由区分符）：64 比特的路由标识符，附加在 IPv4 地址前，构成 VPNv4 地址，来保持路由前缀的唯一性，能够有效地处理重叠的地址空间。PE 从 CE 接收的是标准的 IPv4 路由，如果需要传递给其他的 PE 路由器，就需要为这条路由附加一个 RD。在 IPv4 地址加上 RD 之后，就变成 VPNv4 地址族。VPNv4 地址仅用于服务供应商网络内部。接收端 PE 可以根据 RD 确定路由所属的 VPN，从而把路由安装到正确的 VRF 中。
④ RT（Route Target，路由目标）：64 比特的路由标识符，实现路由的选择性导入和导出，让接收路由的 PE 端知道要把路由导入哪些 VRF。RT 是一个扩展的 BGP 的团体属性，VPNv4 路由在传递时是要带上这个属性的。RT 值有两个，一个是 export RT，一个是 import RT。当 PE 发布路由时，将使用路由所属 VRF 的 RT 的 export 值，直接发送给其他的 PE 设备。PE 在接收路由时，首先接收所有的路由，并根据每个 VRF 配置的 RT 的 import 值进行检查，如果与路由中的 RT 值匹配，则将该路由加入到相应的 VRF 中。也就是说，发送端 PE 的 export 必须和接收端 PE 的 import 匹配，接收端 PE 才会把路由加入 VRF 路由表。

VRF 可以脱离 MPLS 单独应用。VRF Lite（Multi-VRF）就是典型例子，VRF Lite 可在 CE 设备上支持 VRF。

6.3 路由重分布

6.3.1 实验1：路由重分布基本配置

1. 实验目的

通过本实验可以掌握：
① 种子度量值的配置方法。
② 路由重分布参数的含义。
③ 直连和静态路由的重分布方法。
④ IS-IS 和 EIGRP 的重分布方法。
⑤ EIGRP 和 OSPF 的重分布方法。
⑥ 重分布路由的查看和调试方法。

2. 实验拓扑

路由重分布基本配置实验拓扑如图 6-1 所示。

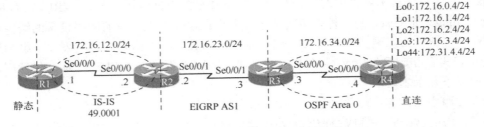

图 6-1　路由重分布基本配置实验拓扑

3. 实验步骤

（1）配置路由器 R1

```
R1(config)#router isis cisco
R1(config-router)#net 49.0001.1111.1111.1111.00
R1(config-router)#is-type level-2-only
R1(config-router)# redistribute static ip metric 15    //重分布静态路由
R1(config)#interface Serial0/0/0
R1(config-if)#ip router isis cisco
R1(config)#ip route 192.168.1.0 255.255.255.0 null0
```

（2）配置路由器 R2

```
R2(config)#router eigrp 1
R2(config-router)#network 172.16.23.2 0.0.0.0
R2(config-router)# redistribute isis cisco level-2 metric 10000 100 255 1 1500
```

```
            //将 IS-IS 路由重分布到 EIGRP 进程中
            R2(config-router)#redistribute connected metric 10000 100 255 1 1500
            R2(config)#router isis cisco
            R2(config-router)# net 49.0001.2222.2222.2222.00
            R2(config-router)# is-type level-2-only
            R2(config-router)#redistribute eigrp 1
            //将 EIGRP 重分布到 IS-IS 中
            R2(config)#interface Serial0/0/0
            R2(config-if)#ip router isis cisco
```

【技术要点】

① 因为 EIGRP 的度量相对复杂，所以重分布时需要分别指定带宽、延迟、可靠性、负载以及 MTU 参数的值。

② 在重分布其他路由进入 IS-IS 路由进程时，默认种子度量值默认为 0。

③ 其他路由协议在重分布 IS-IS 路由时，只重分布 IS-IS 路由本身，而运行 IS-IS 的接口的直连路由不会一起重分布，需要单独重分布直连路由。

④ 对于 RIP、EIGRP 和 OSPF 路由进程，默认情况下，在重分布静态和直连路由时，不用指定种子度量值，路由器使用默认的种子度量值，如在将直连或静态路由重分布进 RIP 时，度量值为 1。

（3）配置路由器 R3

```
            R3(config)#router eigrp 1
            R3(config-router)#network 172.16.23.3 0.0.0.0
            R3(config-router)#redistribute ospf 1 metric 10000 100 255 1 1500
            //将 OSPF 路由重分布到 EIGRP 进程中
            R3(config-router)#distance eigrp 90 150
            //配置 EIGRP 默认管理距离，内部路由为 90，外部路由为 150
            R3(config)#router ospf 1
            R3(config-router)#router-id 3.3.3.3
            R3(config-router)#network 172.16.34.3 0.0.0.0 area 0
            R3(config-router)#redistribute eigrp 1 metric 30 metric-type 1 subnets
            //将 EIGRP 路由重分布到 OSPF 进程中，subnets 参数指定重分布子网路由，否则只重分布主类网络的路由
            R3(config-router)#redistribute maximum-prefix 100 50
```

【技术要点】

命令 redistribute maximum-prefix *maximum* [*threshold*] [warning-only]用于限制重分布到 OSPF 进程中的最大路由条目的数量，各参数含义如下。

① *maximum*：重分布到 OSPF 中的最大路由条目数，达到最大路由条目数后，将不再重分布路由信息，提示的信息类似%IPRT-4-REDIST_MAX_PFX: Redistribution **prefix limit has been reached** "ospf 1" – 100 prefixes。

② *threshold*：当超过最大路由条目数量的百分比之后，路由器将发出告警信息，默认为 75%，告警信息类似%IPRT-4-REDIST_THR_PFX: Redistribution **prefix threshold has been** reached "ospf 1" – 50 prefixes。

③ **warning-only**：当超过了最大路由条目数之后，只显示告警信息，不会限制重分布的路由条目的数量。

（4）配置路由器 R4

```
R4(config)#access-list 1 permit 172.16.0.0 0.0.254.0
//该 ACL 用于匹配 172.16.X.0，X 代表偶数
R4(config)#access-list 2 permit 172.16.1.0 0.0.254.0
//该 ACL 用于匹配 172.16.X.0，X 代表奇数
R4(config)#route-map CONN permit 10        //配置路由映射表
R4(config-route-map)#match ip address 1    //匹配 ACL 1
R4(config-route-map)#set metric 200        //设置度量值
R4(config-route-map)#exit
R4(config)#route-map CONN permit 20
R4(config-route-map)#match ip address 2
R4(config-route-map)#set metric 100
R4(config-route-map)#set metric-type type-1
//设置 OSPF 路由类型为 OE1，默认为 OE2
R4(config-route-map)#exit
R4(config)#route-map CONN permit 30
```
//以上路由映射表 CONN 是为在 OSPF 进程重分布直连路由时调用的，其含义是：对于 172.16 开头的第三位是偶数的路由条目，设置度量值为 100，路由类型为 OE2（默认不用配置），对于 172.16 开头的第三位是奇数的路由条目，设置度量值为 200，路由类型为 OE1，而其他路由条目采用默认设置，即度量值为 20，路由类型为 OE2。
```
R4(config)#router ospf 1
R4(config-router)#router-id 4.4.4.4
R4(config-router)#network 172.16.34.4 0.0.0.0 area 0
R4(config-router)#redistribute connected subnets route-map CONN
//将直连路由重分布到 OSPF 进程中，重分布时调用路由映射表
```

4. 实验调试

（1）查看路由表

以下输出均省略路由代码部分。

① R1#**show ip route**
```
172.16.0.0/16 is variably subnetted, 8 subnets, 2 masks
i L2     172.16.0.0/24 [115/10] via 172.16.12.2, 01:49:07, Serial0/0/0
i L2     172.16.1.0/24 [115/10] via 172.16.12.2, 01:49:07, Serial0/0/0
i L2     172.16.2.0/24 [115/10] via 172.16.12.2, 01:49:07, Serial0/0/0
i L2     172.16.3.0/24 [115/10] via 172.16.12.2, 01:49:07, Serial0/0/0
C        172.16.12.0/24 is directly connected, Serial0/0/0
L        172.16.12.1/32 is directly connected, Serial0/0/0
i L2     172.16.23.0/24 [115/10] via 172.16.12.2, 01:49:07, Serial0/0/0
i L2     172.16.34.0/24 [115/10] via 172.16.12.2, 01:49:07, Serial0/0/0
     172.31.0.0/24 is subnetted, 1 subnets
i L2     172.31.4.0 [115/10] via 172.16.12.2, 01:49:07, Serial0/0/0
S        192.168.1.0/24 is directly connected, Null0
```

以上输出表明路由器 R1 通过 IS-IS 学到从路由器 R2 重分布进 IS-IS 进程中的 L2 路由。

```
② R2#show ip route
     172.16.0.0/16 is variably subnetted, 9 subnets, 2 masks
D EX    172.16.0.0/24 [170/2195456] via 172.16.23.3, 02:03:21, Serial0/0/1
D EX    172.16.1.0/24 [170/2195456] via 172.16.23.3, 02:03:21, Serial0/0/1
D EX    172.16.2.0/24 [170/2195456] via 172.16.23.3, 02:03:21, Serial0/0/1
D EX    172.16.3.0/24 [170/2195456] via 172.16.23.3, 02:03:21, Serial0/0/1
C       172.16.12.0/24 is directly connected, Serial0/0/0
L       172.16.12.2/32 is directly connected, Serial0/0/0
C       172.16.23.0/24 is directly connected, Serial0/0/1
L       172.16.23.2/32 is directly connected, Serial0/0/1
D EX    172.16.34.0/24 [170/2195456] via 172.16.23.3, 02:03:21, Serial0/0/1
     172.31.0.0/24 is subnetted, 1 subnets
D EX    172.31.4.0 [170/2195456] via 172.16.23.3, 02:03:21, Serial0/0/1
i L2    192.168.1.0/24 [115/25] via 172.16.12.1, 01:37:48, Serial0/0/0
```

以上输出表明从路由器 R1 上重分布进 IS-IS 进程中的静态路由被路由器 R2 学习到，路由代码为 **i L2**，度量值为 25（R1 上的种子度量值 15 加上 R2 接口的开销 10）；从路由器 R3 上重分布进 EIGRP 进程中的 OSPF 路由也被路由器 R2 学习到，路由代码为 **D EX**，管理距离为 170，这也说明 EIGRP 能够识别内部路由和外部路由，默认时，EIGRP 内部路由的管理距离是 90，外部路由的管理距离是 170。

```
③ R3#show ip route
     172.16.0.0/16 is variably subnetted, 9 subnets, 2 masks
O E2    172.16.0.0/24 [110/200] via 172.16.34.4, 02:27:49, Serial0/0/0
O E1    172.16.1.0/24 [110/164] via 172.16.34.4, 02:27:49, Serial0/0/0
O E2    172.16.2.0/24 [110/200] via 172.16.34.4, 02:27:49, Serial0/0/0
O E1    172.16.3.0/24 [110/164] via 172.16.34.4, 02:27:49, Serial0/0/0
D EX    172.16.12.0/24 [150/2195456] via 172.16.23.2, 01:51:02, Serial0/0/1
C       172.16.23.0/24 is directly connected, Serial0/0/1
L       172.16.23.3/32 is directly connected, Serial0/0/1
C       172.16.34.0/24 is directly connected, Serial0/0/0
L       172.16.34.3/32 is directly connected, Serial0/0/0
     172.31.0.0/24 is subnetted, 1 subnets
O E2    172.31.4.0 [110/20] via 172.16.34.4, 02:27:49, Serial0/0/0
D EX    192.168.1.0/24 [150/2195456] via 172.16.23.2, 01:53:47, Serial0/0/1
```

以上输出表明，从路由器 R2 上重分布到 EIGRP 进程中的 IS-IS L2 路由被路由器 R3 学习到，同时 EIGRP 外部路由的管理距离被修改为 150；从路由器 R4 上重分布到 OSPF 进程的直连路由被路由器 R3 学习到，OSPF 路由的度量值及路由类型同 R4 上定义的路由映射表 CONN 设置的是一致的。

```
④ R4#show ip route
     172.16.0.0/16 is variably subnetted, 12 subnets, 2 masks
O E1    172.16.12.0/24 [110/94] via 172.16.34.3, 01:53:54, Serial0/0/0
O E1    172.16.23.0/24 [110/94] via 172.16.34.3, 02:30:52, Serial0/0/0
O E1    192.168.1.0/24 [110/94] via 172.16.34.3, 01:56:39, Serial0/0/0
（此处省略直连路由和本地路由的输出）
```

以上输出表明，从路由器 R3 上重分布到 OSPF 进程中的路由被路由器 R4 学习到，路由

代码为 **O E1**。

（2）查看路由协议相关信息

```
R3#show ip protocols | begin Routing Protocol is "eigrp 1"
Routing Protocol is "eigrp 1"        //运行 AS 为 1 的 EIGRP 进程
  Outgoing update filter list for all interfaces is not set
  Incoming update filter list for all interfaces is not set
  Default networks flagged in outgoing updates
  Default networks accepted from incoming updates
  EIGRP metric weight K1=1, K2=0, K3=1, K4=0, K5=0
  EIGRP maximum hopcount 100
  EIGRP maximum metric variance 1
  Redistributing: eigrp 1, ospf 1        //OSPF 进程 1 被重分布到 EIGRP 进程中
  EIGRP NSF-aware route hold timer is 240s
  Automatic network summarization is not in effect
  Maximum path: 4
  Routing for Networks:
    172.16.23.3/32
  Routing Information Sources:
    Gateway          Distance       Last Update
    172.16.23.2         90          00:06:28
  Distance: internal 90 external 150        //管理距离

Routing Protocol is "ospf 1"        //运行 OSPF 进程，进程号为 1
  Outgoing update filter list for all interfaces is not set
  Incoming update filter list for all interfaces is not set
  Router ID 3.3.3.3
  It is an autonomous system boundary router        //R3 是 ASBR 路由器
  Redistributing External Routes from,
    eigrp 1 with metric mapped to 30, includes subnets in redistribution
//EIGRP 进程 1 的路由被重分布到 OSPF 进程中，种子度量值为 30，能够重分布子网路由
  Number of areas in this router is 1. 1 normal 0 stub 0 nssa
  Maximum path: 4
  Routing for Networks:
    172.16.34.3 0.0.0.0 area 0
  Reference bandwidth unit is 100 mbps
  Routing Information Sources:
    Gateway          Distance       Last Update
    4.4.4.4            110          00:12:13
  Distance: (default is 110)
```

以上输出表明路由器 R3 运行 EIGRP 和 OSPF 两种路由协议，而且实现了双向重分布。

（3）查看 route-map 信息

```
R4#show route-map CONN
route-map CONN, permit, sequence 10
  Match clauses:        //匹配条件
    ip address (access-lists): 1        //IP 地址匹配 ACL 1
  Set clauses:        //执行行为
    metric 200        //设置度量值
```

```
    Policy routing matches: 0 packets, 0 bytes    //策略路由匹配的数据包个数和字节数
route-map CONN, permit, sequence 20
    Match clauses:
        ip address (access-lists): 2    //IP 地址匹配 ACL 1
    Set clauses:
        metric 100    //设置度量值
        metric-type type-1    //设置 OSPF 外部路由的类型
    Policy routing matches: 0 packets, 0 bytes
route-map CONN, permit, sequence 30
    Match clauses:    //空表示匹配所有
    Set clauses:        //空表示不执行任何行为
    Policy routing matches: 0 packets, 0 bytes
```

6.3.2 实验 2：路由重分布中次优路由和路由环路问题及其解决方案

1. 实验目的

通过本实验可以掌握：
① 次优路由问题和路由环路的产生及其解决方案。
② 路由重分布过程中修改管理距离的方法。
③ 路由重分布过程中使用路由映射表的方法。

2. 实验拓扑

路由重分布次优路由和路由环路的解决方案实验拓扑如图 6-2 所示。

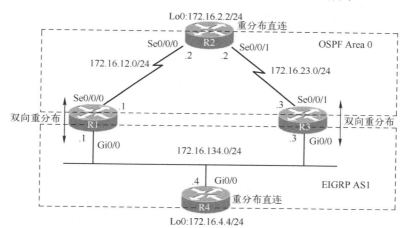

图 6-2 路由重分布次优路由和路由环路的解决方案实验拓扑

3. 实验步骤

（1）配置路由器 R1

```
R1(config)#router eigrp 1
R1(config-router)#network 172.16.134.1 0.0.0.0
```

```
R1(config-router)#redistribute ospf 1 metric 10000 100 255 1 1500
R1(config)#router ospf 1
R1(config-router)#router-id 1.1.1.1
R1(config-router)#network 172.16.12.1 0.0.0.0 area 0
R1(config-router)#redistribute eigrp 1 subnets
```

（2）配置路由器 R2

```
R2(config)#router ospf 1
R2(config-router)#router-id 2.2.2.2
R2(config-router)#redistribute connected subnets
R2(config-router)#network 172.16.12.2 0.0.0.0 area 0
R2(config-router)#network 172.16.23.2 0.0.0.0 area 0
```

（3）配置路由器 R3

```
R3(config)#router eigrp 1
R3(config-router)#network 172.16.134.3 0.0.0.0
R3(config-router)#redistribute ospf 1 metric 10000 100 255 1 1500
R3(config)#router ospf 1
R3(config-router)#router-id 3.3.3.3
R3(config-router)#network 172.16.23.3 0.0.0.0 area 0
R3(config-router)#redistribute eigrp 1 subnets
```

（4）配置路由器 R4

```
R4(config)#router eigrp 1
R4(config-router)#network 172.16.134.4 0.0.0.0
R4(config-router)#redistribute connected metric 10000 100 255 1 1500
```

4. 实验调试

（1）查看路由表

以下路由表输出均省略路由代码部分。

```
① R1#show ip route
（此处省略直连路由和本地路由）
     172.16.0.0/16 is variably subnetted, 7 subnets, 2 masks
O E2    172.16.2.0/24 [110/20] via 172.16.12.2, 00:01:43, Serial0/0/0
O E2    172.16.4.0/24 [110/20] via 172.16.12.2, 00:00:35, Serial0/0/0
O       172.16.23.0/24 [110/128] via 172.16.12.2, 00:01:43, Serial0/0/0
```

本实验中，在路由器 R3 上首先执行 OSPF 和 EIGRP 双向重分布。在路由器 R1 的路由表中，发现到 172.16.4.0 的下一跳地址是 172.16.12.2，也就是经过路由器 R2 和 R3 到达，而没有选择直接到 R4，这就是次优路由，很明显这是不合理的。接下来分析一下次优路由产生的原因：由于 172.16.4.0 是通过 R4 重分布进入 EIGRP 的，所以默认管理距离是 170，在路由器 R1 和 R3 上没有执行双向重分布之前，172.16.4.0 会以 **D EX** 代码出现在路由器 R1 和 R3 的路由表中，默认管理距离是 170。在 R3 上首先执行了 OSPF 和 EIGRP 双向重分布，外部

路由条目 **172.16.4.0** 进入 OSPF，该条目通过 5 类 LSA 传递给路由器 R1，R1 通过比较管理距离，发现从路由器 R2 收到该路由条目的管理距离为 110，而从路由器 R4 收到的路由条目管理距离为 170，所以 R1 更新路由表，选择路由条目管理距离低的路径，即下一跳指向 R2，因而造成了次优路由。这一点可以通过查看 R1 的 OSPF 数据库和 EIGRP 拓扑表得到证实：

```
R1#show ip ospf database
        OSPF Router with ID (1.1.1.1) (Process ID 1)
            Router Link States (Area 0)
Link ID      ADV Router     Age       Seq#           Checksum      Link count
1.1.1.1      1.1.1.1        187       0x80000001     0x0024DD      2
2.2.2.2      2.2.2.2        188       0x80000008     0x00D8C6      4
3.3.3.3      3.3.3.3        121       0x80000002     0x0023B3      2
            Type-5 AS External Link States
Link ID      ADV Router     Age       Seq#           Checksum      Tag
172.16.2.0   2.2.2.2        342       0x80000001     0x0010C9      0
172.16.4.0   3.3.3.3        121       0x80000001     0x00DBF7      0
172.16.134.0 3.3.3.3        121       0x80000001     0x004011      0

R1#show ip eigrp topology
EIGRP-IPv4 Topology Table for AS(1)/ID(172.16.134.1)
Codes: P - Passive, A - Active, U - Update, Q - Query, R - Reply,
       r - reply Status, s - sia Status
P 172.16.2.0/24, 0 successors, FD is Infinity
        via 172.16.134.3 (284160/281600), GigabitEthernet0/0
P 172.16.12.0/24, 0 successors, FD is Infinity
        via 172.16.134.3 (284160/281600), GigabitEthernet0/0
P 172.16.134.0/24, 1 successors, FD is 28160
        via Connected, GigabitEthernet0/0
P 172.16.4.0/24, 0 successors, FD is Infinity
        via 172.16.134.4 (284160/281600), GigabitEthernet0/0
P 172.16.23.0/24, 0 successors, FD is Infinity
        via 172.16.134.3 (284160/281600), GigabitEthernet0/0
```

以上输出说明路由器 R1 的 OSPF 数据库和 EIGRP 拓扑表中都包含 **172.16.4.0**，但是管理距离小的被放入路由表中。可以用 **traceroute** 命令查看如何从路由器 R1 到达 **172.16.4.4**：

```
R1#traceroute 172.16.4.4
Type escape sequence to abort.
Tracing the route to 172.16.4.4
   1 172.16.12.2 8 msec 8 msec 8 msec
   2 172.16.23.3 28 msec 16 msec 12 msec
   3 172.16.134.4 8 msec *   4 msec
```

以上输出信息说明路由器 R1 经过 R2 和 R3 到达 R4 的环回接口 0。次优路由问题可以通过很多方式解决。

② R2#show ip route ospf
```
         172.16.0.0/16 is variably subnetted, 8 subnets, 2 masks
O E2     172.16.4.0/24 [110/20] via 172.16.23.3, 00:05:13, Serial0/0/1
O E2     172.16.134.0/24 [110/20] via 172.16.23.3, 00:08:28, Serial0/0/1
```

[110/20] via 172.16.12.1, 00:00:40, Serial0/0/0

从以上的输出看出，路由条目 **172.16.4.0** 在路由器 R2 的路由表中并没有出现等价路径，因为该路由条目在 R1 的路由表中是 **O E2** 路由。而实验希望的结果是路由条目 **172.16.4.0** 在路由器 R2 的路由表中出现等价路径。

```
③ R3#show ip route
        172.16.0.0/16 is variably subnetted, 7 subnets, 2 masks
O E2      172.16.2.0/24 [110/20] via 172.16.23.2, 00:11:56, Serial0/0/1
D EX      172.16.4.0/24    [170/284160] via 172.16.134.4, 00:01:27, GigabitEthernet0/0
                           [170/284160] via 172.16.134.1, 00:01:27, GigabitEthernet0/0
O         172.16.12.0/24 [110/128] via 172.16.23.2, 00:11:56, Serial0/0/1
C         172.16.23.0/24 is directly connected, Serial0/0/1
L         172.16.23.3/32 is directly connected, Serial0/0/1
C         172.16.134.0/24 is directly connected, GigabitEthernet0/0
L         172.16.134.3/32 is directly connected, GigabitEthernet0/0
```

在路由器 R3 的路由表中，发现到 **172.16.4.0** 的路径是等价路径，下一跳地址分别是 **172.16.134.4** 和 **172.16.134.1**，这是因为在路由器 R1 和 R4 的 EIGRP 进程中执行重分布时 metric 参数设置相同，都是 **10000 100 255 1 1500**，其中经过路由器 R1 的路径明显不是最优路径，因此这也是不合理的。

```
④ R4#show ip route eigrp
         172.16.0.0/16 is variably subnetted, 7 subnets, 2 masks
D EX     172.16.2.0/24    [170/284160] via 172.16.134.3, 00:06:14, GigabitEthernet0/0
                          [170/284160] via 172.16.134.1, 00:06:14, GigabitEthernet0/0
D EX     172.16.12.0/24 [170/284160] via 172.16.134.3, 00:06:14, GigabitEthernet0/0
                          [170/284160] via 172.16.134.1, 00:06:14, GigabitEthernet0/0
D EX     172.16.23.0/24 [170/284160] via 172.16.134.3, 00:06:14, GigabitEthernet0/0
                          [170/284160] via 172.16.134.1, 00:06:14, GigabitEthernet0/0
```

在路由器 R4 的路由表中，发现所有的 EIGRP 条目都是等价路径，下一跳地址分别是 **172.16.134.3** 和 **172.16.134.1**，这是因为在路由器 R1 和 R3 的 EIGRP 进程中执行重分布 OSPF 时 metric 参数设置相同，都是 **10000 100 255 1 1500**，而实际上到达 **172.16.12.0** 和 **172.16.23.0** 的路径根本不是等价路径，其中一条存在次优路由问题。实验希望到达 **172.16.12.0** 和 **172.16.23.0** 选择最优路径，而到达 **172.16.2.0** 的路径是等价路径。

（2）产生路由环路

在路由器 R4 上通过 **offset-list** 命令，增加从接口 **Gi0/0** 发送路由更新信息的度量值，配置如下：

```
R4(config)#router eigrp 1
R4(config-router)#offset-list 1 out 20000 GigabitEthernet0/0
```

【技术要点】

偏移列表的配置命令如下：**offset-list** *{access-list-number | access-list-name}* **{in |out}** *offset [interface-type interface-number]*，各参数含义如下。

● *access-list-number | access-list-name*：标准 ACL 的表号或者名称；

- **in |out**: 偏移列表作用的方向;
- *offset*: 度量值的偏移量,如果为 0,表示没有偏移;
- *interface-type interface-number*: 指出将偏移列表作用的接口类型和编号。

此时查看路由表:

```
① R1#show ip route
(此处省略直连路由和本地路由)
       172.16.0.0/16 is variably subnetted, 7 subnets, 2 masks
O E2      172.16.2.0/24 [110/20] via 172.16.12.2, 00:22:20, Serial0/0/0
O E2      172.16.4.0/24 [110/20] via 172.16.12.2, 00:21:12, Serial0/0/0
O         172.16.23.0/24 [110/128] via 172.16.12.2, 00:22:20, Serial0/0/0
② R2#show ip route
     O E2      172.16.4.0/24 [110/20] via 172.16.23.3, 00:22:16, Serial0/0/1
O E2      172.16.134.0/24 [110/20] via 172.16.23.3, 00:22:16, Serial0/0/1
                          [110/20] via 172.16.12.1, 00:14:28, Serial0/0/0
③ R3#show ip route
(此处省略直连路由和本地路由)
172.16.0.0/16 is variably subnetted, 7 subnets, 2 masks
O E2      172.16.2.0/24 [110/20] via 172.16.23.2, 00:25:52, Serial0/0/1
D EX      172.16.4.0/24 [170/284160] via 172.16.134.1, 00:15:23, GigabitEthernet0/0
O         172.16.12.0/24 [110/128] via 172.16.23.2, 00:25:52, Serial0/0/1
```

以上输出关注三台路由器上路由条目 **172.16.4.0** 的下一跳,发现已经产生路由环路,可以用 **traceroute** 命令在 R1 上查看:

```
R1#traceroute 172.16.4.4
Type escape sequence to abort.
Tracing the route to 172.16.4.4
  1 172.16.12.2 8 msec 8 msec 8 msec
  2 172.16.23.3 16 msec 12 msec 16 msec
  3 172.16.134.1 12 msec 16 msec 16 msec
  4 172.16.12.2 12 msec 12 msec 12 msec
  5 172.16.23.3 20 msec 20 msec 20 msec
  6 172.16.134.1 16 msec 16 msec 20 msec
  7 172.16.12.2 20 msec 16 msec 16 msec
  8 172.16.23.3 24 msec 24 msec 24 msec
  9 172.16.134.1 24 msec 24 msec 24 msec
  ……
```

(3)通过修改管理距离和进行路由过滤解决次优路由和路由环路的问题

解决次优路由和路由环路问题的方案是多种的,本实验给出其中一种。对于 172.16.4.0 路由条目,只要在 OSPF 进程中重分布 EIGRP 路由的时候,把管理距离设置的比 170 大即可。R1 和 R3 上配置如下:

```
R1(config)#access-list 1 permit 172.16.4.0
R1(config)#router ospf 1
R1(config-router)#distance 180 0.0.0.0 255.255.255.255 1
//从任何更新源收到符合 ACL 1 的路由,管理距离为 180
R3(config)#access-list 1 permit 172.16.4.0
```

```
R3(config)#router ospf 1
R3(config-router)#distance 180 0.0.0.0 255.255.255.255 1
```

为了防止在路由器 R4 的路由表中出现 **172.16.12.0** 和 **172.16.23.0** 是等价路径的现象，可以通过路由映射表在重分布路由时进行路由区分或者过滤，方案如下。

① 方案 1：该方案的思路考虑到了备份的需求，重分布不同的路由，采用不同的度量值加以区分，路由器 R4 会选择度量值小的放入路由表，配置如下。

```
R1(config)#access-list 2 permit 172.16.23.0
R1(config)#route-map REDIS permit 10
R1(config-route-map)#match ip address 2
R1(config-route-map)#set metric 100 1000 255 1 1500
R1(config)#route-map REDIS permit 20
R1(config-route-map)#set metric 1000 100 255 1 1500
R1(config)#router eigrp 1
R1(config-router)#redistribute ospf 1 route-map REDIS
R3(config)#access-list 2 permit 172.16.12.0
R3(config)#route-map REDIS permit 10
R3(config-route-map)#match ip address 2
R3(config-route-map)#set metric 100 1000 255 1 1500
R3(config)#route-map REDIS permit 20
R3(config-route-map)#set metric 1000 100 255 1 1500
R3(config)#router eigrp 1
R3(config-router)#redistribute ospf 1 route-map REDIS
```

② 方案 2：该方案的思路是完成重分布时对路由条目进行过滤，不考虑备份需求的配置如下。

```
R1(config)#access-list 2 permit 172.16.23.0
R1(config)#route-map REDIS deny 10      //注意这里是 deny
R1(config-route-map)#match ip address 2
R1(config)#route-map REDIS permit 20
R1(config)#router eigrp 1
R1(config-router)#redistribute ospf 1 route-map REDIS
R3(config)#access-list 2 permit 172.16.12.0
R3(config)#route-map REDIS deny 10
R3(config-route-map)#match ip address 2
R3(config)#route-map REDIS permit 20
R3(config)#router eigrp 1
R3(config-router)#redistribute ospf 1 route-map REDIS
```

在没有网络故障的情况下，上述两种方案的结果相同，各路由器的路由表如下。

```
① R1#show ip route | include O|D
  O E2      172.16.2.0/24 [110/20] via 172.16.12.2, 00:07:24, Serial0/0/0
  D EX      172.16.4.0/24 [170/304160] via 172.16.134.4, 00:06:42, GigabitEthernet0/0
  O         172.16.23.0/24 [110/128] via 172.16.12.2, 00:07:24, Serial0/0/0
② R2#show ip route ospf
     172.16.0.0/16 is variably subnetted, 8 subnets, 2 masks
  O E2      172.16.134.0 [110/20] via 172.16.23.3, 02:21:50, Serial0/0/1
                        [110/20] via 172.16.12.1, 02:21:06, Serial0/0/0
```

```
O E2       172.16.4.0 [110/20] via 172.16.23.3, 00:56:30, Serial0/0/1
                      [110/20] via 172.16.12.1, 00:56:52, Serial0/0/0
③ R3#show ip route     | include O|D
       172.16.0.0/16 is variably subnetted, 7 subnets, 2 masks
O E2       172.16.2.0/24 [110/20] via 172.16.23.2, 00:09:28, Serial0/0/1
D EX       172.16.4.0/24 [170/304160] via 172.16.134.4, 00:09:28, GigabitEthernet0/0
O          172.16.12.0/24 [110/128] via 172.16.23.2, 00:09:28, Serial0/0/1
④ R4#show ip route eigrp
       172.16.0.0/16 is variably subnetted, 7 subnets, 2 masks
D EX       172.16.2.0/24   [170/284160] via 172.16.134.3, 00:11:07, GigabitEthernet0/0
                           [170/284160] via 172.16.134.1, 00:11:07, GigabitEthernet0/0
D EX       172.16.12.0/24 [170/284160] via 172.16.134.1, 00:07:56, GigabitEthernet0/0
D EX       172.16.23.0/24 [170/284160] via 172.16.134.3, 00:09:16, GigabitEthernet0/0
```

6.4 控制路由更新

6.4.1 实验 3：配置被动接口和分布列表控制路由更新

1. 实验目的

通过本实验可以掌握：
① 被动接口的配置。
② 分布列表的配置和调试方法。

2. 实验拓扑

配置分布列表控制路由更新实验拓扑如图 6-3 所示。

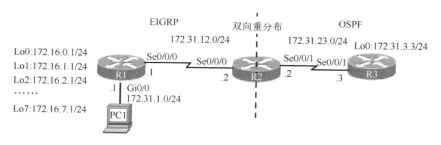

图 6-3　配置分布列表控制路由更新实验拓扑

3. 实验步骤

（1）配置路由器 R1

```
R1(config)#access-list 1 deny 172.16.1.0 0.0.254.0
//拒绝 172.16 开头的第三位为奇数的路由
R1(config)#access-list 1 permit any
R1(config)#router eigrp 1
```

```
R1(config-router)#network 172.16.0.0 0.0.7.255
R1(config-router)#network 172.31.1.1 0.0.0.0
R1(config-router)#network 172.31.12.1 0.0.0.0
R1(config-router)#passive-interface default              //默认所有接口为被动接口
R1(config-router)#no passive-interface Serial0/0/0       //关闭被动接口
R1(config-router)#distribute-list 1 out Serial0/0/0      //出方向配置分布控制列表
```

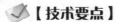【技术要点】

distribute-list 命令可以全局地在一个出或入方向的路由更新过程中过滤路由，也可以为一个路由进程所涉及的每一个接口的入方向或出方向设置路由过滤。在出站方向配置分布列表的命令如下：

distribute-list [*access-list-number* | *name*] | [route-map *map-tag*] | [prefix *prefix-list-name*] out [*interface-type interface-number* | *routing-process*]

在入站方向配置分布列表的命令如下：

distribute-list [*access-list-number* | *name*] | [route-map *map-tag*] | [prefix *prefix-list-name*] in [*interface-type interface-number*]

以上两条命令的参数含义如下。

① *access-list-number* | *name*：访问控制列表表号或名称。
② route-map *map-tag*：使用路由映射的名称，只有 OSPF 和 EIGRP 支持该参数。
③ prefix *prefix-list-name*：使用前缀列表的名称。
④ in 或 out：分布列表作用的方向。
⑤ *interface-type interface-number*：接口类型和编号。

（2）配置路由器 R2

```
R2(config)#access-list 2 permit 172.31.0.0
R2(config)#access-list 2 permit 172.16.0.0 0.0.252.0
//允许 172.16 开头的第三位被 4 整除的路由
R2(config)#interface Serial0/0/0
R2(config-if)#ip summary-address eigrp 1 172.31.0.0 255.255.0.0
//汇总的目的是为了验证重分布的路由回馈问题
R2(config)#router eigrp 1
R2(config-router)#network 172.31.12.2 0.0.0.0
R2(config-router)#redistribute ospf 1 metric 10000 100 255 1 1500
R2(config)#router ospf 1
R2(config-router)#router-id 2.2.2.2
R2(config-router)#network 172.31.23.2 0.0.0.0 area 0
R2(config-router)#redistribute eigrp 1 subnets
R2(config-router)#distribute-list 2 out eigrp 1
//对重分布的 EIGRP 进程的路由进行过滤
```

【提示】

在链路状态协议配置分布列表时，关键字 **out** 不能与接口联合使用，因为不像距离矢量路由协议,链路状态协议不是从自身的路由表中通告路由的,而是通过 LSA 进行路由更新的,如果配置 **out** 与接口联合使用,会提示下面的信息**% Interface not allowed with OUT for OSPF**。

(3) 配置路由器 R3

```
R3(config)#access-list 1 permit 172.31.0.0
R3(config)#access-list 1 permit 172.16.0.0
R3(config)#router ospf 1
R3(config-router)#router-id 3.3.3.3
R3(config-router)#network 172.31.3.3 0.0.0.0 area 0
R3(config-router)#network 172.31.23.3 0.0.0.0 area 0
R3(config-router)#distribute-list 1 in Serial0/0/1
//对入方向 OSPF 路由进行过滤,该过滤指 OSPF 从链路状态数据库中提取特定路由条目进入路由表
```

4. 实验调试

(1) 查看路由表

```
① R1#show ip route eigrp
      172.31.0.0/16 is variably subnetted, 5 subnets, 3 masks
D        172.31.0.0/16 [90/2195456] via 172.31.12.2, 00:01:22, Serial0/0/0
② R2#show ip route | include O|D
D     172.16.0.0 [90/2297856] via 172.31.12.1, 00:02:38, Serial0/0/0
D     172.16.2.0 [90/2297856] via 172.31.12.1, 00:02:38, Serial0/0/0
D     172.16.4.0 [90/2297856] via 172.31.12.1, 00:02:38, Serial0/0/0
D     172.16.6.0 [90/2297856] via 172.31.12.1, 00:02:38, Serial0/0/0
//上面四行说明 172.16 开头的第三位为奇数的路由被过滤
D     172.31.0.0/16 is a summary, 00:04:57, Null0
D     172.31.1.0/24 [90/2172416] via 172.31.12.1, 00:02:38, Serial0/0/0
O     172.31.3.3/32 [110/65] via 172.31.23.3, 00:01:31, Serial0/0/1
③ R3#show ip route ospf
      172.16.0.0/24 is subnetted, 1 subnets
O E2    172.16.0.0 [110/20] via 172.31.23.2, 00:08:17, Serial0/0/1
      172.31.0.0/16 is variably subnetted, 5 subnets, 3 masks
O E2    172.31.0.0/16 [110/20] via 172.31.23.2, 00:08:17, Serial0/0/1
//该路由在路由器 R2 上做了汇总之后,生成了一条指向 NULL0 接口的汇总路由,这样,在把 EIGRP
重分布进 OSPF 时,这条路由又被重分布进来。这种现象称为路由回馈问题
```

以上输出表明了路由器 R1、R2 和 R3 的路由表都按照需求通过分布列表实现了路由过滤。

(2) 查看路由协议相关信息

```
① R1#show ip protocols    | begin Routing Protocol is "eigrp 1"
Routing Protocol is "eigrp 1"
  Outgoing update filter list for all interfaces is not set   //全局下没有设置分布列表
    Serial0/0/0 filtered by 1(per-user), default is not set
// Serial0/0/0 接口的出方向设置了分布列表 1
  Incoming update filter list for all interfaces is not set
(此处省略部分输出)
```

② R2#**show ip protocols | begin Routing Protocol is "ospf 1"**
Routing Protocol is "ospf 1"
　　Outgoing update filter list for all interfaces is not set
　　　Redistributed eigrp 1 filtered by 2
　　　　//出方向设置了分布列表 2，对重分布进 eigrp 进程 1 的路由进行过滤
　　Incoming update filter list for all interfaces is not set
（此处省略部分输出）

6.4.2　实验 4：配置前缀列表和路由映射表控制路由更新

1. 实验目的

通过本实验可以掌握：
① 前缀列表的使用方法和含义。
② 在路由映射表中前缀列表匹配的方法。
③ 用路由映射表控制路由更新的方法。
④ 分布列表中调用路由映射表的方法。

2. 实验拓扑

配置前缀列表和路由映射表控制路由更新实验拓扑如图 6-4 所示。

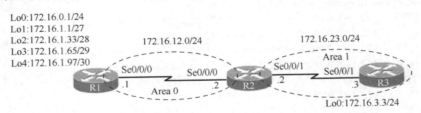

图 6-4　配置前缀列表和路由映射表控制路由更新实验拓扑

3. 实验步骤

（1）配置路由器 R1

```
R1(config)#ip prefix-list CONN1 seq 5 permit 172.16.0.0/16 ge 24 le 28
R1(config)#ip prefix-list CONN2 seq 5 permit 172.16.0.0/16 ge 29
R1(config)#route-map CONN permit 10
R1(config-route-map)#match ip address prefix-list CONN1    //匹配前缀列表
R1(config-route-map)#set metric 100
R1(config-route-map)#set metric-type type-1
R1(config)#route-map CONN permit 20
R1(config-route-map)#match ip address prefix-list CONN2
R1(config-route-map)#set metric 200
R1(config-route-map)#set tag 200
R1(config)#route-map CONN permit 30
R1(config)#router ospf 1
R1(config-router)#router-id 1.1.1.1
```

R1(config-router)#**network 172.16.12.1 0.0.0.0 area 0**
R1(config-router)#**redistribute connected subnets route-map CONN**　　//重分布直连路由

【技术要点】

配置前缀列表的命令如下：

ip prefix-list *list-name* **[seq** *seq-value*] {**deny|permit**} *network/length* [**ge** *ge-value*] [**le** *le-value*]，各参数的含义如下。

① *list name*：前缀列表名，注意列表名区分大小写。

② **seq** *seq-value*：32 比特序号，用于确定语句被处理的次序。默认序号以 5 递增，5，10，15 等。

③ **deny|permit**：匹配条目时所要采取的行为，如果前缀不与前缀列表中任何条目匹配，将被拒绝。

④ *network/length*：前缀和前缀长度。

⑤ **ge** *ge-value*：匹配的前缀长度的范围。当然 ge 和 le 为可选参数，对于前缀长度匹配范围，要满足下列条件，**length<=*ge-value*<=*le-value*<=32**。如果只指定了 **ge** 参数，掩码长度的匹配范围为 ge-value<=掩码长度<=32。

⑥ **le** *le-vlaue*：匹配的前缀长度的范围。如果只定义了 le，掩码长度的匹配范围为 length<=掩码长度<=*le-value*，如果同时定义了 ge 和 le，掩码长度的匹配范围为 *ge-value*<=掩码长度<=*le-value*；如果既没有定义 ge 也没有定义 le，掩码长度的匹配范围只能是 length，也就是精确匹配 network/ length 的路由条目。

下面是前缀列表匹配的实例：

① ip prefix-list test1 seq 5 permit 0.0.0.0/0 ge 32　　//匹配所有主机路由

② ip prefix-list test2 seq 10 permit 0.0.0.0/0 le 32　　//匹配所有路由

③ ip prefix-list test3 seq 15 permit 0.0.0.0/0 ge 1　　//匹配默认路由以外的所有路由

④ ip prefix-list test4 seq 20 permit 0.0.0.0/1 ge 8 le 8　　//匹配 A 类地址

⑤ ip prefix-list test5 seq 25 permit 128.0.0.0/2 ge 16 le 16　　//匹配 B 类地址

⑥ ip prefix-list test6 seq 30　permit 192.0.0.0/3 ge 24 le 24　　//匹配 C 类地址

⑦ ip prefix-list test7 seq 35 permit 192.168.0.0/16 le 20　　//匹配以 192.168 开头的，掩码长度在 16～20 之间（包括 16 和 20）的所有路由

⑧ ip prefix-list test8 seq 40 permit 192.168.0.0/16 ge 20　　//匹配以 192.168 开头的，掩码长度在 20～32 之间（包括 20 和 32）的所有路由，比如 192.168.0.0/16、192.168.128.0/18 不能匹配，但 192.168.64.0/24 的路由能匹配

（2）配置路由器 R2

R2(config)#**router ospf 1**
R2(config-router)#**router-id 2.2.2.2**
R2(config-router)#**network 172.16.12.2 0.0.0.0 area 0**
R2(config-router)#**network 172.16.23.2 0.0.0.0 area 1**

（3）配置路由器 R3

R3(config)#**route-map TAG permit 10**

```
R3(config-route-map)#match tag 200
R3(config)#router ospf 1
R3(config-router)#router-id 3.3.3.3
R3(config-router)#network 172.16.3.3 0.0.0.0 area 1
R3(config-router)#network 172.16.23.3 0.0.0.0 area 1
R3(config-router)#distribute-list route-map TAG in
//R3 通过分布列表在入向进行过滤，只把 TAG 为 200 的路由从数据库中提取到路由表中
```

4. 实验调试

（1）查看路由表

① R2#show ip route ospf
```
      172.16.0.0/16 is variably subnetted, 10 subnets, 6 masks
O E1     172.16.0.0/24 [110/164] via 172.16.12.1, 00:02:22, Serial0/0/0
O E1     172.16.1.0/27 [110/164] via 172.16.12.1, 00:02:22, Serial0/0/0
O E1     172.16.1.32/28 [110/164] via 172.16.12.1, 00:02:22, Serial0/0/0
```
//以上三条路由条目是路由映射表 CONN 序号 10 匹配前缀列表 CONN1 的设置
```
O E2     172.16.1.64/29 [110/200] via 172.16.12.1, 00:02:22, Serial0/0/0
O E2     172.16.1.96/30 [110/200] via 172.16.12.1, 00:02:22, Serial0/0/0
```
//以上两条路由条目是路由映射表 CONN 序号 20 匹配前缀列表 CONN2 的设置
```
O        172.16.3.3/32 [110/65] via 172.16.23.3, 00:01:12, Serial0/0/1
```

② R3#show ip route ospf
```
      172.16.0.0/16 is variably subnetted, 6 subnets, 4 masks
O E2     172.16.1.96/30 [110/200] via 172.16.23.2, 00:08:25, Serial0/0/1
O E2     172.16.1.64/29 [110/200] via 172.16.23.2, 00:08:25, Serial0/0/1
```
//以上两条路由条目是路由映射表 TAG 序号 10 匹配 TAG 为 200 的设置

（2）查看 DSPF 链路状态数据库

R3#show ip ospf database

            OSPF Router with ID (3.3.3.3) (Process ID 1)

                Router Link States (Area 1)

Link ID	ADV Router	Age	Seq#	Checksum	Link count
2.2.2.2	2.2.2.2	846	0x80000003	0x006FCC	2
3.3.3.3	3.3.3.3	845	0x80000008	0x00E37B	3

                Summary Net Link States (Area 1)

Link ID	ADV Router	Age	Seq#	Checksum
172.16.12.0	2.2.2.2	842	0x80000003	0x00DC7D

                Summary ASB Link States (Area 1)

Link ID	ADV Router	Age	Seq#	Checksum
1.1.1.1	2.2.2.2	847	0x80000001	0x00AE71

                Type-5 AS External Link States

Link ID	ADV Router	Age	Seq#	Checksum	Tag
172.16.0.0	1.1.1.1	1613	0x80000002	0x00E12D	0
172.16.1.0	1.1.1.1	1613	0x80000002	0x001C11	0
172.16.1.32	1.1.1.1	1616	0x80000002	0x003BC1	0
172.16.1.64	**1.1.1.1**	**1506**	**0x80000005**	**0x00B172**	**200**
172.16.1.96	**1.1.1.1**	**1506**	**0x80000005**	**0x008877**	**200**

从路由器 R3 的链路状态数据库中可以看出，从路由器 R1 重分布直连进入 OSPF 的 5 条条目都有，但是由于针对路由器 R3 的 Se0/0/1 接口的入向的分布列表要求匹配 Tag 值为 200 的路由，所以 OSPF 只把 Tag 值为 200 的放入路由表中。

（3）查看前缀列表信息

```
R1#show ip prefix-list
ip prefix-list CONN1: 1 entries
    seq 5 permit 172.16.0.0/16 ge 24 le 28
ip prefix-list CONN2: 1 entries
    seq 5 permit 172.16.0.0/16 ge 29
```

（4）查看 route-map 信息

```
R1#show route-map
route-map CONN, permit, sequence 10
  Match clauses:
    ip address prefix-lists: CONN1
//注意 prefix-lists 参数，如果没有这个参数，表示匹配的是命名的 ACL
  Set clauses:
    metric 100
    metric-type type-1
  Policy routing matches: 0 packets, 0 bytes
route-map CONN, permit, sequence 20
  Match clauses:
    ip address prefix-lists: CONN2
  Set clauses:
    metric 200
    tag 200
  Policy routing matches: 0 packets, 0 bytes
route-map CONN, permit, sequence 30
  Match clauses:
  Set clauses:
  Policy routing matches: 0 packets, 0 bytes
```

6.4.3 实验 5：配置 Cisco IP SLA 控制路径选择

1. 实验目的

通过本实验可以掌握：
① IP SLA 的工作原理和特征。
② IP SLA 的配置和调试方法。

2. 实验拓扑

配置 IP SLA 控制路径选择实验拓扑如图 6-5 所示。本实验中的两台服务器的地址在路由器上用环回接口模拟。

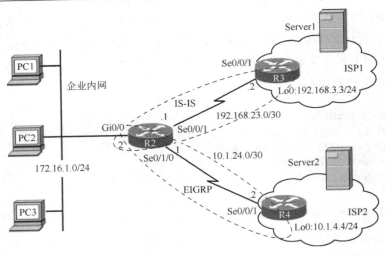

图 6-5 配置 IP SLA 控制路径选择实验拓扑

3. 实验步骤

（1）配置路由器 R2

```
R2(config)#router eigrp 1
R2(config-router)#network 10.1.24.1 0.0.0.0
R2(config-router)#network 172.16.1.2 0.0.0.0
R2(config)#router isis cisco
R2(config-router)#net 49.0001.2222.2222.2222.00
R2(config)#interface gigabitEthernet0/0
R2(config-if)#ip router isis cisco
R2(config)#interface serial0/0/1
R2(config-if)#ip router isis cisco
//R2 上运行 IS-IS 和 EIGEP 路由协议的主要目的是为了获得监控对象的可达性
R2(config)#ip sla 23
//定义一个 IP SLA 操作，23 是操作的编号，编号范围为 1～2147483647
R2(config-ip-sla)#icmp-echo 192.168.3.3 source-interface gigabitEthernet0/0
//定义操作类型为 icmp-echo，用于连通性测试，指定目标地址和源接口，并进入 SLA 的 echo 配置模式
R2(config-ip-sla-echo)#frequency 10    //定义 IP SLA 操作的时间频率，默认为 60 秒
R2(config-ip-sla-echo)#timeout 6000    //定义等待响应的时间，默认为 5000 毫秒
R2(config-ip-sla-echo)#owner test      //定义 owner
R2(config-ip-sla-echo)#exit
R2(config)#ip sla schedule 23 life 7200 start-time now ageout 7200
//配置 IP SLA 调度参数，life 默认为 3600 秒，forever 表示一直采集数据，ageout 定义停止信息采集后，在内存保留时间，默认是 0，即一直保留。start-time 定义开始采集的时间，now 参数表示立即采集数据，也可以用 after 参数定义多长时间后开始采集数据
R2(config)#ip sla 24
R2(config-ip-sla)#icmp-echo 10.1.4.4 source-interface gigabitEthernet0/0
R2(config-ip-sla-echo)#frequency 10
R2(config-ip-sla-echo)#exit
R2(config)#ip sla schedule 24 start-time now
R2(config)#track 1 ip sla 23 reachability
//定义跟踪对象，跟踪 ip sla 23 配置的目标地址是否可达
R2(config-track)#delay up 10 down 10
```

//定义 up 和 down 的延迟时间，单位为秒，可以避免路由抖动
R2(config)#**track 2 ip sla 24 reachability**
R2(config-track)#**delay up 10 down 10**
R2(config)#**ip route 0.0.0.0 0.0.0.0 Serial0/0/1 10 track 1**
//配置跟踪对象的默认路由，管理距离为 10
R2(config)#**ip route 0.0.0.0 0.0.0.0 Serial0/1/0 20 track 2**
//配置跟踪对象的默认路由，管理距离为 20，浮动静态路由使得 R2 首选 ISP1

（2）配置路由器 R3

    R3(config)#**router isis cisco**
    R3(config-router)#**net 49.0001.3333.3333.3333.00**
    R3(config)#**interface serial0/0/1**
    R3(config-if)#**ip router isis cisco**
    R3(config)#**interface loopback0**
    R3(config-if)#**ip router isis cisco**

（3）配置路由器 R4

    R4(config)#**router eigrp 1**
    R4(config-router)#**network 10.1.4.4 0.0.0.0**
    R4(config-router)#**network 10.1.24.2 0.0.0.0**

4．实验调试

（1）查看 IP SLA 的配置信息

R2#show ip sla configuration
IP SLAs Infrastructure Engine-III
Entry number: 23　　//IP SLA 操作编号
Owner: test
Tag:　　//用户指定的 IP SLA 的标识
Operation timeout (milliseconds): 6000　　//等待 IP SLA 操作的请求数据包的响应时间
Type of operation to perform: icmp-echo　　//IP SLA 操作类型
Target address/Source interface: 192.168.3.3/FastEthernet0/0　　//监控的目标地址和源接口
Type Of Service parameter: 0x0　　//服务类型参数
Request size (ARR data portion): 28　　//IP SLA 操作的请求数据包的 payload 大小
Data pattern: 0xABCDABCD　　//实现 IP SLA 操作的请求数据包的数据填充
Verify data: No　　//没有验证 IP SLA 响应的数据
Vrf Name:
Schedule:　　//IP SLA 调度
   Operation frequency (seconds): 10　　(not considered if randomly scheduled)　　//操作频率
   Next Scheduled Start Time: Start Time already passed　　//启动时间已过，证明不是周期性启动
   Group Scheduled : FALSE
   Randomly Scheduled : FALSE
   Life (seconds): 7200　　//IP SLA 主动收集信息的时间，单位为秒
   Entry Ageout (seconds): 7200　　//停止收集信息后，信息保留在内存的时间，单位为秒
   Recurring (Starting Everyday): FALSE
//Recurring 指的是每天在指定时间自动执行，并持续 life 指定的时间，本实验没有启用该参数
   Status of entry (SNMP RowStatus): Active
Threshold (milliseconds): 5000 (not considered if react RTT is configured)
//计算收集信息统计的门限值

Distribution Statistics:
    Number of statistic hours kept: 2
    Number of statistic distribution buckets kept: 1
    Statistic distribution interval (milliseconds): 20
History Statistics:
    Number of history Lives kept: 0
    Number of history Buckets kept: 15
    History Filter Type: None
Enhanced History:

Entry number: 24
Owner:
Tag:
Type of operation to perform: icmp-echo
Target address/Source interface: 10.1.4.4/FastEthernet0/0
Type Of Service parameter: 0x0
Request size (ARR data portion): 28
Operation timeout (milliseconds): 5000
Verify data: No
Vrf Name:
Schedule:
    Operation frequency (seconds): 10   (not considered if randomly scheduled)
    Next Scheduled Start Time: Start Time already passed
    Group Scheduled : FALSE
    Randomly Scheduled : FALSE
    Life (seconds): 3600
    Entry Ageout (seconds): never
    Recurring (Starting Everyday): FALSE
    Status of entry (SNMP RowStatus): Active
Threshold (milliseconds): 5000 (not considered if react RTT is configured)
Distribution Statistics:
    Number of statistic hours kept: 2
    Number of statistic distribution buckets kept: 1
    Statistic distribution interval (milliseconds): 20
History Statistics:
    Number of history Lives kept: 0
    Number of history Buckets kept: 15
    History Filter Type: None
Enhanced History:

（2）查看 IP SAL 的统计信息

```
R2#show ip sla statistics
IPSLAs Latest Operation Statistics     //IP SLA 最近的统计信息
IPSLA operation id: 23    //IP SLA 操作编号
Type of operation: icmp-echo   //操作类型
        Latest RTT: 1 milliseconds   //发送 IP SLA 请求数据包最近的往返时间（Round-Trip Time，RTT）
Latest operation start time: 01:28:45 UTC Sat Sep 15 2018   //最后一次探测时间
Latest operation return code: OK   //最后操作返回代码
Number of successes: 62   //探测成功的次数
Number of failures: 55   //探测失败的次数
Operation time to live: 6036 sec   //还有 6036 秒可以收集数据
```

```
IPSLA operation id: 24
Type of operation: icmp-echo
        Latest RTT: 8 milliseconds
Latest operation start time: 01:28:38 UTC Sat Sep 15 2018
Latest operation return code: OK    //最后操作返回代码
Number of successes: 55
Number of failures: 0
Operation time to live: 3052 sec
```

（3）查看跟踪信息

```
R2#show track
Track 1
    IP SLA 23 reachability
    Reachability is Up    //可达性为 Up
      1 change, last change 00:16:28
    Delay up 10 secs, down 10 secs    //up 和 down 的延迟时间
    Latest operation return code: 1
    Tracked by:
        STATIC-IP-ROUTING 0    //跟踪结果与静态路由联动
Track 2
    IP SLA 24 reachability
    Reachability is Up
      1 change, last change 00:15:27
    Delay up 10 secs, down 10 secs
    Latest operation return code: OK
    Latest RTT (millisecs) 1
    Tracked by:
        STATIC-IP-ROUTING 0
```

（4）查看路由表

```
R2#show ip route
（此处省略路由代码、直连路由和本地路由部分）
Gateway of last resort is 0.0.0.0 to network 0.0.0.0
S*      0.0.0.0/0 is directly connected, Serial0/0/1
        10.0.0.0/8 is variably subnetted, 3 subnets, 3 masks
D       10.1.4.0/24 [90/2297856] via 10.1.24.2, 00:23:17, Serial0/1/0
        172.16.0.0/16 is variably subnetted, 2 subnets, 2 masks
i L1    192.168.3.0/24 [115/20] via 192.168.23.2, 00:25:31, Serial0/0/1
```

现在将路由器 R3 的环回接口 0 **shutdown**，路由器 R2 上会出现如下日志信息，提示 IP SLA 23 跟踪的目标的可达性状态由 Up 变为 Down：

```
05:01:58: %TRACKING-5-STATE: 1 ip sla 23 reachability Up->Down
```

此时在路由器 R2 上查看路由表：

```
R2#show ip route
（此处省略路由代码、直连路由和本地路由部分）
Gateway of last resort is 0.0.0.0 to network 0.0.0.0
```

```
S*       0.0.0.0/0 is directly connected, Serial0/1/0
         10.0.0.0/8 is variably subnetted, 3 subnets, 3 masks
D           10.1.4.0/24 [90/2297856] via 10.1.24.2, 00:30:28, Serial0/1/0
```

以上输出信息表明，当 IP SLA 23 检测 192.168.3.3 发生故障后，IP SLA 24 检测 10.1.4.4 可达，此时把出接口为 Se0/1/0 的默认路由加入到路由表中。此时将 R3 的环回接口 **no shutdown**，路由器 R2 上会出现如下日志信息，提示 IP SLA 23 跟踪的目标的可达性状态由 Down 变为 Up：

05:01:58: %TRACK-6-STATE: 1 ip sla 23 reachability Down -> Up

此时在路由器 R2 上查看路由表：

```
R2#show ip route
（此处省略路由代码、直连路由和本地路由部分）
Gateway of last resort is 0.0.0.0 to network 0.0.0.0

S*       0.0.0.0/0 is directly connected, Serial0/0/1
         10.0.0.0/8 is variably subnetted, 3 subnets, 3 masks
D           10.1.4.0/24 [90/2297856] via 10.1.24.2, 00:35:05, Serial0/1/0
         172.16.0.0/16 is variably subnetted, 2 subnets, 2 masks
i L1        192.168.3.0/24 [115/20] via 192.168.23.2, 00:02:30, Serial0/0/1
```

以上输出信息表明，当 IP SLA 23 检测 192.168.3.3 可达后，把出接口为 Se0/0/1 的默认路由重新加入到路由表中。因为此条静态路由条目的管理距离低。

6.5 策略路由

6.5.1 实验 6：配置基于源 IP 地址的策略路由

1. 实验目的

通过本实验可以掌握：
① 用 route-map 定义路由策略的方法。
② 在接口下应用路由策略的方法。
③ 基于源 IP 地址的策略路由的配置和调试方法。

2. 实验拓扑

配置基于源 IP 地址的策略路由实验拓扑如图 6-6 所示。

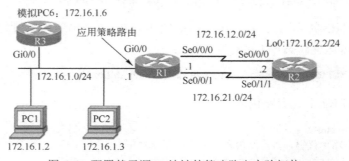

图 6-6 配置基于源 IP 地址的策略路由实验拓扑

3. 实验步骤

实验设计如下：在路由器 R1 的 Gi0/0 接口应用 IP 策略路由 CCNA，将从主机 PC1 来的数据的下一跳地址设置为 172.16.12.2；从主机 PC2 来的数据的下一跳地址设置为 172.16.21.2，所有其他数据包正常转发，整个网络运行 EIGRP 路由协议（本实验 EIGRP 配置省略）。

```
R1(config)#access-list 1 permit 172.16.1.2
R1(config)#access-list 2 permit 172.16.1.3
R1(config)#route-map CCNA permit 10
R1(config-route-map)#match ip address 1
R1(config-route-map)#set ip next-hop 172.16.12.2    //设置下一跳
R1(config)#route-map CCNA permit 20
R1(config-route-map)#match ip address 2
R1(config-route-map)#set ip next-hop 172.16.21.2
R1(config)#interface gigabitEthernet0/0
R1(config-if)#ip policy route-map CCNA
//接口应用路由策略，一个接口只能应用一个路由策略
```

4. 实验调试

（1）执行路由策略调试操作

```
R1#debug ip policy
```

① 在主机 PC1 上 ping 地址 172.16.2.2，路由器 R1 上显示的调试信息如下。

```
01:15:48: IP: s=172.16.1.2 (GigabitEthernet0/0), d=172.16.2.2, len 100, FIB policy match
01:15:48: IP: s=172.16.1.2 (GigabitEthernet0/0), d=172.16.2.2, len 100, PBR Counted
01:15:48: IP: s=172.16.1.2 (GigabitEthernet0/0), d=172.16.2.2, g=172.16.12.2, len 100, FIB policy routed
```

以上输出信息表明源地址为 172.16.1.2 的主机发送给目的地址为 172.16.2.2 的数据包在接口 Gi0/0 匹配路由策略，执行策略路由，设置数据包下一跳地址为 172.16.12.2。

② 在主机 PC2 上 ping 地址 172.16.2.2，路由器 R1 上显示的调试信息如下。

```
01:18:06: IP: s=172.16.1.3 (GigabitEthernet0/0), d=172.16.2.2, len 100, FIB policy match
01:18:06: IP: s=172.16.1.3 (GigabitEthernet0/0), d=172.16.2.2, len 100, PBR Counted
01:18:06: IP: s=172.16.1.3 (GigabitEthernet0/0), d=172.16.2.2, g=172.16.21.2, len 100, FIB policy routed
```

以上输出信息表明源地址为 172.16.1.3 的主机发送给目的地址为 172.16.2.2 的数据包在接口 Gi0/0 匹配路由策略，执行策略路由，设置数据包下一跳地址为 172.16.21.2。

③ 在主机 PC6 上 ping 地址 172.16.2.2，路由器 R1 上显示的调试信息如下。

```
01:20:03: IP: s=172.16.1.6 (GigabitEthernet0/0), d=172.16.2.2, len 100, FIB policy rejected(no match) - normal forwarding
```

以上输出信息表明源地址为 172.16.1.6 的主机发送到目的主机 172.16.2.2 的数据包在接口 Gi0/0 不匹配路由策略，数据包正常转发。

（2）查看路由策略应用情况

R1#show ip policy
Interface   Route map
Gi0/0   CCNA

以上输出信息表明在路由器 R1 的 Gi0/0 接口应用了路由策略 CCNA 来执行策略路由。

（3）查看定义的路由策略及与路由策略匹配的情况

R1#show route-map CCNA
route-map CCNA, permit, sequence 10
  Match clauses:
    ip address (access-lists): 1
  Set clauses:
    ip next-hop 172.16.12.2
  Policy routing matches: 10 packets, 1140 bytes
route-map CCNA, permit, sequence 20
  Match clauses:
    ip address (access-lists): 2
  Set clauses:
    ip next-hop 172.16.21.2
  Policy routing matches: 5 packets, 570 bytes

以上输出信息表明匹配策略路由的每个策略的数据包的数量和字节数。

6.5.2　实验 7：配置基于数据包长度的策略路由

1. 实验目的

通过本实验可以掌握：
① 用 route-map 定义路由策略的方法。
② 在接口下应用路由策略的方法。
③ 基于数据包长度的策略路由的配置和调试方法。

2. 实验拓扑

配置基于数据包长度的策略路由实验拓扑如图 6-7 所示。

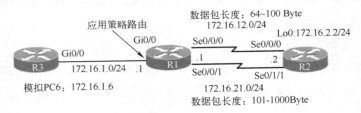

图 6-7　配置基于数据包长度的策略路由实验拓扑

3. 实验步骤

实验设计如下：在路由器 R1 的 Gi0/0 接口应用 IP 策略路由 CCNP，将大小为 64～100

字节的数据包的出接口设置为 Se0/0/0；大小为 101～1000 字节的数据包的出接口设置为 Se0/0/1，所有其他的数据包正常转发，整个网络运行 EIGRP 路由协议（本实验 EIGRP 配置省略）。

```
R1(config)#route-map CCNP permit 10
R1(config-route-map)#match length 64 100
R1(config-route-map)#set interface serial0/0/0
R1(config)#route-map CCNP permit 20
R1(config-route-map)#match length 101 1000
R1(config-route-map)#set interface serial0/0/1
R1(config)#interface gigabitEthernet0/0
R1(config-if)#ip policy route-map CCNP
R1(config)#ip local policy route-map CCNP
```
//配置本地策略路由，需要注意：在接口下应用的策略路由对路由器自身产生的数据包不起作用。如果需要对本地路由器产生的数据包执行策略路由，要用 **ip local policy route-map** 命令配置本地策略路由。路由器只能应用一个本地路由策略

4．实验调试

（1）执行扩展 ping 命令

① 在主机 PC6 上执行扩展 ping 命令，数据包的长度为 90 字节，路由器 R1 上显示的调试信息如下。

```
R3#ping 172.16.2.2 repeat 1 size 90
01:28:48: IP: s=172.16.1.6 (GigabitEthernet0/0), d=172.16.2.2, len 90, FIB policy match
01:28:48: IP: s=172.16.1.6 (GigabitEthernet0/0), d=172.16.2.2, len 90, PBR Counted
01:28:48: IP: s=172.16.1.6 (GigabitEthernet0/0), d=172.16.2.2 (Serial0/0/0), len 90, FIB policy routed
```

以上输出信息表明数据包长度为 90 字节的数据包在接口 Gi0/0 匹配路由策略，执行策略路由，设置数据包的出接口为 Se0/0/0。

② 在主机 PC6 上执行扩展 ping 命令，数据包的长度为 300 字节，路由器 R1 上显示的调试信息如下。

```
R3# ping 172.16.2.2 repeat 1 size 300
01:31:46: IP: s=172.16.1.6 (GigabitEthernet0/0), d=172.16.2.2, len 300, FIB policy match
01:31:46: IP: s=172.16.1.6 (GigabitEthernet0/0), d=172.16.2.2, len 300, PBR Counted
01:31:46: IP: s=172.16.1.6 (GigabitEthernet0/0), d=172.16.2.2 (Serial0/0/1), len 300, FIB policy routed
```

以上输出信息表明数据包长度为 300 字节的数据包在接口 Gi0/0 匹配路由策略，执行策略路由，设置数据包的出接口为 Se0/0/1。

③ 在主机 PC6 上执行扩展 ping 命令，数据包的长度为 1200 字节，路由器 R1 上显示的调试信息如下。

```
R3# ping 172.16.2.2 repeat 1 size 1200
01:35:55: IP: s=172.16.1.6 (GigabitEthernet0/0), d=172.16.2.2, len 1200, FIB policy rejected(no match) - normal forwarding
```

以上输出信息表明数据包长度为 1200 字节的数据包在接口 Gi0/0 不匹配路由策略，数据

包正常转发。

（2）查看路由策略应用情况

```
R1#show ip policy
Interface        Route map
local            CCNP
Gi0/0            CCNP
```

以上输出信息表明在接口 Gi0/0 和本地（local）应用了路由策略 CCNP 来执行策略路由。

6.5.3 实验 8：配置基于应用的策略路由

1. 实验目的

通过本实验可以掌握：
① 用 route-map 定义路由策略的方法。
② 在接口下应用路由策略的方法。
③ 基于应用的策略路由的配置和调试方法。

2. 实验拓扑

基于应用的策略路由配置实验拓扑如图 6-8 所示。

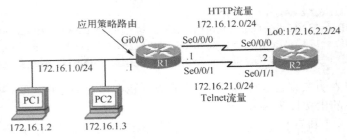

图 6-8　基于应用的策略路由配置实验拓扑

3. 实验步骤

实验设计如下：在路由器 R1 的 Gi0/0 接口应用 IP 策略路由 CCIE，将 HTTP 数据包的下一跳地址设置为 172.16.12.2，并且设置 IP 数据包优先级为 flash；将 Telnet 数据包的下一跳地址设置为 172.16.21.2，并且设置 IP 数据包优先级为 critical，所有其他的数据包正常转发，整个网络运行 EIGRP 路由协议（本实验 EIGRP 配置省略）。路由器 R2 需要开启 HTTP 和 Telnet 服务。

```
R1(config)#ip access-list extended HTTP
R1(config-ext-nacl)#permit tcp any any eq 80
R1(config)#ip access-list extended TELNET
R1(config-ext-nacl)#permit tcp any any eq 23
R1(config)#route-map CCIE permit 10
R1(config-route-map)#match ip address HTTP
```

```
R1(config-route-map)#set ip precedence flash
R1(config-route-map)#set ip next-hop 172.16.12.2
R1(config)#route-map CCIE permit 20
R1(config-route-map)#match ip address TELNET
R1(config-route-map)#set ip precedence critical
R1(config-route-map)#set ip next-hop 172.16.21.2
R1(config)#interface gigabitEthernet0/0
R1(config-if)#ip policy route-map CCIE
R1(config)#ip local policy route-map CCIE
```

4．实验调试

① 在主机 PC1 上访问 172.16.2.2 的 Web 服务，路由器 R1 上显示的调试信息如下。

```
02:22:29: IP: s=172.16.1.2 (GigabitEthernet0/0), d=172.16.2.2, len 44, FIB policy match
02:22:29: IP: s=172.16.1.2 (GigabitEthernet0/0), d=172.16.2.2, len 44, PBR Counted
02:22:29: IP: s=172.16.1.2 (GigabitEthernet0/0), d=172.16.2.2, g=172.16.12.2, len 44, FIB policy routed
```

以上输出信息表明 HTTP 的数据包在接口 Gi0/0 匹配路由策略，执行策略路由，设置数据包的下一跳地址为 172.16.12.2。

② 在主机 PC1 上访问 172.16.2.2 的 Telnet 服务，路由器 R1 上显示的调试信息如下。

```
02:23:49: IP: s=172.16.1.2 (GigabitEthernet0/0), d=172.16.2.2, len 44, FIB policy match
02:23:49: IP: s=172.16.1.2 (GigabitEthernet0/0), d=172.16.2.2, len 44, PBR Counted
02:23:49: IP: s=172.16.1.2 (GigabitEthernet0/0), d=172.16.2.2, g=172.16.21.2, len 44, FIB policy routed
```

以上输出信息表明 Telnet 的数据包在接口 Gi0/0 匹配路由策略，执行策略路由，设置数据包下一跳地址为 172.16.21.2。

③ 在主机 PC1 上 ping 172.16.2.2，路由器 R1 上显示的调试信息如下。

```
02:27:28: IP: s=172.16.1.2 (GigabitEthernet0/0), d=172.16.2.2, len 100, FIB policy rejected(no match) - normal forwarding
```

以上输出信息表明 ping 的数据包在接口 Gi0/0 不匹配路由策略，数据包正常转发。

6.6　VRF Lite

6.6.1　实验 9：配置 VRF Lite

1．实验目的

通过本实验可以掌握：
① VRF Lite 的概念和工作原理。
② VRF Lite 的配置和调试方法。

2．实验拓扑

配置 VRF Lite 实验拓扑如图 6-9 所示。

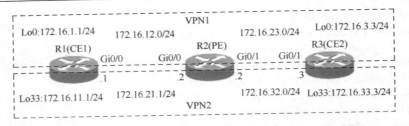

图 6-9　配置 VRF Lite 实验拓扑

3. 实验步骤

实验设计如下：通过 VRF 实现 VPN1 和 VPN2 通信完全隔离，两个 VRF 实例分别运行 OSPF 进程。路由器 R2 模拟 ISP 的网络。三台路由器分别使用子接口，地址分配方案如表 6-3 所示。

表 6-3　地址分配方案

	Gi0/0.1	Gi0/0.2	Gi0/1.1	Gi0/1.2
路由器 R1	172.16.12.1 (VLAN 10)	172.16.21.1 (VLAN 20)		
路由器 R2	172.16.12.2 (VLAN 10)	172.16.21.2 (VLAN 20)	172.16.23.2 (VLAN 10)	172.16.32.2 (VLAN 20)
路由器 R3			172.16.23.3 (VLAN 10)	172.16.32.3 (VLAN 20)

（1）配置路由器 R1

```
R1(config)#ip vrf vpn1              //创建 VRF
R1(config-vrf)#rd 1:1               //配置路由区分符
R1(config)#ip vrf vpn2
R1(config-vrf)#rd 2:2
R1(config)#interface Loopback0
R1(config-if)#ip vrf forwarding vpn1    //接口加入 VRF
R1(config-if)#ip address 172.16.1.1 255.255.255.0
R1(config)#interface Loopback11
R1(config-if)#ip vrf forwarding vpn2
R1(config-if)#ip address 172.16.11.1 255.255.255.0
R1(config)#interface GigabitEthernet0/0
R1(config-if)#no shutdown
R1(config)#interface GigabitEthernet0/0.1
R1(config-subif)#encapsulation dot1Q 10
R1(config-subif)#ip vrf forwarding vpn1   //接口加入 VRF
R1(config-subif)#ip address 172.16.12.1 255.255.255.0
R1(config)#interface GigabitEthernet0/0.2
R1(config-subif)#encapsulation dot1Q 20
R1(config-subif)#ip vrf forwarding vpn2
R1(config-subif)#ip address 172.16.21.1 255.255.255.0
R1(config)#router ospf 1 vrf vpn1    //启动与 VRF 关联的 OSPF 进程
R1(config-router)#router-id 1.1.1.1
R1(config-router)#network 172.16.1.1 0.0.0.0 area 0
R1(config-router)#network 172.16.12.1 0.0.0.0 area 0
```

```
R1(config)#router ospf 11 vrf vpn2
R1(config-router)#router-id 11.11.11.11
R1(config-router)#network 172.16.11.1 0.0.0.0 area 0
R1(config-router)#network 172.16.21.1 0.0.0.0 area 0
```

（2）配置路由器 R2

```
R2(config)#ip vrf v1
R2(config-vrf)#rd 11:11
R2(config-vrf)#route-target export 11:2      //配置 RT export 值
R2(config-vrf)#route-target import 11:2      //配置 RT import 值
```

> 【提示】
>
> route-target 的 export 和 import 值用于在 PE 之间实现路由的选择性导入和导出，由于本实验中 PE 路由器只有一台，所以可以不配置。可以通过 show ip vrf detail 命令查看每个 VRF 的 RT 值。

```
R2(config)#ip vrf v2
R2(config-vrf)#rd 22:22
R2(config-vrf)#route-target export 22:2
R2(config-vrf)#route-target import 22:2
R2(config)#interface GigabitEthernet0/0
R2(config-if)#no shutdown
R2(config)#interface GigabitEthernet0/0.1
R2(config-subif)#encapsulation dot1Q 10
R2(config-subif)#ip vrf forwarding v1
R2(config-subif)#ip address 172.16.12.2 255.255.255.0
R2(config)#interface GigabitEthernet0/0.2
R2(config-subif)#encapsulation dot1Q 20
R2(config-subif)#ip vrf forwarding v2
R2(config-subif)#ip address 172.16.21.2 255.255.255.0
R2(config)#interface GigabitEthernet0/1
R2(config-if)#no shutdown
R2(config)#interface GigabitEthernet0/1.1
R2(config-subif)#encapsulation dot1Q 10
R2(config-subif)#ip vrf forwarding v1
R2(config-subif)#ip address 172.16.23.2 255.255.255.0
R2(config)#interface GigabitEthernet0/1.2
R2(config-subif)#encapsulation dot1Q 20
R2(config-subif)#ip vrf forwarding v2
R2(config-subif)#ip address 172.16.32.2 255.255.255.0
R2(config)#router ospf 1 vrf v1
R2(config-router)#router-id 2.2.2.2
R2(config-router)#network 172.16.12.2 0.0.0.0 area 0
R2(config-router)#network 172.16.23.2 0.0.0.0 area 0
R2(config)#router ospf 2 vrf v2
R2(config-router)#router-id 22.22.22.22
R2(config-router)#network 172.16.21.2 0.0.0.0 area 0
R2(config-router)#network 172.16.32.2 0.0.0.0 area 0
```

（3）配置路由器 R3

```
R3(config)#ip vrf vpn1
R3(config-vrf)#rd 3:3
R3(config)#ip vrf vpn2
R3(config-vrf)#rd 4:4
R3(config)#interface Loopback0
R3(config-if)#ip vrf forwarding vpn1
R3(config-if)#ip address 172.16.3.3 255.255.255.0
R3(config)#interface Loopback33
R3(config-if)#ip vrf forwarding vpn2
R3(config-if)#ip address 172.16.33.3 255.255.255.0
R3(config)#interface GigabitEthernet0/1
R3(config-if)#no shutdown
R3(config)#interface GigabitEthernet0/1.1
R3(config-subif)#encapsulation dot1Q 10
R3(config-subif)#ip vrf forwarding vpn1
R3(config-subif)#ip address 172.16.23.3 255.255.255.0
R3(config)#interface GigabitEthernet0/1.2
R3(config-subif)#encapsulation dot1Q 20
R3(config-subif)#ip vrf forwarding vpn2
R3(config-subif)#ip address 172.16.32.3 255.255.255.0
R3(config)#router ospf 1 vrf vpn1
R3(config-router)#router-id 3.3.3.3
R3(config-router)#network 172.16.3.3 0.0.0.0 area 0
R3(config-router)#network 172.16.23.3 0.0.0.0 area 0
R3(config)#router ospf 2 vrf vpn2
R3(config-router)#router-id 33.33.33.33
R3(config-router)#network 172.16.32.3 0.0.0.0 area 0
R3(config-router)#network 172.16.33.3 0.0.0.0 area 0
```

4. 实验调试

（1）查看创建的 VRF 以及每个 VRF 的 RD 和加入到该 VRF 的接口

```
R1#show ip vrf
  Name                             Default RD        Interfaces
  vpn1                             1:1               Lo0
                                                     Gi0/0.1
  vpn2                             2:2               Gi0/0.2
                                                     Lo11
```

（2）查看加入 VRF 的接口、接口地址以及接口状态

```
R1#show ip vrf interfaces vpn1
  Interface         IP-Address        VRF                         Protocol
  Lo0               172.16.1.1        vpn1                        up
  Gi0/0.1           172.16.12.1       vpn1                        up
```

（3）查看路由表

以下输出省略路由代码部分。

① R1#**show ip route vrf vpn1 ospf**   //查看 vpn1 的虚拟路由表中 OSPF 路由
172.16.0.0/16 is variably subnetted, 6 subnets, 2 masks
O   172.16.3.3/32   [110/3] via 172.16.12.2, 00:05:21, GigabitEthernet0/0.1
O   172.16.23.0/24 [110/2] via 172.16.12.2, 00:05:21, GigabitEthernet0/0.1
② R1#**show ip route vrf vpn2 ospf**   //查看 vpn2 的虚拟路由表中 OSPF 路由
172.16.0.0/16 is variably subnetted, 6 subnets, 2 masks
O   172.16.32.0/24 [110/2] via 172.16.21.2, 00:07:05, GigabitEthernet0/0.2
O   172.16.33.3/32 [110/3] via 172.16.21.2, 00:07:05, GigabitEthernet0/0.2
③ R1#**show ip route**   //查看 R1 主路由表

④ R2#**show ip route vrf v1 ospf**   //查看 v1 的虚拟路由表中 OSPF 路由
172.16.0.0/16 is variably subnetted, 6 subnets, 2 masks
O   172.16.1.1/32 [110/2] via 172.16.12.1, 00:09:55, GigabitEthernet0/0.1
O   172.16.3.3/32 [110/2] via 172.16.23.3, 00:09:55, GigabitEthernet0/1.1
⑤ R2#**show ip route vrf v2 ospf**   //查看 v2 的虚拟路由表中 OSPF 路由

O   172.16.11.1/32 [110/2] via 172.16.21.1, 00:15:13, GigabitEthernet0/0.2
O   172.16.33.3/32 [110/2] via 172.16.32.3, 00:15:13, GigabitEthernet0/1.2
⑥ R2#**show ip route**   //查看 R2 主路由表

⑦ R3#**show ip route vrf vpn1 ospf**   //查看 vpn1 的虚拟路由表中 OSPF 路由
172.16.0.0/16 is variably subnetted, 6 subnets, 2 masks
O   172.16.1.1/32   [110/3] via 172.16.23.2, 00:11:32, GigabitEthernet0/1.1
O   172.16.12.0/24 [110/2] via 172.16.23.2, 00:11:32, GigabitEthernet0/1.1
⑧ R3#**show ip route vrf vpn2 ospf**   //查看 vpn2 的虚拟路由表中 OSPF 路由
172.16.0.0/16 is variably subnetted, 6 subnets, 2 masks
O   172.16.11.1/32 [110/3] via 172.16.32.2, 00:12:18, GigabitEthernet0/1.2
O   172.16.21.0/24 [110/2] via 172.16.32.2, 00:12:18, GigabitEthernet0/1.2
⑨ R3#**show ip route**   //查看 R3 主路由表

以上①～⑨输出显示了路由器 R1、R2 和 R3 的 VRF 的虚拟路由表以及主路由表，主路由表为空，虚拟路由表完全独立，互相隔离，从而实现 vpn1 和 vpn2 之间数据流的分离。

（4）执行 ping 命令

① R1#**ping vrf vpn1 172.16.3.3 source 172.16.1.1**
Type escape sequence to abort.
Sending 5, 100-byte ICMP Echos to 172.16.3.3, timeout is 2 seconds:
Packet sent with a source address of 172.16.1.1
!!!!!
Success rate is 100 percent (5/5), round-trip min/avg/max = 1/1/4 ms
② R1#**ping vrf vpn2 172.16.33.3 source 172.16.11.1**
Type escape sequence to abort.
Sending 5, 100-byte ICMP Echos to 172.16.33.3, timeout is 2 seconds:
Packet sent with a source address of 172.16.11.1
!!!!!
Success rate is 100 percent (5/5), round-trip min/avg/max = 1/1/4 ms

本 章 小 结

路由重分布实现了不同路由协议之间的路由信息的共享，而路径控制对提高网络的稳定性、安全性和收敛速度等意义重大。本章介绍了路由重分布和路径控制的多种方式，包括被动接口、路由映射表、分布列表、前缀列表、Cisco IOS IP SLA 和策略路由，最后介绍了 VRF 的概念，并用实验演示和验证了路由重分布基本配置、路由重分布中次优路由和路由环路问题的解决方法、被动接口和分布列表控制路由更新、前缀列表和路由映射表控制路由更新、IP SLA 控制路径选择、策略路由配置和 VRF Lite 配置。

第 7 章 BGP

通常可以将路由协议分为 IGP（内部网关协议）和 EGP（外部网关协议）两类。EGP 主要用于在 ISP 之间交换路由信息。目前使用最为广泛的 EGP 是 BGP 版本 4，它是第一个支持 CIDR 和路由汇总的 BGP 版本。而 RFC 4760 中定义的 BGP4+是对 BGP4 的扩展，支持包括 IPv6 在内的多种协议。

7.1 BGP 概述

7.1.1 BGP 特征

BGP 被称为是基于策略的路径向量路由协议，它的任务是在自治系统之间交换路由信息，同时确保没有路由环路，其特征如下。

① 用属性（Attribute）描述路径，丰富的属性特征方便实现基于策略的路由控制。
② 使用 TCP（端口 179）作为传输协议，继承了 TCP 的可靠性和面向连接的特性。
③ 通过 Keepalive 消息来检验 TCP 的连接。
④ 拥有自己的 BGP 邻居表、BGP 表和路由表。
⑤ 支持 VLSM 和 CIDR。
⑥ 支持 MD5 身份验证。
⑦ 采用增量更新和触发更新方式。
⑧ 适合在大型网路中使用。

当网络满足下列一个或者多个条件时，建议使用 BGP。

① 自治系统允许数据包穿越它前往其他自治系统。
② 自治系统有多条到达其他自治系统的连接。
③ 必须对进入或者是离开自治系统的数据包进行路由策略控制。

当网络符合下面的条件时，不建议使用 BGP。

① 与 Internet 或者另一个自治系统只有单一连接。
② 路由器没有足够的 CPU 处理能力和足够的内存处理 BGP 进程。
③ 网管人员对 BGP 路由操纵理解有限，无法预计启动 BGP 后的结果。

7.1.2 BGP 术语

① 对等体（Peer）：当两台 BGP 路由器之间建立了一条基于 TCP 的连接后，就称它们为邻居或对等体；
② 自治系统（AS）：是一组处于统一管理控制和策略下的路由器或主机，它们使用内部网关路由协议决定如何在自治系统内部路由数据包，并使用自治系统间路由协议决定如何把

数据包路由到其他自治系统。AS 号由因特网注册机构分配，长度为 16 比特（在 RFC 4893 中描述了长度为 32 位的扩展 AS 号），范围为 1～65535，其中 64512～65535 为私有使用。

③ IBGP：当 BGP 在一个 AS 内运行时，被称为内部 BGP（IBGP）。

④ EBGP：当 BGP 运行在 AS 之间时，被称为外部 BGP（EBGP）。

⑤ NLRI（网络层可达性信息）：是 BGP 更新报文的一部分，用于列出通过该路径可到达的目的地的集合。

⑥ 同步：在 BGP 能够通告路由之前，该路由必须存在于当前的 IP 路由表中。也就是说，BGP 和 IGP 必须在网络能被通告前同步。Cisco 允许通过命令 no synchronization 来关闭同步，这也是高版本 IOS 的默认配置。

⑦ IBGP 水平分割：通过 IBGP 学到的路由信息不能通告给其他的 IBGP 邻居。

⑧ 对等体组（Peer Group）：是一组采用相同更新策略的 BGP 邻居。当一个对等体加入对等体组中时，此对等体将获得与所在对等体组相同的配置；当对等体组的配置改变时，组内成员的配置也相应改变。

7.1.3 BGP 属性

BGP 具有丰富的属性，为路由控制带来很大的方便，BGP 路径属性分为以下 4 类。

（1）公认必遵（Well-Known Mandatory）

公认必遵属性是 BGP 更新信息中必须包含的信息，并且是必须被所有 BGP 厂商设备所能识别的属性，包括 ORIGIN、AS_PATH 和 Next_HOP 三个属性。

① ORIGIN（起源）：该属性说明了路由信息的来源，有 IGP、EGP 和 INCOMPLETE 三个可能的源。路由器在多个路由选择的处理过程中使用该信息。路由器选择具有最低 ORIGIN 类型的路径。ORIGIN 类型从低到高的顺序为 IGP<EGP<INCOMPLETE。

② AS_PATH（AS 路径）：包含在 Update 中的路由信息所经过的自治系统的序列。

③ Next_HOP（下一跳）：路由器所获得的 BGP 路由的下一跳。对 EBGP 会话来说，下一跳就是通告该路由的邻居路由器的源地址。对于 IBGP 会话，有两种情况，一是起源 AS 内部路由的下一跳通告该路由的邻居路由器的源地址；二是由 EBGP 注入 AS 的路由，它的下一跳会不变地被带入 IBGP 中。

（2）公认自决（Well-Known Discretionary）

公认自决指属性必须被所有 BGP 设备所识别，但是在 BGP 更新信息中可以发送，也可以不发送该属性，包括 LOCAL_PREF 和 ATOMIC_AGGREGATE 两个属性。

① LOCAL_PREF（本地优先级）：用于告诉自治系统内的路由器当有多条路径时，怎样离开自治系统。本地优先级越高，路由优先级越高。该属性仅仅在 IBGP 邻居之间传递。

② ATOMIC_AGGREGATE（原子聚合）：指出已被丢失了的信息。当路由聚合时会导致信息的丢失，因为聚合来自具有不同属性的不同源。如果一台路由器发送了导致信息丢失的聚合，路由器被要求将原子聚合属性附加到该路由上。

（3）可选过渡（Optional Transitive）

可选过渡属性并不要求所有的 BGP 实现都支持。如果该属性不能被 BGP 进程识别，它就会去看过渡标志。如果过渡标志被设置了，BGP 进程会接受这个属性并将它不加改变地传送，包括 AGGREGATOR 和 COMMUNITY。

① AGGREGATOR（聚合者）：标明实施路由聚合的 BGP 路由器 ID 和聚合路由的路由器的 AS 号。

② COMMUNITY（团体）：指共享一个公共属性的一组路由器。

（4）可选非过渡（Optional Nontransitive）

可选非过渡属性并不要求所有的 BGP 都支持。如果这些属性被发送到不能对其识别的路由器，这些属性将会被丢弃，不能传送给 BGP 邻居，包括 MED、ORIGINATOR_ID 和 CLUSTER_LIST。

① MED（多出口区分）：通知 AS 外的路由器采用哪一条路径到达 AS。它也被认为是路由的外部度量，低的 MED 值表示高的优先级。MED 属性在自治系统间交换，但 MED 属性不能传递到第三方 AS。默认情况下，仅当路径来自同一个自治系统的不同邻居时，路由器才比较它们的 MED 属性。

② ORIGINATOR_ID（起源 ID）：路由反射器会附加到这个属性上，它携带本 AS 源路由器的路由器 ID，用以防止环路。

③ CLUSTER_LIST（簇列表）：此属性显示了采用的反射路径。

7.1.4　BGP 消息类型及格式

1. BGP 包头

BGP 消息类型主要包括 Open、Update、Notification、Keepalive 和 Route-refresh。这些消息具有相同的包头，长度为 19 字节。BGP 包头格式如图 7-1 所示，各字段含义如下。

图 7-1　BGP 包头格式

① 标记：16 字节，用来检测对等体之间同步的丢失情况，以及在支持验证功能时用于验证消息。当不使用验证时所有比特均为"1"。

② 长度：2 字节，BGP 消息总长度（包括包头在内），以字节为单位，范围为 19～4096。

③ 类型：1 字节，BGP 消息的类型。其取值为 1～5，分别表示 Open、Update、Notification、Keepalive 和 Route-refresh 消息。

2. BGP Open 消息

Open 消息是 TCP 连接建立后发送的第一个消息，用于建立 BGP 对等体之间的连接关系。BGP Open 消息格式如图 7-2 所示，各字段含义如下。

图 7-2　BGP Open 消息格式

① 版本：1 字节，BGP 的版本号，对于 BGP4 来说其值为 4。

② 我的自治系统：2 字节，BGP 邻居建立发起者的 AS 号。用来决定双方是 IBGP 邻居，还是 EBGP 邻居。

③ 保持时间：2 字节，是设备收到一个 Keepalive 消息之前允许等待的最长时间。如果在这个时间内未收到对端发来的 Keepalive 消息，则认为 BGP 连接中断。这个时间为 0 秒则不发送 Keepalive 消息，如果不为 0，则至少是 3 秒。在建立对等体关系时两端要协商保持时间，在协商时，采用 Open 消息中较小端的保持时间作为双方的保持时间。Cisco 默认保持时间是 180 秒。

④ BGP 标识符：4 字节，发送者的 BGP 路由器 ID，以 IP 地址的形式表示。BGP 路由器 ID 的确定方法和 OSPF 路由器 ID 确定方法相同。

⑤ 可选参数长度：1 字节，表示可选参数的长度。如果为 0 则没有可选参数。

⑥ 可选参数：可变长度，用于 BGP 验证或多协议扩展等功能。包括一个可选参数列表，每个参数由 1 字节类型字段、1 字节长度字段和一个包含参数值的可变长度字段来确定，即 TLV（Type-Length-Value）方式。

3．BGP Update 消息

Update 消息用于在对等体之间交换路由信息。它既可以发布可达路由信息，也可以撤销不可达的路由信息。BGP Update 消息格式如图 7-3 所示，各字段含义如下。

① 不可用路由长度：2 字节，撤销路由字段的整体长度，如果为 0，说明没有路由被撤销，并且在该消息中没有撤销路由的字段。

② 撤销路由：可变长度，包含不可达路由的列表。

图 7-3　BGP Update 消息格式

③ 全部路径属性长度：2 字节，路径属性字段的长度。如果为 0 则说明没有路径属性字段。

④ 路径属性：可变长度，列出与 NLRI 相关的所有路径属性列表，包括 AS_PATH、本地优先级和起源等，每个路径属性由一个 TLV（Type-Length-Value）三元组构成。路径属性是 BGP 用以进行路由控制和决策的重要信息。

⑤ 网络层可达信息：可变长度，是可达路由的前缀和前缀长度二元组。

4．BGP Notification 消息

当 BGP 检测到错误状态时，就向对等体发出 Notification 消息，之后 BGP 连接会立即中断。BGP Notification 消息格式如图 7-4 所示，各字段含义如下：

① 错误编码：1 字节，错误类型。
② 错误子码：1 字节，错误类型更详细的信息。
③ 数据：可变长度，用于诊断错误的原因，它的内容依赖于具体的错误编码和错误子码。

5．BGP Keepalive 消息

BGP 会周期性（Cisco 默认为 60 秒）地向对等体发出 Keepalive 消息，用来保持连接的有效性。其消息格式中只包含 BGP 包头，没有附加其他任何字段。Keepalive 消息的发送周

期是保持时间的 1/3，但该时间不能低于 1 秒。如果协商后的保持时间为 0，则不发送 Keepalive 消息。

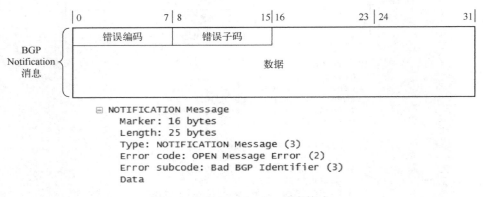

图 7-4 BGP Notification 消息格式

6. Route-refresh 消息

Route-refresh 消息用来要求对等体重新发送指定地址族的路由信息，BGP Route-refresh 消息格式如图 7-5 所示，地址族标识可以是 IPv4 或 IPv6 等，子地址族标识可以是单播或组播路由等。

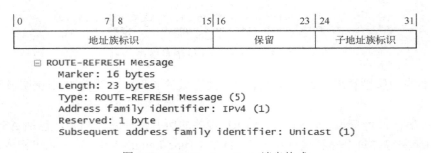

图 7-5 BGP Route-refresh 消息格式

7.1.5 BGP 路由决策

BGP 使用很多属性描述路由特性，这些属性和每一条路由一起在 BGP 更新消息中被发送。路由器使用这些属性去选择到目的地的最佳路由。理解 BGP 路由判定的过程非常重要，下面按优先顺序给出了路由器在 BGP 路径选择中的判定过程：

① 如果下一跳不可达，则不考虑该路由。
② 优先选取具有最大权重（Weight）值的路径，权重是 Cisco 专有属性。
③ 优先选取具有最高本地优先级的路由。
④ 优先选取源自于本路由器（即下一跳为 0.0.0.0）上 BGP 的路由。
⑤ 优先选取具有最短 AS 路径的路由。
⑥ 优先选取有最低起源代码（IGP<EGP<INCOMPLETE）的路由。
⑦ 优先选取具有最低 MED 值的路径。

⑧ 在 EBGP 路由和联盟 EBGP 路由中，首选 EBGP 路由，在联盟 EBGP 路由和 IBGP 路由中，首选联盟 EBGP 路由。
⑨ 优先选取离 IGP 邻居最近的路径。
⑩ 优先选取最老的 EBGP 路径。
⑪ 优先选取具有最低 BGP 路由器 ID 的路径。
⑫ 优先选取邻居 IP 地址最小的路径。

7.1.6 BGP 路由抑制

BGP 路由抑制（Route Dampening）用来解决路由不稳定的问题。路由不稳定的主要表现形式是路由翻转（Route Flaps），即路由表中的某条路由反复消失和重现。当发生路由翻转时，路由协议就会向邻居发布路由更新信息，收到更新信息的路由器需要重新计算路由并修改路由表。所以频繁的路由翻转会消耗大量的带宽资源和 CPU 资源，严重时会影响网络正常工作。BGP 使用路由抑制来防止持续的路由翻转带来的不利影响。BGP 路由抑制的工作原理如图 7-6 所示。

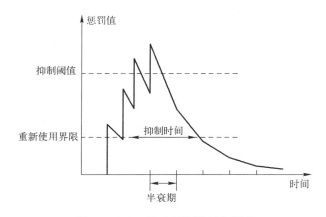

图 7-6 BGP 路由抑制的工作原理

下面介绍几个术语。
① 半衰期：单位为分钟，每经过半衰期的时间，抑制值就会减半，Cisco 默认为 15 分钟。
② 重新使用界限：当一条被抑制路由的惩罚值低于该值后，该路由可重新使用。Cisco 默认值为 750。
③ 抑制阈值：超过该值路由被抑制，不再向其他 BGP 对等体发布更新报文。Cisco 默认值为 2000。
④ 最大抑制时间：被抑制路由的抑制时间超过该值后，不管惩罚值为多少，都会重新使用，Cisco 默认为 60 分钟，即 4 倍的半衰期时间。

当一条路由出现翻转时，这条路由会被加上一个 1000 的惩罚值，这个惩罚值会每 5 秒钟递减，等到 15 分钟时，惩罚值会减少到一半，也就是 500。

如果这条路由连续出现翻转，每一次翻转这条路由的惩罚值都会加 1000，例如，一条路由连续出现了 3 次翻转，那么这时候惩罚值就接近 3000。当惩罚值大于 2000 时，这条路由就会在 BGP 表中被标为 Suppressed。经过了 15 分钟后（没有再发生翻转），这条路由的惩罚值会被减少一半到 1500，这条路由还是被标为 Suppressed。再经过 15 分钟后，惩罚值减为 750，这条路由才可以重新被使用。也就是说经过 30 分钟，路由器认为这条路由是稳定的，才会重新使用它。如果这条路由不停地翻转，好比说 15 次，这时候惩罚值会接近 15000。每过 15 分钟惩罚值会减少一半，可是到了 60 分钟时，惩罚值还是大于 750，这时因为已经超过了最大抑制时间，所以这条路由会被重新使用。需要注意的是，BGP 路由抑制只对通过 EBGP 学到的路由起作用，对 IBGP 学到的路由不起作用。

7.1.7 BGP 邻居状态

BGP 建立邻居有限状态机如图 7-7 所示，共有如下 6 种状态。

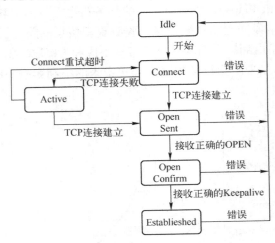

图 7-7 BGP 建立邻居有限状态机

（1）Idle（空闲）状态

这是初始状态，BGP 进程检查是否有前往指定邻居的路由。如果没有路由，则保持空闲状态；如果有路由，则进入连接状态。

（2）Connect（连接）状态

该状态下，BGP 等待完成 TCP 连接。若连接成功，则向对等体发送 Open 消息，然后状态进入 OpenSent 状态；如果连接失败，则继续侦听是否有对等体启动连接并进入 Active 状态。

（3）Active（激活）状态

该状态下，BGP 试图建立 TCP 连接，如果连接成功，则向对等体发送 Open 消息，并转至 OpenSent 状态。

（4）OpenSent（打开发送）状态

该状态下，BGP 等待对等体的 Open 消息。收到 Open 消息后对其进行检查，如果发现错误，本地发送 Notification 消息给对等体并进入空闲状态；如果消息没有错误，BGP 发送 Keepalive 消息并进入 OpenConfirm 状态。

（5）OpenConfirm（打开确认）状态

该状态下，BGP 等待 Keepalive 消息或 Notification 消息。如果收到 Keepalive 消息，则进入 Established 状态；如果收到 Notification 消息，则进入空闲状态。

（6）Established（已建立）

该状态下，BGP 可以和其他对等体交换 Update、Notification 和 Keepalive 消息，开始路由选择。如果收到了正确的 Update 和 Keepalive 消息，就认为对端处于正常运行状态，本地重置保持时间计时器；如果收到 Notification 消息，则进入到空闲状态；如果 TCP 连接中断，则关闭 BGP 连接并回到空闲状态。

7.2 配置基本 BGP

7.2.1 实验 1：配置 IBGP 和 EBGP

1. 实验目的

通过本实验可以掌握：
① 启动 BGP 路由进程。
② BGP 进程中通告网络的方法。
③ IBGP 邻居和 EBGP 邻居配置。
④ BGP 路由更新源和 next-hop-self 配置。
⑤ BGP 路由汇总配置。
⑥ 对等体组配置。
⑦ BGP 路由调试方法。

2. 实验拓扑

配置 IBGP 和 EBGP 实验拓扑如图 7-8 所示。

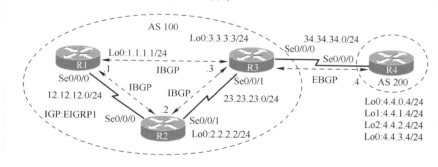

图 7-8　配置 IBGP 和 EBGP 实验拓扑

3. 实验步骤

因为本实验中 IBGP 路由器（R1，R2 和 R3）形成全互连的邻居关系，所以路由器 R1、R2 和 R3 均关闭同步。AS 100 内部路由器之间运行的 IGP 是 EIGRP，实现网络的连通性，为 BGP 邻居关系建立提供 TCP 连接。

（1）配置路由器 R1

```
R1(config)#router eigrp 1
R1(config-router)#network 1.1.1.1 0.0.0.0
R1(config-router)#network 12.12.12.1 0.0.0.0
R1(config)#router bgp 100                              //启动 BGP 进程
R1(config-router)#no synchronization                   //关闭同步，高版本 IOS 默认配置
R1(config-router)#bgp router-id 1.1.1.1
//配置 BGP 路由器 ID，如果建立邻居关系的两台路由器的 BGP 路由器 ID 相同，会出现类似如下
的信息：04:53:11: %BGP-3-NOTIFICATION: received from neighbor 3.3.3.3 2/3 (BGP identifier wrong) 4 bytes
03030303，提示 BGP 标识符错误，不能建立邻居关系
R1(config-router)#neighbor 2.2.2.2 remote-as 100
//指定 BGP 邻居路由器及所在的 AS
R1(config-router)#neighbor 2.2.2.2 update-source Loopback0    //指定 BGP 更新源
R1(config-router)#neighbor 3.3.3.3 remote-as 100
R1(config-router)#neighbor 3.3.3.3 update-source Loopback0
R1(config-router)#network 1.1.1.0 mask 255.255.255.0          //通告网络
R1(config-router)#no auto-summary                             //关闭自动汇总，高版本 IOS 默认配置
```

（2）配置路由器 R2

```
R2(config)#router eigrp 1
R2(config-router)#network 2.2.2.2 0.0.0.0
R2(config-router)#network 12.12.12.2 0.0.0.0
R2(config-router)#network 23.23.23.2 0.0.0.0
R2(config)#router bgp 100
R2(config-router)#bgp router-id 2.2.2.2
R2(config-router)#neighbor 1.1.1.1 remote-as 100
R2(config-router)#neighbor 1.1.1.1 update-source Loopback0
R2(config-router)#neighbor 3.3.3.3 remote-as 100
R2(config-router)#neighbor 3.3.3.3 update-source Loopback0
```

（3）配置路由器 R3

```
R3(config)#router eigrp 1
R3(config-router)#network 3.3.3.3 0.0.0.0
R3(config-router)#network 23.23.23.3 0.0.0.0
R3(config)#router bgp 100
R3(config-router)#bgp router-id 3.3.3.3
R3(config-router)#neighbor 1.1.1.1 remote-as 100
R3(config-router)#neighbor 1.1.1.1 update-source Loopback0
R3(config-router)#neighbor 1.1.1.1 next-hop-self
//配置下一跳自我，即对从 EBGP 邻居传入的路由，在通告给 IBGP 邻居时，强迫路由器通告自己
是发送 BGP 更新信息的下一跳，而不是 EBGP 邻居
```

```
R3(config-router)#neighbor 2.2.2.2 remote-as 100
R3(config-router)#neighbor 2.2.2.2 update-source Loopback0
R3(config-router)#neighbor 2.2.2.2 next-hop-self
```

【提示】

上述 6 行配置也可以通过对等体组来实现，替代配置如下：

```
R3(config-router)#neighbor cisco peer-group              //创建对等体组
R3(config-router)#neighbor cisco remote-as 100           //定义对等体组策略
R3(config-router)#neighbor cisco update-source Loopback0
R3(config-router)#neighbor cisco next-hop-self
R3(config-router)#neighbor 1.1.1.1 peer-group cisco      //加入对等体组
R3(config-router)#neighbor 2.2.2.2 peer-group cisco
```

可以通过命令 **show ip bgp peer-group** 查看对等体组的信息：

```
R3#show ip bgp peer-group cisco
BGP peer-group is cisco, remote AS 100                   //对等体组名称及 remote AS 号
  BGP version 4
  Default minimum time between advertisement runs is 0 seconds
 For address family: IPv4 Unicast                        //地址族
  BGP neighbor is cisco, peer-group internal, members:   //对等体组成员
  1.1.1.1 2.2.2.2
  Index 0, Offset 0, Mask 0x0
  NEXT_HOP is always this router                         //下一跳一直是本路由器
  Update messages formatted 0, replicated 0
  Number of NLRIs in the update sent: max 0, min 0
R3(config-router)#neighbor 34.34.34.4 remote-as 200
```

（4）配置路由器 R4

```
R4(config)#ip route 4.4.0.0 255.255.252.0 null0
//在 IGP 路由表中构造该汇总路由，否则不能在 BGP 中用 network 命令通告
R4(config)#router bgp 200
R4(config-router)#bgp router-id 4.4.4.4
R4(config-router)#neighbor 34.34.34.3 remote-as 100
R4(config-router)#network 4.4.0.0 mask 255.255.255.0
R4(config-router)#network 4.4.1.0 mask 255.255.255.0
R4(config-router)#network 4.4.2.0 mask 255.255.255.0
R4(config-router)#network 4.4.3.0 mask 255.255.255.0
R4(config-router)#network 4.4.0.0 mask 255.255.252.0
```
//用 network 命令进行路由汇总通告，做这条配置是为了说明在 BGP 中，network 命令不仅可以通告直连路由，还可以通告 IGP 路由表中的其他路由条目，从功能上讲，该条路由可以取代上面通告的四条直连子网路由。本实验中汇总和明细路由都被通告，实际应用中不需要

【技术要点】

① 一台路由器只能启动一个 BGP 进程，当配置多个进程时会提示如下的错误信息：BGP is **already running;** AS is 100。

② 命令 **neighbor** 后边跟的是邻居路由器 BGP 路由更新源的地址。

③ BGP 中的 **network** 命令与 IGP 不同，它只是将 IGP 中存在的路由条目（可以是直连、静态路由或动态路由）在 BGP 中通告，同时 **network** 命令使用参数 **mask** 来通告单独的子网。如果 BGP 的自动汇总功能没有关闭，而且在 IGP 路由表中存在子网路由，在 BGP 中可以用 **network** 命令通告主类网络，当然也可以通过参数 **mask** 来通告单独的子网；如果 BGP 的自动汇总功能关闭，则通告必须通过参数 **mask** 严格匹配掩码长度。

④ 在命令 **neighbor** 后边跟 **update-source** 参数，是用来指定 BGP 更新源的。如果网络中有多条路径，那么用环回接口建立 TCP 连接，并作为 BGP 路由的更新源，会增加 BGP 的稳定性。

⑤ 在命令 **neighbor** 后边跟 **next-hop-self** 参数是为了解决下一跳可达性问题，因为当路由通过 EBGP 注入到 AS 时，从 EBGP 获得的下一跳地址会被不变地在 IBGP 中传递，**next-hop-self** 参数使得路由器会把自己作为发送 BGP 更新信息的下一跳通告给 IBGP 邻居。

⑥ BGP 的下一跳是指 BGP 表中路由条目的下一跳，也就是相应 **neighbor** 命令所指的地址。

4．实验调试

（1）查看 TCP 连接信息摘要

```
R3#show tcp brief
TCB         Local Address           Foreign Address         (state)
64752BAC    3.3.3.3.11002           1.1.1.1.179             ESTAB
64753B5C    3.3.3.3.11000           2.2.2.2.179             ESTAB
6472708     34.34.34.3.11001        34.34.34.4.179          ESTAB
```

以上输出表明路由器 R3 与路由器 R1、R2 和 R4 的 179 端口建立了 TCP 连接。建立 TCP 连接的双方使用 BGP 路由更新源的地址。只要两台路由器之间建立了一条 TCP 连接，就可以形成 BGP 邻居关系。

（2）查看 BGP 邻居的详细信息

```
R3#show ip bgp neighbors 34.34.34.4
BGP neighbor is 34.34.34.4, remote AS 200, external link
//BGP 邻居的地址和所在 AS，external 表示建立的是 EBGP 邻居关系
  BGP version 4, remote router ID 4.4.4.4        //BGP 版本和远程邻居的 BGP 路由器 ID
  BGP state = Established, up for 00:50:29       //BGP 邻居关系的状态以及建立的时间
  Last read 00:00:21, hold time is 180, keepalive interval is 60 seconds
  //默认保持时间和 keepalive 发送周期，可以通过命令 timers bgp keepalive holdtime 来调整，该命
令对所有邻居生效，如果想针对某个邻居调整，命令为 neighbor ip-address timers keepalive holdtime，调整
之后，在进行 BGP 连接建立时，会协商使用小的值
            （此处省略部分输出）
```

以上输出表明路由器有一个外部 BGP 邻居路由器 R4（34.34.34.4）在 AS 200 中。此邻居路由器 ID 是 **4.4.4.4**。命令 **show ip bgp neighbors** 显示出的信息最重要的一部分是 BGP state= 那一行。此行给出了 BGP 连接的状态。**Established** 状态表示 BGP 对等体间的会话正在运行，可以交换 BGP 路由信息。

(3) 查看 BGP 的摘要信息

```
R3#show ip bgp summary
BGP router identifier 3.3.3.3, local AS number 100      //BGP 路由器 ID 及本地 AS
BGP table version is 8, main routing table version 8
//BGP 表的版本号（BGP 表变化时该号会逐次加 1）和注入主路由表的最后版本号
6 network entries using 792 bytes of memory
6 path entries using 312 bytes of memory
3/2 BGP path/bestpath attribute entries using 504 bytes of memory
1 BGP AS-PATH entries using 24 bytes of memory
0 BGP route-map cache entries using 0 bytes of memory
0 BGP filter-list cache entries using 0 bytes of memory
Bitfield cache entries: current 2 (at peak 2) using 64 bytes of memory
BGP using 1696 total bytes of memory
//以上八行显示了 BGP 使用内存的情况
BGP activity 6/0 prefixes, 6/0 paths, scan interval 60 secs
//BGP 活动的前缀、路径和扫描间隔

Neighbor      V    AS     MsgRcvd  MsgSent   TblVer   InQ  OutQ  Up/Down    State/PfxRcd
1.1.1.1       4    100    58       58        8        0    0     00:44:01   1
2.2.2.2       4    100    51       49        8        0    0     00:43:54   0
34.34.34.4    4    200    19       19        8        0    0     00:15:21   5
//以上输出的邻居表的各个字段的含义如下
① Neighbor：BGP 邻居的路由器 ID
② V：BGP 的版本
③ AS：邻居所在的 AS 号
④ MsgRcvd：接收的 BGP 数据包数量
⑤ MsgSent：发送的 BGP 数据包数量
⑥ InQ/OutQ：入站队列或出站队列中等待处理的数据包数量
⑦ TblVer：发送给该邻居的最后一个 BGP 表的版本号
⑧ Up/Down：保持邻居关系的时间
⑨ State/PfxRcd：BGP 连接状态或者收到的路由前缀数量
```

【技术要点】

为了确保能够建立 BGP 邻居关系，**neighbor** 命令指定的邻居地址必须可达（但是两端不能全都通过默认路由实现可达性，因为用默认路由不可以主动发起 BGP 连接），同时要确保发送方路由器的更新源地址（BGP 路由器默认以到达邻居的出接口为更新源）和接收方路由器 **neighbor** 命令所指定的地址相同。BGP 邻居无法建立的可能原因如下所述。

如果 BGP 邻居关系一直停在 **Idle** 状态，可能的原因如下。

① 没有去往邻居的路由。
② **neighbor** 命令指的邻居的地址不正确。
③ BGP 路由器 ID 相同。

如果 BGP 邻居关系一直停在 **Active** 状态，可能的原因如下。

① 邻居没有更新源的路由。
② 邻居的 **neighbor** 命令指的地址不正确。
③ 邻居没有配置 **neighbor** 命令。
④ 邻居的 **neighbor** 命令配置的 AS 号不正确。

⑤ 双方配置的 BGP 更新源不匹配。

⑥ 两端全都通过默认路由实现更新源可达性。

（4）查看 BGP 表的信息

```
R3#show ip bgp
BGP table version is 11, local router ID is 3.3.3.3   //BGP 表的版本号和路由器的 BGP 路由器 ID
Status codes: s suppressed, d damped, h history, * valid, > best, i - internal,
              r RIB-failure, S Stale                  //BGP 路由状态代码区
Origin codes: i - IGP, e - EGP, ? – incomplete        //BGP 路由起源代码区
   Network          Next Hop            Metric  LocPrf  Weight  Path
r>i 1.1.1.0/24      1.1.1.1             0       100     0       i
*>  4.4.0.0/24      34.34.34.4          0               0       200 i
*>  4.4.0.0/22      34.34.34.4          0               0       200 i
*>  4.4.1.0/24      34.34.34.4          0               0       200 i
*>  4.4.2.0/24      34.34.34.4          0               0       200 i
*>  4.4.3.0/24      34.34.34.4          0               0       200 i
```

以上输出中，路由条目表项的状态代码（Status Codes）的含义解释如下。

① s：表示路由条目被抑制。

② d：表示路由条目由于被惩罚而受到抑制，从而阻止了不稳定路由的发布。

③ h：表示该路由正在被惩罚，但还未达到抑制阈值而使它被抑制。

④ *：表示该路由条目有效。

⑤ >：表示该路由条目最优，可以被传递，达到最优的重要前提是下一跳可达。

⑥ i：表示该路由条目是从 IBGP 邻居学到的。

⑦ r：表示将 BGP 表中的路由条目安装到 IP 路由表中失败，可以通过命令 **show ip bgp rib-failure** 显示没有安装到路由表的 BGP 路由以及没有装入的原因，如下所示。

```
R3#show ip bgp rib-failure
Network          Next Hop         RIB-failure              RIB-NH Matches
1.1.1.0/24       1.1.1.1          Higher admin distance    n/a
//路由条目没有被安装到路由表的原因是应为管理距离大，因为通过 EIGRP 学到该路由的管理距
离是 90，而通过 IBGP 学到该路由的管理距离是 200
```

⑧ S：表示该路由条目过期，用于支持 NSF 的路由器中。

以上输出中，起源代码（Origin codes）中 i 的含义表示路由条目来源为 IGP，e 的含义表示路由条目来源为 EGP，? 的含义表示路由条目来源不清楚，通常是从 IGP 重分布到 BGP 的路由条目。

下面具体地解释 BGP 路由条目的含义。

```
r>i 1.1.1.0/24          1.1.1.1        0     100     0 i
```

① r：因为路由器 R3 通过 EIGRP 学到 **1.1.1.0/24** 路由条目，其管理距离为 90，而通过 IBGP 学到 **1.1.1.0/24** 路由条目的管理距离是 200，而且关闭了同步，BGP 表中的路由条目放入到 IP 路由表中失败，所以出现代码 **r**。

② >：表示该路由条目最优，可以被传递。

③ i：紧跟>的 i，表示该路由条目是从 IBGP 邻居学到的。

④ **1.1.1.1**：表示该 BGP 路由的下一跳，即邻居的 BGP 路由更新源。
⑤ **0**（标题栏对应 Metric）：表示该路由外部度量值即 MED 值为 0。
⑥ **100**：表示该路由本地优先级为 100。
⑦ **0**（标题栏对应 Weight）：表示该路由的权重值为 0，如果是本地产生的，默认权重值是 32768；如果是从 BGP 邻居学来的，默认权重值为 0。
⑧ 由于该路由是通过相同 AS 的 IBGP 邻居传递而来的，所以 PATH 字段为空。
⑨ **i**：最后的 **i**，表示路由条目来源为 IGP，它是路由器 R1 用 **network** 命令通告的，而不是通过 EGP 或者重分布学到的。

（5）查看路由表

```
① R1#show ip route bgp
      4.0.0.0/8 is variably subnetted, 5 subnets, 2 masks
B        4.4.0.0/24 [200/0] via 3.3.3.3, 00:22:39
B        4.4.0.0/22 [200/0] via 3.3.3.3, 00:22:39
B        4.4.1.0/24 [200/0] via 3.3.3.3, 00:22:40
B        4.4.2.0/24 [200/0] via 3.3.3.3, 00:22:40
B        4.4.3.0/24 [200/0] via 3.3.3.3, 00:22:40
② R3#show ip route bgp
      4.0.0.0/8 is variably subnetted, 5 subnets, 2 masks
B        4.4.0.0/24 [20/0] via 34.34.34.4, 01:11:28
B        4.4.0.0/22 [20/0] via 34.34.34.4, 01:11:28
B        4.4.1.0/24 [20/0] via 34.34.34.4, 01:11:28
B        4.4.2.0/24 [20/0] via 34.34.34.4, 01:11:28
B        4.4.3.0/24 [20/0] via 34.34.34.4, 01:12:53
```

以上输出表明 IBGP 的管理距离是 200，EBGP 的管理距离是 20。由于在路由器 R3 的 IBGP 邻居配置了 **next-hop-self** 参数，所以看到路由器 R1 的 BGP 路由条目的下一跳为 **3.3.3.3**，即 R3 的 BGP 的更新源。

（6）采用 ping 命令测试

在路由器 R4 上 ping 1.1.1.1，结果是不通的，原因很简单，路由器 R1 和 R2 的路由表中没有 34.34.34.0 的路由。此时如果能顺利执行扩展 ping 命令，就是通的，测试结果如下。

```
① R4#ping 1.1.1.1
Type escape sequence to abort.
Sending 5, 100-byte ICMP Echos to 1.1.1.1, timeout is 2 seconds:
.....
Success rate is 0 percent (0/5)
② R4#ping 1.1.1.1 source 4.4.0.4
Type escape sequence to abort.
Sending 5, 100-byte ICMP Echos to 1.1.1.1, timeout is 2 seconds:
Packet sent with a source address of 4.4.0.4
!!!!!
```

如果一定采用标准 ping 命令，无非就是让路由器 R1 和 R2 学到 34.34.34.0 的路由，方法很多，比如在路由器 R3 的 EIGRP 进程中重分布直连路由。

（7）验证 BGP 同步

在 R1 上打开 BGP 同步，然后查看 BGP 表。

```
R1(config)#router bgp 100
R1(config-router)#synchronization          //打开同步
R1#clear ip bgp *                          //重置 BGP 连接
R1#show ip bgp
BGP table version is 1, local router ID is 1.1.1.1
Status codes: s suppressed, d damped, h history, * valid, > best, i - internal,
              r RIB-failure, S Stale
Origin codes: i - IGP, e - EGP, ? - incomplete
   Network          Next Hop         Metric    LocPrf    Weight    Path
*> 1.1.1.0/24       0.0.0.0          0                   32768     i
*  i4.4.0.0/24      3.3.3.3          0         100       0         200 i
*  i4.4.0.0/22      3.3.3.3          0         100       0         200 i
*  i4.4.1.0/24      3.3.3.3          0         100       0         200 i
*  i4.4.2.0/24      3.3.3.3          0         100       0         200 i
*  i4.4.3.0/24      3.3.3.3          0         100       0         200 i
```

以上输出表明 R1 从 AS 200 学到的 BGP 路由不是被优化的，因为 R1 的 IGP 路由表中并没有这些路由条目。

（8）查看 BGP 发送和接收的更新信息

```
R4#debug ip bgp updates
08:44:34: BGP(0): 34.34.34.3 rcvd UPDATE w/ attr: nexthop 34.34.34.3, origin i, path 100
08:44:34: BGP(0): 34.34.34.3 rcvd 1.1.1.0/24
08:44:34: BGP(0): Revise route installing 1 of 1 routes for 1.1.1.0/24 -> 34.34.34.3(main) to main IP table
//以上三行表明收到携带属性的更新信息并放入路由表中
08:44:34: BGP(0): nettable_walker 4.4.0.0/24 route sourced locally
08:44:34: BGP(0): nettable_walker 4.4.0.0/22 route sourced locally
08:44:34: BGP(0): nettable_walker 4.4.1.0/24 route sourced locally
08:44:34: BGP(0): nettable_walker 4.4.2.0/24 route sourced locally
08:44:34: BGP(0): nettable_walker 4.4.3.0/24 route sourced locally
//以上五行表明本地产生的路由
08:44:34: BGP(0): 34.34.34.3 send UPDATE (format) 4.4.3.0/24, next 34.34.34.4, metric 0, path Local
08:44:34: BGP(0): 34.34.34.3 send UPDATE (prepend, chgflags: 0x820) 4.4.2.0/24, next 34.34.34.4, metric 0, path Local, extended community
08:44:34: BGP(0): 34.34.34.3 send UPDATE (prepend, chgflags: 0x820) 4.4.1.0/24, next 34.34.34.4, metric 0, path Local, extended community
08:44:34: BGP(0): 34.34.34.3 send UPDATE (prepend, chgflags: 0x820) 4.4.0.0/22, next 34.34.34.4, metric 0, path Local, extended community
08:44:34: BGP(0): 34.34.34.3 send UPDATE (prepend, chgflags: 0x820) 4.4.0.0/24, next 34.34.34.4, metric 0, path Local, extended community
//以上十行表明发送的 BGP 路由更新信息，每个路由条目都携带了相应的属性
```

（9）验证 IBGP 水平分割

删除路由器 R1 和 R3 之间的邻居关系，保持路由器 R1 和 R2 建立邻居关系，路由器 R2 和 R3 建立邻居关系，配置如下。

```
R1(config)#router bgp 100
R1(config-router)#no synchronization
R1(config-router)#no neighbor 3.3.3.3
R3(config)#router bgp 100
R3(config-router)#no neighbor 1.1.1.1
```

在路由器 R1 和 R2 上查看 BGP 表：

① R1#**show ip bgp**
BGP table version is 2, local router ID is 1.1.1.1
Status codes: s suppressed, d damped, h history, * valid, > best, i - internal,
              r RIB-failure, S Stale
Origin codes: i - IGP, e - EGP, ? - incomplete

Network	Next Hop	Metric	LocPrf	Weight	Path
*> 1.1.1.0/24	0.0.0.0	0		32768	i

② R2#**show ip bgp**
BGP table version is 8, local router ID is 2.2.2.2
Status codes: s suppressed, d damped, h history, * valid, > best, i - internal,
              r RIB-failure, S Stale
Origin codes: i - IGP, e - EGP, ? - incomplete

Network	Next Hop	Metric	LocPrf	Weight	Path
r>i1.1.1.0/24	1.1.1.1	0	100	0	i
*>i4.4.0.0/24	3.3.3.3	0	100	0	200 i
*>i4.4.0.0/22	3.3.3.3	0	100	0	200 i
*>i4.4.1.0/24	3.3.3.3	0	100	0	200 i
*>i4.4.2.0/24	3.3.3.3	0	100	0	200 i
*>i4.4.3.0/24	3.3.3.3	0	100	0	200 i

以上①和②输出表明路由器 R2 并没有将路由器 R3 通告的 BGP 路由通告给路由器 R1，这也进一步验证了 IBGP 水平分割的基本原理：通过 IBGP 学到的路由不能通告给相同 AS 内的其他 IBGP 邻居。通常的解决办法有两个：IBGP 形成全互连邻居关系或使用路由反射器。

7.2.2 实验 2：配置 BGP 验证、路由抑制和 EBGP 多跳

1. 实验目的

通过本实验可以掌握：
① BGP 邻居 MD5 验证的配置和调试方法。
② BGP 路由抑制的配置和调试方法。
③ EBGP 多跳的原理、配置和调试方法。

2. 实验拓扑

配置 BGP 验证、路由抑制和 EBGP 多跳实验拓扑如图 7-9 所示。

3. 实验步骤

本实验中，路由器 R1 和路由器 R2 用环回接口 0 建立 EBGP 邻居关系，并通过静态路由实现环回接口 0 的可达性，同时通过关闭和开启路由器 R1 的环回接口 11 完成路由抑制调试。

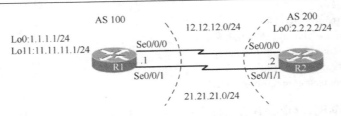

图 7-9 配置 BGP 验证、路由抑制和 EBGP 多跳实验拓扑

（1）配置路由器 R1

```
R1(config)#ip route 2.2.2.0 255.255.255.0 Serial0/0/0
R1(config)#ip route 2.2.2.0 255.255.255.0 Serial0/0/1
//配置静态路由实现 BGP 更新源可达性
R1(config)#router bgp 100
R1(config-router)#bgp router-id 1.1.1.1
R1(config-router)#neighbor 2.2.2.2 remote-as 200
R1(config-router)#neighbor 2.2.2.2 password cisco123      //配置 BGP 邻居身份验证
R1(config-router)#neighbor 2.2.2.2 ebgp-multihop 2        //配置 EBGP 多跳
R1(config-router)#neighbor 2.2.2.2 update-source Loopback0
R1(config-router)#network 11.11.11.0 mask 255.255.255.0
```

（2）配置路由器 R2

```
R2(config)#ip route 1.1.1.0 255.255.255.0 Serial0/0/0
R2(config)#ip route 1.1.1.0 255.255.255.0 Serial0/1/1
R2(config)#ip prefix-list DAMP seq 5 permit 11.11.11.0/24
R2(config)#route-map DAMP permit 10
R2(config-route-map)#match ip address prefix-list DAMP
R2(config-route-map)#set dampening 15 750 2000 60         //设置 dampening 参数
R2(config)#router bgp 200
R2(config-router)#bgp router-id 2.2.2.2
R2(config-router)#neighbor 1.1.1.1 remote-as 100
R2(config-router)#neighbor 1.1.1.1 password cisco123
R2(config-router)#neighbor 1.1.1.1 ebgp-multihop 2
R2(config-router)#neighbor 1.1.1.1 update-source Loopback0
R2(config-router)#bgp dampening route-map DAMP            //配置 BGP dampening
```

【技术要点】

① 当 EBGP 对等体之间有多条物理链路时，建议使用环回接口来建立邻居关系，这样可以增加 BGP 连接的稳定性。但是默认情况下，EBGP 在建立邻居关系时，使用直连的物理接口，即 TTL 为 1。由于环回接口不是直连的，所以必须通过命令 neighbor {*ip-address* | *peer- group-name*} ebgp-multihop [*ttl*]启用多跳 EBGP。如果没有配置 *ttl* 可选项，默认将 TTL 设置为 255。

② BGP 对等体之间采用的是 MD5 身份验证。

③ dampening 功能可以在路由模式下针对全部的 BGP 路由配置，配置举例如下：

```
R2(config-router)#dampening 15 750 2000 60
```

也可以针对具体的路由条目配置，本实验针对 11.11.11.0/24 路由条目配置 dampening。

4．实验调试

（1）BGP 邻居 MD5 验证调试

① 如果路由器 R1 配置 BGP 邻居 MD5 身份验证，R2 没有配置 BGP 邻居 MD5 身份验证，则 R1 上的提示信息如下。

```
%TCP-6-BADAUTH: No MD5 digest from 2.2.2.2(17898) to 1.1.1.1(179) tableid - 0
```

② 如果路由器 R1 和 R2 都配置 MD5 身份验证，但是配置的密码不同，则 R1 上的提示信息如下。

```
%TCP-6-BADAUTH: Invalid MD5 digest from 2.2.2.2(49845) to 1.1.1.1(179) tableid - 0
```

（2）进行 dampening 调试

① 查看 dampening 的参数及其值。

```
R2#show ip bgp dampening parameters
 dampening 15 750 2000 60 (route-map DAMP 10)    //在路由映射表 DAMP 的序号 10 中配置
  Half-life time         : 15 mins     Decay Time          : 2320 secs
  Max suppress penalty   : 12000       Max suppress time   : 60 mins
  Suppress penalty       : 2000        Reuse penalty       : 750
```

② 在路由器 R1 上将环回接口 11 关闭和开启各一次，在路由器 R2 查看信息。

```
R2#show ip bgp dampening flap-statistics              //查看 dampening 状态和统计信息
BGP table version is 11, local router ID is 2.2.2.2
Status codes: s suppressed, d damped, h history, * valid, > best, i - internal,
              r RIB-failure, S Stale
Origin codes: i - IGP, e - EGP, ? - incomplete

   Network          From           Flaps Duration Reuse   Path
h 11.11.11.0/24    1.1.1.1           1    00:00:13         100
//BGP 路由条目 11.11.11.0 的状态为 h，翻转 1 次
```

③ 在路由器 R1 上将环回接口 11 关闭和开启多次，在路由器 R2 查看信息。

```
R2#show ip bgp dampening flap-statistics
BGP table version is 15, local router ID is 2.2.2.2
Status codes: s suppressed, d damped, h history, * valid, > best, i - internal,
              r RIB-failure, S Stale
Origin codes: i - IGP, e - EGP, ? - incomplete

   Network          From           Flaps Duration Reuse    Path
d 11.11.11.0/24    1.1.1.1           3    00:05:50 00:01:13 100
//BGP 路由条目 11.11.11.0 的状态为 d，即正在被惩罚，翻转 3 次
```

④ 查看 BGP 路由信息。

```
R2#show ip route   bgp
```

以上输出表明 BGP 路由条目 **11.11.11.0** 没有出现在路由表中，因为处于 dampening 惩罚状态的路由不会被放入 IP 路由表中，同时也不被 BGP 传递。

继续查看 BGP 表中有关 **11.11.11.0** 的详细信息：

```
R2#show ip bgp 11.11.11.0
BGP routing table entry for 11.11.11.0/24, version 15
Paths: (1 available, no best path)          //不是最优路径
  Not advertised to any peer                //通告给任何对等体
  Refresh Epoch 1
  100, (suppressed due to dampening)        //由于 dampening 被抑制
    1.1.1.1 from 1.1.1.1 (1.1.1.1)
      Origin IGP, metric 0, localpref 100, external
      Dampinfo: penalty 1855, flapped 3 times in 00:13:13, reuse in 00:08:34
```

以上输出显示了处于 dampening 惩罚状态的 BGP 路由条目更为详细的信息，包括属性、惩罚值、翻转次数和重新使用时间等。

⑤ 清除被抑制 BGP 路由的状态。

```
R2#clear ip bgp dampening
```

该命令手动清除被抑制 BGP 路由的状态，使之可以被重新使用。

7.3 配置高级 BGP

7.3.1 实验 3：配置 BGP 地址聚合

1．实验目的

通过本实验可以掌握：
① 启动 BGP 路由进程的方法。
② EBGP 邻居的配置方法。
③ BGP 中通告网络的方法。
④ BGP 地址聚合配置和调试方法。
⑤ 地址聚合中参数 as-set、summary-only 和 suppress-map 的含义。

2．实验拓扑

配置 BGP 地址聚合实验拓扑如图 7-10 所示。

3．实验步骤

本实验实现在路由器 R2 上将路由器 R1 和路由器 R3 通告的环回接口的路由进行地址聚合，并通告给路由器 R4。

（1）配置路由器 R1

```
R1(config)#router bgp 100
R1(config-router)#bgp router-id 1.1.1.1
R1(config-router)#neighbor 12.12.12.2 remote-as 200
```

```
R1(config-router)#network 1.1.0.0 mask 255.255.255.0
R1(config-router)#network 1.1.1.0 mask 255.255.255.0
```

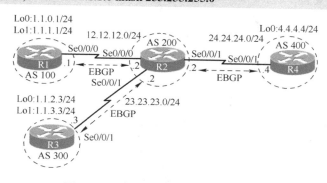

图 7-10　配置 BGP 地址聚合实验拓扑

（2）配置路由器 R2

```
R2(config)#router bgp 200
R2(config-router)#bgp router-id 2.2.2.2
R2(config-router)#neighbor 12.12.12.1 remote-as 100
R2(config-router)#neighbor 23.23.23.3 remote-as 300
R2(config-router)#neighbor 24.24.24.4 remote-as 400
R2(config-router)#aggregate-address 1.1.0.0 255.255.252.0     //配置地址聚合
```

（3）配置路由器 R3

```
R3(config)#router bgp 300
R3(config-router)#bgp router-id 3.3.3.3
R3(config-router)#neighbor 23.23.23.2 remote-as 200
R3(config-router)#network 1.1.2.0 mask 255.255.255.0
R3(config-router)#network 1.1.3.0 mask 255.255.255.0
```

（4）配置路由器 R4

```
R4(config)#router bgp 400
R4(config-router)#bgp router-id 4.4.4.4
R4(config-router)#neighbor 24.24.24.2 remote-as 200
R4(config-router)#network 4.4.4.0 mask 255.255.255.0
```

4．实验调试

1）在路由器 R1、R2、R3 和 R4 上查看 BGP 表，并在路由器 R2 上查看路由表。此处省略 BGP 表中状态代码部分、起源代码部分和路由表中路由代码部分。

① R1#show ip bgp

Network	Next Hop	Metric	LocPrf	Weight	Path
*> 1.1.0.0/24	0.0.0.0	0		32768	i
*> 1.1.0.0/22	**12.12.12.2**	**0**		**0**	**200** i
*> 1.1.1.0/24	0.0.0.0	0		32768	i
*> 1.1.2.0/24	12.12.12.2			0	200 300 i
*> 1.1.3.0/24	12.12.12.2			0	200 300 i

*> 4.4.4.0/24	12.12.12.2	0		200 400 i

② R2#**show ip bgp**

Network	Next Hop	Metric LocPrf	Weight	Path
*> 1.1.0.0/24	12.12.12.1	0	0	100 i
***> 1.1.0.0/22**	**0.0.0.0**		32768	**i**
*> 1.1.1.0/24	12.12.12.1	0	0	100 i
*> 1.1.2.0/24	23.23.23.3	0	0	300 i
*> 1.1.3.0/24	23.23.23.3	0	0	300 i
*> 4.4.4.0/24	24.24.24.4	0	0	400 i

③ R2#**show ip route bgp**
```
     1.0.0.0/8 is variably subnetted, 5 subnets, 2 masks
B       1.1.0.0/24 [20/0] via 12.12.12.1, 02:15:59
B       1.1.0.0/22 [200/0] via 0.0.0.0, 00:00:32, Null0
B       1.1.1.0/24 [20/0] via 12.12.12.1, 02:15:59
B       1.1.2.0/24 [20/0] via 23.23.23.3, 02:15:59
B       1.1.3.0/24 [20/0] via 23.23.23.3, 02:15:59
     4.0.0.0/24 is subnetted, 1 subnets
B       4.4.4.0 [20/0] via 24.24.24.4, 02:15:59
```

④ R3#**show ip bgp**

Network	Next Hop	Metric LocPrf	Weight	Path
***> 1.1.0.0/24**	23.23.23.2		0	200 100 i
***> 1.1.0.0/22**	**23.23.23.2**	**0**	**0**	**200 i**
***> 1.1.1.0/24**	23.23.23.2		0	200 100 i
*> 1.1.2.0/24	0.0.0.0	0	32768	i
*> 1.1.3.0/24	0.0.0.0	0	32768	i
*> 4.4.4.0/24	23.23.23.2		0	200 400 i

⑤ R4#**show ip bgp**

Network	Next Hop	Metric LocPrf	Weight	Path
***> 1.1.0.0/24**	24.24.24.2		0	200 100 i
***> 1.1.0.0/22**	**24.24.24.2**	**0**	**0**	**200 i**
***> 1.1.1.0/24**	24.24.24.2		0	200 100 i
*> 1.1.2.0/24	24.24.24.2		0	200 300 i
*> 1.1.3.0/24	24.24.24.2		0	200 300 i
*> 4.4.4.0/24	**0.0.0.0**	0	32768	i

以上输出表明：

① 路由器 R1、R3 和 R4 收到 **1.1.0.0/22** 聚合路由，通过 AS-PATH 属性可以看出，执行地址聚合的路由器 R2 成为聚合路由的创造者，原来的 AS-PATH 属性丢失。

② 路由器 R4 同时收到 4 条明细路由，在显示的 AS-PATH 属性中，路由的始发 AS 号在列表的末端（右侧），当传递给其他 AS 时，BGP 对等体会把它自己的 AS 号追加在列表的开头（左侧）。

③ BGP 路由器下一跳为 0.0.0.0，表示该 BGP 路由起源本地，Weight 值为 32768。

④ 因为所有 BGP 路由条目的代码为"*>"，所以所有 BGP 路由条目都为最优。

⑤ 由于配置了地址聚合，所以在路由器 R2 的路由表中产生一条指向 Null0 的汇总路由 1.1.0.0/22，主要是为了避免路由环路。

2）as-set 参数可以使 BGP 聚合路由不丢失原来的 AS-PATH 属性，从而避免路由环路，可在路由器 R2 上进行如下操作。

第 7 章 BGP

```
R2(config-router)#aggregate-address 1.1.0.0 255.255.252.0 as-set
```

在路由器 R1、R3 和 R4 上再次查看 BGP 表。

① R1#**show ip bgp**

Network	Next Hop	Metric	LocPrf	Weight	Path
*> 1.1.0.0/24	0.0.0.0	0		32768	i
*> 1.1.1.0/24	0.0.0.0	0		32768	i
*> 1.1.2.0/24	12.12.12.2			0	200 300 i
*> 1.1.3.0/24	12.12.12.2			0	200 300 i
*> 4.4.4.0/24	12.12.12.2			0	200 400 i

② R3#**show ip bgp**

Network	Next Hop	Metric	LocPrf	Weight	Path
*> 1.1.0.0/24	23.23.23.2			0	200 100 i
*> 1.1.1.0/24	23.23.23.2			0	200 100 i
*> 1.1.2.0/24	0.0.0.0	0		32768	i
*> 1.1.3.0/24	0.0.0.0	0		32768	i
*> 4.4.4.0/24	23.23.23.2			0	200 400 i

③ R4#**show ip bgp**

Network	Next Hop	Metric	LocPrf	Weight	Path
*> 1.1.0.0/24	24.24.24.2			0	200 100 i
*> **1.1.0.0/22**	**24.24.24.2**	**0**		**0**	**200 {100,300} i**
*> 1.1.1.0/24	24.24.24.2			0	200 100 i
*> 1.1.2.0/24	24.24.24.2			0	200 300 i
*> 1.1.3.0/24	24.24.24.2			0	200 300 i
*> 4.4.4.0/24	0.0.0.0	0		32768	i

以上输出表明：

① 在路由器 R4 上收到的汇总路由 **1.1.0.0/22** 中，AS-PATH 包含了被聚合路由中所有的明细路由所在的 AS 号的集合**{100,300}**。

② 聚合路由正是由于携带了所有明细路由的 AS-PATH 属性，所以该聚合路由在路由器 R1 和 R3 的 BGP 表中没有出现，从而可以有效地避免路由环路。但是命令 **neighbor** *ip-address* **allowas-in** 可以允许在 AS-PATH 属性中出现的本 AS 的 BGP 路由条目进入本 AS，当然这样有环路的风险。

【技术要点】

BGP 使用 AS-PATH 属性作为路由更新的一部分来确保没有路由环路。因为在 BGP 对等体之间传递的每条路由都携带它所经过的 AS 号序列表，如果该路由被通告给它始发的 AS，该 AS 路由器将在 AS 号序列表中看到自己的 AS 号，它将不接受该路由。以下的输出充分说明了这一点。

R2#**show ip bgp neighbor 12.12.12.1 advertised-routes**

Network	Next Hop	Metric	LocPrf	Weight	Path
*> **1.1.0.0/24**	12.12.12.1	0		0	100 i
*> **1.1.0.0/22**	0.0.0.0		100	32768	{100,300} i
*> **1.1.1.0/24**	12.12.12.1	0		0	100 i
*> 1.1.2.0/24	23.23.23.3	0		0	300 i
*> 1.1.3.0/24	23.23.23.3	0		0	300 i

*> 4.4.4.0/24	24.24.24.4	0		0	400	i

Total number of prefixes 6

以上输出表明路由器 R2 向邻居 **12.12.12.1** 通告聚合路由 **1.1.0.0/22** 路由条目。

R1#**show ip bgp neighbors 12.12.12.2 received-routes**

Network	Next Hop	Metric	LocPrf	Weight	Path
*> 1.1.2.0/24	12.12.12.2			0	200 300 i
*> 1.1.3.0/24	12.12.12.2			0	200 300 i
*> 4.4.4.0/24	12.12.12.2			0	200 400 i

以上输出表明路由器 R1 没有接收聚合路由 **1.1.0.0/22** 条目，因为它发现该路由条目中的 AS-PATH 属性列表中包含自己的 AS 号 100，所以不接收。

【提示】

要执行 **show ip bgp neighbors 12.12.12.2 received-routes** 命令，必须执行下面这条命令：

R1(config-router)#**neighbor 12.12.12.2 soft-reconfiguration inbound**

3）如果在路由器 R4 上只想看到汇总路由，没有明细路由，通过配置 **summary-only** 参数可以实现，该参数在路由器 R2 上的配置如下。

R2(config-router)#**aggregate-address 1.1.0.0 255.255.252.0 as-set summary-only**

在路由器 R2、R4 上查看 BGP 表。

① R2#**show ip bgp**

Network	Next Hop	Metric	LocPrf	Weight	Path
s> 1.1.0.0/24	12.12.12.1	0		0	100 i
*> 1.1.0.0/22	0.0.0.0	100		32768	{100,300} i
s> 1.1.1.0/24	12.12.12.1	0		0	100 i
s> 1.1.2.0/24	23.23.23.3	0		0	300 i
s> 1.1.3.0/24	23.23.23.3	0		0	300 i
*> 4.4.4.0/24	24.24.24.4	0		0	400 i

② R4#**show ip bgp**

Network	Next Hop	Metric	LocPrf	Weight	Path
*> 1.1.0.0/22	24.24.24.2	0		0	200 {100,300} i
*> 4.4.4.0/24	0.0.0.0			32768	i

以上输出表明：

① 路由器 R2 上所有被聚合的明细路由被标记为 s，表示被抑制，不被发送。

② 路由器 R4 只收到一条聚合路由 **1.1.0.0/22**。如果不配置 **as-set** 参数，则路由器 R1、R3 也会收到该聚合路由。

4）如果有特殊的需求，在聚合后只抑制部分明细路由条目，通过配置参数 **suppress-map** 可以完成。本实验要求路由器 R2 完成地址聚合后，路由器 R1 的两条明细路由被抑制，而路由器 R3 的明细路由将被传递给路由器 R4，路由器 R2 配置步骤如下。

R2(config)#**ip prefix-list 1 seq 5 permit 1.1.0.0/24**   //匹配路由条目
R2(config)#**ip prefix-list 1 seq 10 permit 1.1.1.0/24**

```
R2(config)#route-map SUP permit 10
R2(config-route-map)#match ip address prefix-list 1
R2(config)#router bgp 200
R2(config-router)#aggregate-address 1.1.0.0 255.255.252.0 as-set suppress-map SUP
```

分别在四台路由器查看 BGP 表。

① R1#**show ip bgp**

Network	Next Hop	Metric	LocPrf	Weight	Path
*> 1.1.0.0/24	0.0.0.0	0		32768	i
*> 1.1.1.0/24	0.0.0.0	0		32768	i
*> 1.1.2.0/24	12.12.12.2			0	200 300 i
*> 1.1.3.0/24	12.12.12.2			0	200 300 i
*> 4.4.4.0/24	12.12.12.2			0	200 400 i

② R2#**show ip bgp**

Network	Next Hop	Metric	LocPrf	Weight	Path
s> 1.1.0.0/24	12.12.12.1	0		0	100 i
*> 1.1.0.0/22	0.0.0.0		100	32768	{100,300} i
s> 1.1.1.0/24	12.12.12.1	0		0	100 i
*> 1.1.2.0/24	23.23.23.3	0		0	300 i
*> 1.1.3.0/24	23.23.23.3	0		0	300 i
*> 4.4.4.0/24	24.24.24.4	0		0	400 i

③ R3#**show ip bgp**

Network	Next Hop	Metric	LocPrf	Weight	Path
*> 1.1.2.0/24	0.0.0.0	0		32768	i
*> 1.1.3.0/24	0.0.0.0	0		32768	i
*> 4.4.4.0/24	23.23.23.2			0	200 400 i

④ R4#**show ip bgp**

Network	Next Hop	Metric	LocPrf	Weight	Path
*> 1.1.0.0/22	24.24.24.2	0		0	200 {100,300} i
*> 1.1.2.0/24	24.24.24.2			0	200 300 i
*> 1.1.3.0/24	24.24.24.2			0	200 300 i
*> 4.4.4.0/24	0.0.0.0	0		32768	i

以上输出表明：

① 由于在路由器 R2 上将路由器 R1 的 BGP 明细路由 **1.1.0.0/24** 和 **1.1.1.0/24** 抑制，所以路由器 R3 和 R4 无法收到这两条路由条目。

② 由于在路由器 R2 上没有将 BGP 明细路由 **1.1.2.0/24** 和 **1.1.3.0/24** 抑制，又没有配置 **summary-only** 参数，所以 R1、R2 和 R4 路由器全部收到 **1.1.2.0/24** 和 **1.1.3.0/24** 的 BGP 路由条目。

③ 由于配置了 **as-set** 参数，所以只有 R4 收到 BGP 汇总路由 **1.1.0.0/22**。

7.3.2 实验 4：配置路由反射器（RR）

1．实验目的

通过本实验可以掌握：

① RR 的反射原理和反射规则。

② RR 的配置。
③ ORIGINATOR_ID（起源 ID）属性。
④ CLUSTER_LIST（簇列表）属性。

2. 实验拓扑

配置路由反射器（RR）实验拓扑如图 7-11 所示。本实验中，路由器 R2 作为路由反射器，路由器 R1 和 R3 作为它的客户端。

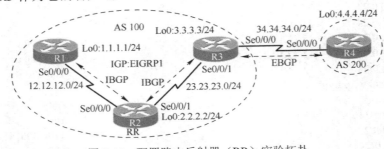

图 7-11　配置路由反射器（RR）实验拓扑

3. 实验步骤

（1）配置路由器 R1

```
R1(config)#router eigrp 1
R1(config-router)#network 1.1.1.1 0.0.0.0
R1(config-router)#network 12.12.12.1 0.0.0.0
R1(config)#router bgp 100
R1(config-router)#bgp router-id 1.1.1.1
R1(config-router)#neighbor 2.2.2.2 remote-as 100
R1(config-router)#neighbor 2.2.2.2 update-source Loopback0
R1(config-router)#network 1.1.1.0 mask 255.255.255.0
```

（2）配置路由器 R2

```
R2(config)#router eigrp 1
R2(config-router)#network 2.2.2.2 0.0.0.0
R2(config-router)#network 12.12.12.2 0.0.0.0
R2(config-router)#network 23.23.23.2 0.0.0.0
R2(config)#router bgp 100
R2(config-router)#bgp router-id 2.2.2.2
R2(config-router)#neighbor 1.1.1.1 remote-as 100
R2(config-router)#neighbor 1.1.1.1 update-source Loopback0
R2(config-router)#neighbor 1.1.1.1 route-reflector-client       //配置 RR 客户端
R2(config-router)#neighbor 3.3.3.3 remote-as 100
R2(config-router)#neighbor 3.3.3.3 update-source Loopback0
R2(config-router)#neighbor 3.3.3.3 route-reflector-client
```

（3）配置路由器 R3

```
R3(config)#router eigrp 1
```

```
R3(config-router)#network 3.3.3.3 0.0.0.0
R3(config-router)#network 23.23.23.3 0.0.0.0
R3(config)#router bgp 100
R3(config-router)#bgp router-id 3.3.3.3
R3(config-router)#neighbor 2.2.2.2 remote-as 100
R3(config-router)#neighbor 2.2.2.2 update-source Loopback0
R3(config-router)#neighbor 2.2.2.2 next-hop-self
R3(config-router)#neighbor 34.34.34.4 remote-as 200
```

（4）配置路由器 R4

```
R4(config)#router bgp 200
R4(config-router)#bgp router-id 4.4.4.4
R4(config-router)#neighbor 34.34.34.3 remote-as 100
R4(config-router)#network 4.4.4.0 mask 255.255.255.0
```

【技术要点】

当一个 AS 包含多个 IBGP 对等体时，路由反射器非常有用。因为 IBGP 客户只需要和路由反射器建立邻居关系，从而降低了 IBGP 的连接数量。路由反射器和它的客户合称为一个簇。路由反射器是克服 IBGP 水平分割的重要解决方案之一。路由反射器的反射规则如下。

① 如果路由是从非客户的 IBGP 邻居学来的，则 RR 只将它反射给客户。

② 如果路由是从客户学来的，RR 会将它反射给所有的非客户和客户（除了发起该路由的客户）。

③ 如果路由是从 EBGP 邻居学来的，RR 会将它反射给所有的非客户和客户。

4．实验调试

① R2#**show ip bgp neighbors 1.1.1.1**
```
BGP neighbor is 1.1.1.1,   remote AS 100, internal link
  BGP version 4, remote router ID 1.1.1.1
  BGP state = Established, up for 00:31:06
  Last read 00:00:07, hold time is 180, keepalive interval is 60 seconds
  （此处省略部分输出）
  For address family: IPv4 Unicast
  BGP table version 4, neighbor version 4
  Index 1, Offset 0, Mask 0x2
  Route-Reflector Client                  //路由反射器的客户端
（此处省略部分输出）
```

以上输出表明邻居 1.1.1.1 是路由反射器的客户端。

② R2#**show ip bgp 4.4.4.0**
```
BGP routing table entry for 4.4.4.0/24, version 4
Paths: (1 available, best #1, table Default-IP-Routing-Table)
  Advertised to non peer-group peers:    //通告给非对等体组的对等体
  1.1.1.1
  200, (Received from a RR-client)       //该 BGP 路由起源 AS 号，从 RR 客户端收到该路由
    3.3.3.3 (metric 2297856) from 3.3.3.3 (3.3.3.3)
```

Origin IGP, metric 0, localpref 100, valid, internal, best

以上输出表明 BGP 路由条目 **4.4.4.0/24** 是从 RR 的客户端收到的，客户端是 3.3.3.3，并且将它反射给 1.1.1.1。注意上面输出中（**3.3.3.3**）指的是 BGP 路由器 ID。

③ R1#**show ip bgp 4.4.4.0**
BGP routing table entry for 4.4.4.0/24, version 3
Paths: (1 available, best #1, table Default-IP-Routing-Table)
　Not advertised to any peer
　200　　//该 BGP 路由起源 AS 号
　　3.3.3.3 (metric 2809856) from 2.2.2.2 (2.2.2.2)
　　　Origin IGP, metric 0, localpref 100, valid, internal, best
　　Originator: 3.3.3.3, Cluster list: 2.2.2.2

以上输出表明在 AS 100 内 BGP 路由条目 4.4.4.0/24 的创造者是 3.3.3.3，簇 ID 是 2.2.2.2。

【术语】

① **ORIGINATOR_ID**（起源 ID）：由路由反射器生成，是本 AS 内路由创造者的路由器 ID。
② **CLUSTER_ID**（簇 ID）：一个 AS 内的每个簇必须用一个唯一的 4 字节的簇 ID 来标识，如果簇内只有一个 RR，那么簇 ID 就是 RR 的路由器 ID。当 RR 收到一条更新信息时，它将检查 CLUSTER_LIST，如果发现在列表中有自己的簇 ID，就知道出现了路由环路，从而可以有效避免环路。

7.3.3　实验 5：配置 BGP 联邦

1．实验目的

通过本实验可以掌握：
① BGP 联邦的含义。
② BGP 联邦的配置。

2．实验拓扑

配置 BGP 联邦实验拓扑如图 7-12 所示。本实验中，联邦的成员为 AS 65001 和 AS 65002，联邦对外 AS 为 100。

3．实验步骤

（1）配置路由器 R1

```
R1(config)#router bgp 65001
R1(config-router)#bgp router-id 1.1.1.1
R1(config-router)#bgp confederation identifier 100          //配置 BGP 联邦 ID
R1(config-router)#bgp confederation peers 65002             //联邦的成员
R1(config-router)#neighbor 12.12.12.2 remote-as 200
R1(config-router)#neighbor 13.13.13.3 remote-as 65002
R1(config-router)#network 1.1.1.0 mask 255.255.255.0
```

第 7 章 BGP

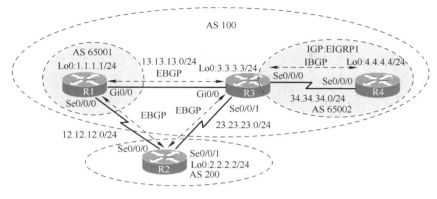

图 7-12 配置 BGP 联邦实验拓扑

（2）配置路由器 R2

```
R2(config)#router bgp 200
R2(config-router)#bgp router-id 2.2.2.2
R2(config-router)#neighbor 12.12.12.1 remote-as 100
R2(config-router)#neighbor 23.23.23.3 remote-as 100
R2(config-router)#network 2.2.2.0 mask 255.255.255.0
```

（3）配置路由器 R3

```
R3(config)#router eigrp 1
R3(config-router)#network 3.3.3.3 0.0.0.0
R3(config-router)#network 34.34.34.3 0.0.0.0
R3(config)#router bgp 65002
R3(config-router)#bgp router-id 3.3.3.3
R3(config-router)#bgp confederation identifier 100
R3(config-router)#bgp confederation peers 65001
R3(config-router)#neighbor 4.4.4.4 remote-as 65002
R3(config-router)#neighbor 4.4.4.4 update-source Loopback0
R3(config-router)#neighbor 4.4.4.4 next-hop-self
R3(config-router)#neighbor 13.13.13.1 remote-as 65001
R3(config-router)#neighbor 13.13.13.1 next-hop-self
R3(config-router)#neighbor 23.23.23.2 remote-as 200
R3(config-router)#network 3.3.3.0 mask 255.255.255.0
```

（4）配置路由器 R4

```
R4(config)#router eigrp 1
R4(config-router)#network 4.4.4.4 0.0.0.0
R4(config-router)#network 34.34.34.4 0.0.0.0
R4(config)#router bgp 65002
R4(config-router)#bgp router-id 4.4.4.4
R4(config-router)#neighbor 3.3.3.3 remote-as 65002
R4(config-router)#neighbor 3.3.3.3 update-source Loopback0
R4(config-router)#network 4.4.4.0 mask 255.255.255.0
```

【技术要点】

BGP 联邦用于将 AS 分割成多个子 AS，是大型网络解决 IBGP 水平分割的另一条途径。而子 AS 被称为成员自治系统。每个联邦都有一个被分配的联邦 ID，对联邦外部来讲，这个联邦 ID 是代表整个联邦的 AS 号。外部看不到联邦内部结构，联邦看起来就是一个 AS，成员自治系统信息被隐藏起来。

4．实验调试

（1）在路由器 R2 上查看 BGP 表

```
R2#show ip bgp
   Network          Next Hop         Metric LocPrf  Weight  Path
*  1.1.1.0/24       23.23.23.3                           0  100 i
*>                  12.12.12.1            0              0  100 i
*> 2.2.2.0/24       0.0.0.0               0          32768  i
*  3.3.3.0/24       12.12.12.1                           0  100 i
*>                  23.23.23.3            0              0  100 i
*  4.4.4.0/24       12.12.12.1                           0  100 i
*>                  23.23.23.3                           0  100 i
```

以上输出表明路由器 R2 学到的 **1.1.1.0/24**、**3.3.3.0/24** 和 **4.4.4.0/24** 网络都有两条路径，而且都是来自 AS 100。由此看出 BGP 联邦内所有成员的信息对外都被隐藏。

（2）在路由器 R3 查看 BGP 表

```
R3#show ip bgp
   Network          Next Hop         Metric LocPrf Weight Path
*> 1.1.1.0/24       13.13.13.1            0    100      0 (65001) i
*  2.2.2.0/24       13.13.13.1            0    100      0 (65001) 200 i
*>                  23.23.23.2            0           200       i
*> 3.3.3.0/24       0.0.0.0               0         32768       i
r>i4.4.4.0/24       4.4.4.4               0    100      0       i
```

以上输出表明，联邦内的 AS-PATH 属性用（ ）括起来。

（3）查看 bgp 2.2.2.0 条目详细信息

```
R3#show ip bgp 2.2.2.0
BGP routing table entry for 2.2.2.0/24, version 6
Paths: (2 available, best #1, table Default-IP-Routing-Table)
  Advertised to update-groups:
        1         2
  200
    23.23.23.2 from 23.23.23.2 (2.2.2.2)
      Origin IGP, metric 0, localpref 100, valid, external, best
  (65001) 200
    12.12.12.2 (inaccessible) from 13.13.13.1 (1.1.1.1)        //该 BGP 路由不是最优的
      Origin IGP, metric 0, localpref 100, valid, confed-external
//外部 BGP 路由通过联邦成员 AS 65001 传递给 R3
```

【技术要点】

在联邦范围内,将成员 AS 加入到 AS-PATH 中,并且用括号括起来,但是并不将它们公布到联邦的范围以外。AS-PATH 中联邦的 AS 号用于避免出现路由环路。

7.3.4 实验 6:配置 BGP 团体

1. 实验目的

通过本实验可以掌握:
① BGP 团体的配置。
② BGP 团体属性 local-AS。
③ BGP 团体属性 no-export。
④ BGP 团体属性 no-advertise。

2. 实验拓扑

实验拓扑如图 7-12 所示。

3. 实验步骤与实验测试

保留实验 5 的所有配置,因本实验完全在实验 5 配置的基础上完成,通过让路由器 R4 上的 4.4.4.0 路由条目携带不同的团体属性来验证团体的各个属性的传递特征。本书重点讨论熟知的团体属性 local-AS、no-export 和 no-advertise。本实验只给出在实验 5 基础上增加的配置。

(1) 在路由器 R4 上配置团体属性 local-AS

```
R4(config)#ip prefix-list 1 permit 4.4.4.0/24              //定义前缀列表
R4(config)#route-map Local_AS permit 10                    //定义 route-map
R4(config-route-map)#match ip address prefix-list 1        //匹配前缀列表
R4(config-route-map)#set community local-AS                //设置团体属性
R4(config-route-map)#router bgp 65002
R4(config-router)#neighbor 3.3.3.3 send-community          //开启发送团体属性的能力
R4(config-router)#neighbor 3.3.3.3 route-map Local_AS out
//在出方向向邻居发送团体属性
R1(config)#router bgp 65001
R1(config-router)#neighbor 12.12.12.2 send-community
R3(config)#router bgp 65002
R3(config-router)#neighbor 13.13.13.1 send-community
R3(config-router)#neighbor 23.23.23.2 send-community
```

【提示】

后面的验证 no-export 和 no-advertise 团体属性的实验当然也需要在路由器 R1 和 R3 上开启发送团体属性的能力,如上五行所示,后面的配置就不重复了。

（2）测试团体属性 local-AS

① 在路由器 R3 上查看 BGP 表。

```
R3#clear ip bgp *
R3#show ip bgp
   Network          Next Hop         Metric LocPrf Weight   Path
*> 1.1.1.0/24       13.13.13.1         0     100     0     (65001)        i
*  2.2.2.0/24       13.13.13.1         0     100     0     (65001) 200    i
*>                  23.23.23.2         0             0     200            i
*> 3.3.3.0/24       0.0.0.0            0           32768                  i
r>i4.4.4.0/24       4.4.4.4            0     100     0                    i
```

② 在路由器 R2 上查看 BGP 表。

```
R2#clear ip bgp *
R2#show ip bgp
   Network          Next Hop         Metric LocPrf Weight   Path
*  1.1.1.0/24       23.23.23.3                       0     100 i
*>                  12.12.12.1         0             0     100 i
*> 2.2.2.0/24       0.0.0.0            0           32768         i
*  3.3.3.0/24       23.23.23.3                       0     100 i
*>                  12.12.12.1                       0     100 i
```

③ 在路由器 R1 上查看 BGP 表。

```
R1#clear ip bgp *
R1#show ip bgp
   Network          Next Hop         Metric LocPrf Weight   Path
*> 1.1.1.0/24       0.0.0.0            0           32768                 i
*  2.2.2.0/24       13.13.13.3         0     100     0     (65002) 200  i
*>                  12.12.12.2         0             0     200           i
*> 3.3.3.0/24       13.13.13.3         0     100     0     (65002)      i
```

以上①、②和③输出表明携带团体 local-AS 属性的条目 **4.4.4.0/24** 只传递给路由器 R3，因为路由器 R3 和 R4 都在 AS 65002 内，并没有传递给路由器 R2 和 R1。由此可见携带 local-AS 团体属性的路由条目只能在本 AS 内传递。

（3）在路由器 R4 上配置团体属性 **no-export**

```
R4(config)#route-map NO-EXPORT permit 10
R4(config-route-map)#match ip address prefix-list 1
R4(config-route-map)#set community no-export                    //设置团体属性
R4(config)#router bgp 65002
R4(config-router)#neighbor 3.3.3.3 send-community
R4(config-router)#neighbor 3.3.3.3 route-map NO-EXPORT out
```

（4）测试团体属性 no-export

① 在路由器 R3 上查看 BGP 表。

```
R3#clear ip bgp *
R3#show ip bgp
```

```
     Network           Next Hop        Metric LocPrf Weight Path
*>   1.1.1.0/24        13.13.13.1         0     100     0 (65001)     i
*    2.2.2.0/24        13.13.13.1         0     100     0 (65001) 200 i
*>                     23.23.23.2         0             0        200 i
*>   3.3.3.0/24        0.0.0.0            0         32768             i
r>i  4.4.4.0/24        4.4.4.4            0     100     0             i
```

② 在路由器 R2 上查看 BGP 表。

```
R2#clear ip bgp *
R2#show ip bgp
     Network           Next Hop        Metric LocPrf Weight Path
*    1.1.1.0/24        23.23.23.3                     0    100 i
*>                     12.12.12.1         0           0    100 i
*>   2.2.2.0/24        0.0.0.0            0       32768         i
*    3.3.3.0/24        23.23.23.3                     0    100 i
*>                     12.12.12.1                     0    100 i
```

③ 在路由器 R1 上查看 BGP 表。

```
R1#clear ip bgp *
R1#show ip bgp
     Network           Next Hop        Metric LocPrf Weight Path
*>   1.1.1.0/24        0.0.0.0            0         32768         i
*    2.2.2.0/24        13.13.13.3         0     100     0 (65002) 200 i
*>                     12.12.12.2         0             0        200 i
*>   3.3.3.0/24        13.13.13.3         0     100     0 (65002)     i
*>   4.4.4.0/24        13.13.13.3         0     100     0 (65002)     i
```

以上①、②和③输出表明携带团体 no-export 属性的条目 4.4.4.0/24 传递给路由器 R3 和 R1，因为路由器 R1、R3 和 R4 都在联邦 AS 100 内，并没有传递给路由器 R2。由此可见携带 no-export 团体属性的路由条目能在联邦的大 AS 内传递，如果没有联邦，则只能在本 AS 内传递。

【提示】

通过命令 **show ip bgp community no-export** 来查看 BGP 表哪些条目携带相应的属性，如下所示：

```
① R1#show ip bgp community no-export
     Network           Next Hop        Metric LocPrf Weight Path
*>   4.4.4.0/24        13.13.13.3         0     100     0 (65002) i
② R3#show ip bgp community no-export
r>i  4.4.4.0/24        4.4.4.4            0     100     0         i
```

以上输出表明路由器 R1 和 R3 的 BGP 表中的 **4.4.4.0/24** 携带了 **no-export** 团体属性。

（5）在路由器 R4 上配置团体属性 no-advertise

```
R4(config)#route-map NO_ADV permit 10
R4(config-route-map)#match ip address prefix-list 1
R4(config-route-map)#set community no-advertise
```

```
R4(config)#router bgp 65002
R4(config-router)#neighbor 3.3.3.3 send-community
R4(config-router)#neighbor 3.3.3.3 route-map NO_ADV out
```

(6)测试团体属性 no-advertise

① 在路由器 R3 上查看 BGP 表

```
R3#clear ip bgp *
R3#show ip bgp
   Network          Next Hop         Metric LocPrf Weight  Path
*> 1.1.1.0/24       13.13.13.1       0      100    0       (65001) i
*  2.2.2.0/24       13.13.13.1       0      100    0       (65001) 200 i
*>                  23.23.23.2       0             0       200 i
*> 3.3.3.0/24       0.0.0.0          0             32768   i
r>i4.4.4.0/24       4.4.4.4          0      100    0       i
```

② 在路由器 R2 上查看 BGP 表

```
R2#clear ip bgp *
R2#show ip bgp
   Network          Next Hop         Metric LocPrf Weight  Path
*  1.1.1.0/24       23.23.23.3                    0        100 i
*>                  12.12.12.1       0            0        100 i
*> 2.2.2.0/24       0.0.0.0          0            32768    i
*  3.3.3.0/24       12.12.12.1                    0        100 i
*>                  23.23.23.3       0            0        100 i
```

③ 在路由器 R1 上查看 BGP 表

```
R1#clear ip bgp *
R1#show ip bgp
   Network          Next Hop         Metric LocPrf Weight  Path
*> 1.1.1.0/24       0.0.0.0          0             32768   i
*  2.2.2.0/24       13.13.13.3       0      100    0       (65002) 200 i
*>                  12.12.12.2       0             0       200 i
*> 3.3.3.0/24       13.13.13.3       0      100    0       (65002) i
```

以上①、②和③输出表明携带团体 no-advertise 属性的条目 **4.4.4.0/24** 只传递给路由器 R3，并没有继续传递给路由器 R2 和 R1。由此可见携带 no-advertise 团体属性的条目被收到后，将不通告给任何 BGP 对等体。

【提示】

可以为一条 BGP 路由设置多个团体属性。举例如下：

```
R4(config-route-map)#set community local-AS no-advertise
```

(7)在路由器 R4 上配置团体属性值

```
R4(config)#route-map COMM permit 10
R4(config-route-map)#match ip address prefix-list 1
```

R4(config-route-map)#**set community 100:200**
R4(config)#**router bgp 65002**
R4(config-router)#**neighbor 3.3.3.3 send-community**
R4(config-router)#**neighbor 3.3.3.3 route-map COMM out**

(8)测试团体属性值

在路由器 R3 上查看 BGP 表。

R3#**clear ip bgp ***
R3#**show ip bgp 4.4.4.0**
BGP routing table entry for 4.4.4.0/24, version 2
Paths: (1 available, best #1, table default, RIB-failure(17))
　Not advertised to any peer
　Refresh Epoch 1
　Local
　　4.4.4.4 (metric 2297856) from 4.4.4.4 (4.4.4.4)
　　　Origin IGP, metric 0, localpref 100, valid, internal, best
　　　Community: 100:200　　//BGP 路由条目携带的团体属性值，十进制值为 6553800
　　　rx pathid: 0, tx pathid: 0x0

【提示】

在默认情况下，Cisco IOS 使用较旧的十进制格式表示和显示团体属性值。如果要以 AA:NN 格式进行配置和显示，则需要配置 **ip bgp-community new-format** 全局命令，其中 AA:NN 格式的 AA 表示 2 字节 AS 号，NN 表示 2 字节编号。

7.4　配置 BGP 属性控制选路

BGP 具有丰富的属性，但本节只研究 ORIGIN、AS-PATH、LOCAL_PREF、Weight 和 MED 属性。本节的实验是一个有机的整体，根据 BGP 路由判定的顺序（优先级别从低到高）设计实验，每个分解实验都用较高的优先级别影响前面分解实验的 BGP 路由选路。实验的拓扑如图 7-11 所示。通过修改 ORIGIN、AS-PATH、LOCAL_PREF、Weight 属性来控制在 AS 100 内路由器 R1、R2 和 R3 对路由器 R4 上通告的 4.4.4.0/24 路由的选路。最后通过在路由器 R2 和 R3 上发布环回接口来控制从路由器 R4 进入 AS 100 的选路。本实验中 IBGP 的路由器（R1、R2 和 R3）形成全互连的邻居关系。IGP 运行 EIGRP 路由协议。在完成每个分解实验后，最好用 **clear ip bgp *** 清除一下 BGP 表，然后再查看结果。整节实验的目的请读者对照 BGP 判定原则仔细研究和体会。

7.4.1　实验 7：配置 BGP ORIGIN 属性控制选路

1. 实验目的

通过本实验可以掌握：
① BGP 路由传递的条件。

② ORIGIN 代码的优先级。
③ 用 ORIGIN 属性选路的原则。

2. 实验拓扑

配置 BGP 属性控制选路实验拓扑如图 7-13 所示。

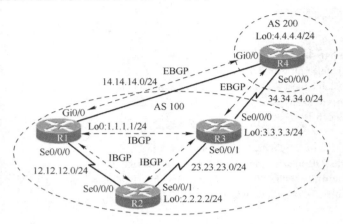

图 7-13 配置 BGP 属性控制选路实验拓扑

3. 实验步骤

本实验中,在路由器 R4 上设置 BGP 路由条目 4.4.4.0/24 的起源代码属性为 EGP,并通过 EBGP 邻居 14.14.14.1 传入 AS 100 内,然后观察路由器 R1、R2 和 R3 对路由器 R4 上通告的 BGP 路由条目 4.4.4.0/24 的选路。

(1) 配置路由器 R1

```
R1(config)#router eigrp 1
R1(config-router)#network 1.1.1.1 0.0.0.0
R1(config-router)#network 12.12.12.1 0.0.0.0
R1(config)#router bgp 100
R1(config-router)#bgp router-id 1.1.1.1
R1(config-router)#neighbor 2.2.2.2 remote-as 100
R1(config-router)#neighbor 2.2.2.2 update-source Loopback0
R1(config-router)#neighbor 2.2.2.2 next-hop-self
R1(config-router)#neighbor 3.3.3.3 remote-as 100
R1(config-router)#neighbor 3.3.3.3 update-source Loopback0
R1(config-router)#neighbor 3.3.3.3 next-hop-self
R1(config-router)#neighbor 14.14.14.4 remote-as 200
```

(2) 配置路由器 R2

```
R2(config)#router eigrp 1
R2(config-router)#network 2.2.2.2 0.0.0.0
R2(config-router)#network 12.12.12.2 0.0.0.0
R2(config-router)#network 23.23.23.2 0.0.0.0
```

第 7 章 BGP

```
R2(config)#router bgp 100
R2(config-router)#bgp router-id 2.2.2.2
R2(config-router)#neighbor 1.1.1.1 remote-as 100
R2(config-router)#neighbor 1.1.1.1 update-source Loopback0
R2(config-router)#neighbor 3.3.3.3 remote-as 100
R2(config-router)#neighbor 3.3.3.3 update-source Loopback0
```

（3）配置路由器 R3

```
R3(config)#router eigrp 1
R3(config-router)#network 3.3.3.3 0.0.0.0
R3(config-router)#network 23.23.23.3 0.0.0.0
R3(config)#router bgp 100
R3(config-router)#bgp router-id 3.3.3.3
R3(config-router)#neighbor 1.1.1.1 remote-as 100
R3(config-router)#neighbor 1.1.1.1 update-source Loopback0
R3(config-router)#neighbor 1.1.1.1 next-hop-self
R3(config-router)#neighbor 2.2.2.2 remote-as 100
R3(config-router)#neighbor 2.2.2.2 update-source Loopback0
R3(config-router)#neighbor 2.2.2.2 next-hop-self
R3(config-router)#neighbor 34.34.34.4 remote-as 200
```

（4）配置路由器 R4

```
R4(config)#ip prefix-list 1 permit 4.4.4.0/24
R4(config)#route-map ORIGIN permit 10
R4(config-route-map)#match ip address prefix-list 1
R4(config-route-map)#set origin egp 900                //设置起源代码
R4(config)#route-map ORIGIN permit 20
R4(config)#router bgp 200
R4(config-router)#bgp router-id 4.4.4.4
R4(config-router)#neighbor 14.14.14.1 remote-as 100
R4(config-router)#neighbor 14.14.14.1 route-map ORIGIN out
//在出方向为去往邻居 14.14.14.1 的路由设置策略
R4(config-router)#neighbor 34.34.34.3 remote-as 100
R4(config-router)#network 4.4.4.0 mask 255.255.255.0
R4#clear ip bgp *
```

4．实验调试

在路由器 R1、R2 和 R3 上查看 BGP 表。

① R1#show ip bgp

Network	Next Hop	Metric	LocPrf	Weight	Path
*>i4.4.4.0/24	3.3.3.3	0	100	0	200 i
*	14.14.14.4	0		0	200 e

② R2#show ip bgp

Network	Next Hop	Metric	LocPrf	Weight	Path
*>i4.4.4.0/24	3.3.3.3	0	100	0	200 i

③ R3#show ip bgp

Network	Next Hop	Metric	LocPrf	Weight	Path

*> 4.4.4.0/24	34.34.34.4	0	0 200 i

以上输出表明路由器 R1 学到两条关于 4.4.4.0/24 的 BGP 路由，但是由于起源代码 i 优先于 e，所以从路由器 R3 学到的 BGP 路由被优化，而从邻居路由器 R4 学到的路由不能被优化（路由代码为*，没有>），不能继续通告给路由器 R2 和 R3，所以路由器 R2 和 R3 只有一条关于 4.4.4.0/24 的路由。

7.4.2　实验 8：配置 BGP AS-PATH 属性控制选路

1．实验目的

通过本实验可以掌握：
① AS-PATH 控制路由环路的原理。
② 配置 AS-PATH 属性的方法。
③ 用 AS-PATH 属性选路的原则。

2．实验拓扑

实验拓扑如图 7-13 所示。

3．实验步骤

路由器 R1、R2 和 R3 上的配置和前面 7.4.1 实验 7 相同，在路由器 R4 上修改 BGP 路由条目 4.4.4.0/24 的 AS_PATH 属性，并通过 EBGP 邻居 34.34.34.3 传入 AS 100 内，然后观察路由器 R1、R2 和 R3 对路由器 R4 上通告的 BGP 路由条目 4.4.4.0/24 的选路，配置如下：

```
R4(config)#ip prefix-list 1 permit 4.4.4.0/24
R4(config)#route-map ASPATH permit 10
R4(config-route-map)#match ip address prefix-list 1
R4(config-route-map)#set as-path prepend 600 700          //为匹配的路由条目追加 AS
R4(config)#route-map ASPATH permit 20
R4(config)#router bgp 200
R4(config-router)#neighbor 34.34.34.3 route-map ASPATH out
R4#clear ip bgp *
```

4．实验调试

在路由器 R1、R2 和 R3 上查看 BGP 表。

```
① R1#show ip bgp
  Network         Next Hop       Metric LocPrf Weight  Path
  *> 4.4.4.0/24   14.14.14.4        0            0     200 e
② R2#show ip bgp
  Network         Next Hop       Metric LocPrf Weight  Path
  *>i4.4.4.0/24   1.1.1.1           0    100     0     200 e
③ R3#show ip bgp
  Network         Next Hop       Metric LocPrf Weight  Path
  *>i4.4.4.0/24   1.1.1.1           0    100     0     200 e
```

| * | 34.34.34.4 | 0 | 0 200 600 700 i |

以上输出表明路由器 R3 学到两条关于 4.4.4.0/24 的路由，但是由于下一跳为 1.1.1.1 的路由的 AS-PATH 属性值比下一跳为 34.34.34.4 的路由的 AS-PATH 属性值短，所以优选下一跳为 1.1.1.1 的路由，而下一跳为 34.34.34.4 的路由不能被优化（路由代码为*，没有>），不能继续通告给路由器 R1 和 R2，所以路由器 R1 和 R2 只有一条 4.4.4.0/24 的 BGP 路由。同时，也说明 BGP 在进行路由判定时，AS-PATH 属性是优于 ORIGIN 属性的。

【提示】

为了完成和上述一样的路径控制目的，也可以在路由器 R3 上进行如下配置：

```
R3(config)#ip prefix-list 1 seq 5 permit 4.4.4.0/24
R3(config)#route-map ASPATH permit 10
R3(config-route-map)#match ip address prefix-list 1
R3(config-route-map)#set as-path prepend 600 700
R3(config-route-map)#route-map ASPATH permit 20
R3(config)#router bgp 100
R3(config-router)#neighbor 34.34.34.4 route-map ASPATH in
```

路由器 R3 的 BGP 表显示如下：

```
R3#show ip bgp
   Network          Next Hop         Metric LocPrf Weight  Path
*>i4.4.4.0/24       1.1.1.1          0      100    0       200 e
*                   34.34.34.4       0             0       600 700 200 i
```

通过对比发现，如果在 R4 上配置 AS_PATH 属性并应用到出向，AS_PATH 属性为 **200 600 700**，如果在 R3 上配置 AS_PATH 属性并应用到入向，AS_PATH 属性为 **600 700 200**。结论：在配置追加 AS_PATH 时，如果应用到出向，是将 AS_PATH 追加到右边，如果应用到入向，是将 AS_PATH 追加到左边。为了保持实验的连续性，请将在路由器 R3 上配置的关于 AP_PATH 属性的修改删除，继续使用 R4 的关于 AS_PATH 属性的配置。

7.4.3 实验 9：配置 BGP LOCAL_PREF 属性控制选路

1. 实验目的

通过本实验可以掌握：
① 配置 LOCAL_PREF 属性的方法。
② 用 LOCAL_PREF 属性选路的原则。

2. 实验拓扑

实验拓扑如图 7-13 所示。

3. 实验步骤

路由器 R1、R2 和 R4 上的配置和 7.4.2 实验 8 相同，修改路由器 R3 从邻居 R4 收到的关于 BGP 路由条目 **4.4.4.0/24** 的本地优先级属性，然后观察路由器 R1、R2 和 R3 对路由器 R4 上通告的 BGP 路由条目 4.4.4.0/24 的选路，配置如下。

```
R3(config)#ip prefix-list 1 permit 4.4.4.0/24
R3(config)#route-map LOCAL_PREF permit 10
R3(config-route-map)#match ip address prefix-list 1
R3(config-route-map)#set local-preference 2000           //修改本地优先级属性
R3(config)#route-map LOCAL_PREF permit 20
R3(config-router)#neighbor 34.34.34.4 route-map LOCAL_PREF in
//对从邻居 34.34.34.4 进入的路由条目设置策略
R3#clear ip bgp *
```

4. 实验调试

在路由器 R1、R2 和 R3 上查看 BGP 表。

① R1#**show ip bgp**

Network	Next Hop	Metric	LocPrf	Weight	Path
* 4.4.4.0/24	14.14.14.4	0		0	200 e
*>i	3.3.3.3	0	2000	0	200 600 700 i

② R2#**show ip bgp**

Network	Next Hop	Metric	LocPrf	Weight	Path
*>i4.4.4.0/24	3.3.3.3	0	2000	0	200 600 700 i

③ R3#**show ip bgp**

Network	Next Hop	Metric	LocPrf	Weight	Path
*> 4.4.4.0/24	34.34.34.4	0	2000	0	200 600 700 i

以上输出表明路由器 R1 学到两条关于 4.4.4.0/24 的 BGP 路由，但是由于下一跳为 3.3.3.3 的路由本地优先级的值比下一跳为 14.14.14.4 的路由的本地优先级的值高，所以优选下一跳为 3.3.3.3 的路由，而下一跳为 14.14.14.4 的路由不能被优化，不能继续通告给路由器 R2 和 R3，所以路由器 R2 和 R3 只有一条关于 4.4.4.0/24 的 BGP 路由。同时也说明 BGP 在进行路由判定时，本地优先级属性是优于 AS-PATH 属性的。

【技术要点】

① 默认情况下，本地优先级的值为 100。
② 本地优先级属性只在 AS 内部传递，不会通告给 EBGP 邻居。
③ 本地优先级属性值越高，路由的优选程度越高。
④ 路由模式下命令 **bgp default local-preference** 也可以修改本地优先级属性，但是用 route-map 设置本地优先级灵活性更大。

7.4.4 实验 10：配置 BGP Weight 属性控制选路

1. 实验目的

通过本实验可以掌握：
① 配置 Weight 属性的方法。

② 用 Weight 属性选路的原则。

2．实验拓扑

实验拓扑如图 7-13 所示。

3．实验步骤

路由器 R2、R3 和 R4 上的配置和 7.4.3 实验 9 相同，修改路由器 R1 从邻居 R2、R3 和 R4 收到的路由条目的 Weight 属性，使得路由器 R1 优选从路由器 R4 学到的路由条目，配置如下。

```
R1(config)#router bgp 100
R1(config-router)#neighbor 2.2.2.2 weight 200
//为从 2.2.2.2 学到路由设置权重值
R1(config-router)#neighbor 3.3.3.3 weight 200
//为从 3.3.3.3 学到路由设置权重值
R1(config-router)#neighbor 14.14.14.4 weight 500
//为从 14.14.14.4 学到路由设置权重值
R1#clear ip bgp *
```

4．实验调试

在路由器 R1、R2 和 R3 上查看 BGP 表。

① R1#show ip bgp

Network	Next Hop	Metric	LocPrf	Weight	Path
*> 4.4.4.0/24	14.14.14.4	0		500	200 e
* i	3.3.3.3	0	2000	200	200 600 700 i

② R2#show ip bgp

Network	Next Hop	Metric	LocPrf	Weight	Path
* i4.4.4.0/24	1.1.1.1	0	100	0	200 e
*>i	3.3.3.3	0	2000	0	200 600 700 i

③ R3#show ip bgp

Network	Next Hop	Metric	LocPrf	Weight	Path
* i4.4.4.0/24	1.1.1.1	0	100	0	200 e
*>	34.34.34.4	0	2000	0	200 600 700 i

以上输出表明路由器 R1 学到两条关于 4.4.4.0/24 的 BGP 路由，但是由于下一跳为 14.14.14.4 的路由的 Weight 属性比下一跳为 3.3.3.3 的路由的 Weight 值高，所以优选下一跳为 14.14.14.4 的路由。因为 Weight 属性只影响路由器 R1 自身选路，所以路由器 R2 和 R3 仍然通过本地优先级选路。路由器 R1 的选路策略说明 BGP 在进行路由判定时 Weight 属性是优于本地优先级属性的。注意 Weight 属性越大，优先级越高。

7.4.5 实验 11：用 MED 属性控制选路

1．实验目的

通过本实验可以掌握：

① 配置 MED 属性的方法。
② 用 MED 属性选路的原则。

2. 实验拓扑

实验拓扑如图 7-13 所示。本实验需要在路由器 R2 上添加一个环回地址 Lo1：20.1.1.2/24，并在 BGP 进程中通告；在路由器 R3 上添加一个环回地址 Lo1：30.1.1.3/24，并在 BGP 进程中通告。通过设置 MED 属性，使得在路由器 R4 上访问 30.1.1.3 时走 R4→R3→R4 路径；在 R4 上访问 20.1.1.2 时走 R4→R1→R2→R1→R4 的路径。

3. 实验步骤

路由器 R4 上的配置和 7.4.4 实验 10 相同，修改路由器 R1、R2、R3 配置如下。

（1）配置路由器 R1

```
R1(config)#ip prefix-list 20 permit 20.1.1.0/24
R1(config)#ip prefix-list 30 permit 30.1.1.0/24
R1(config)#route-map MED permit 10
R1(config-route-map)#match ip address prefix-list 20
R1(config-route-map)#set metric 50                      //设置 MED 值
R1(config)#route-map MED permit 20
R1(config-route-map)#match ip address prefix-list 30
R1(config-route-map)#set metric 100
R1(config)#route-map MED permit 30
R1(config-route-map)#router bgp 100
R1(config-router)#neighbor 14.14.14.4 route-map MED out
```

（2）配置路由器 R2

```
R2(config-if)#router bgp 100
R2(config-router)#network 20.1.1.0 mask 255.255.255.0
R2(config-router)#neighbor 1.1.1.1 weight 1000
R2(config-router)#neighbor 3.3.3.3 weight 500
//以上两行修改 weight 属性的目的是为了控制从 R4 来的数据包按设定路径返回
```

（3）配置路由器 R3

```
R3(config)#ip prefix-list 20    permit 20.1.1.0/24
R3(config)#ip prefix-list 30    permit 30.1.1.0/24
R3(config)#route-map MED permit 10
R3(config-route-map)#match ip address prefix-list 20
R3(config-route-map)#set metric 100
R3(config)#route-map MED permit 20
R3(config-route-map)#match ip address prefix-list 30
R3(config-route-map)#set metric 50
R3(config)#route-map MED permit 30
R3(config)#router bgp 100
R3(config-router)#network 30.1.1.0 mask 255.255.255.0
R3(config-router)#neighbor 34.34.34.4 route-map MED out
```

4．实验调试

（1）在路由器 R4 查看 BGP 表

```
R4#show ip bgp
   Network          Next Hop           Metric LocPrf Weight Path
*> 4.4.4.0/24       0.0.0.0                 0         32768 i
*  20.1.1.0/24      34.34.34.3            100            0 100 i
*>                  14.14.14.1             50            0 100 i
*> 30.1.1.0/24      34.34.34.3             50            0 100 i
*                   14.14.14.1            100            0 100 i
```

以上输出表明路由器 R4 学到的 BGP 路由携带了 MED 值，而且优选 MED 值低的路径。

【技术要点】

① MED 只用来向 EBGP 邻居发送。
② MED 用来影响外部 AS 选路。
③ 一个 AS 中的 MED 属性是不会从这个 AS 中再传递出去的。
④ MED 的值越低，路由的优选程度越高。

（2）用扩展 ping 跟踪路径

① 跟踪路由器 R4 上访问 20.1.1.2 的路径。

```
R4#ping
Protocol [ip]:
Target IP address: 20.1.1.2
Repeat count [5]: 1
Datagram size [100]:
Timeout in seconds [2]:
Extended commands [n]: y
Source address or interface: 4.4.4.4
Type of service [0]:
Set DF bit in IP header? [no]:
Validate reply data? [no]:
Data pattern [0xABCD]:
Loose, Strict, Record, Timestamp, Verbose[none]: r
Number of hops [ 9 ]: 6
Loose, Strict, Record, Timestamp, Verbose[RV]:
Sweep range of sizes [n]:
Type escape sequence to abort.
Sending 1, 100-byte ICMP Echos to 20.1.1.2, timeout is 2 seconds:
Packet sent with a source address of 4.4.4.4
Packet has IP options:    Total option bytes= 27, padded length=28
  Record route: <*>
    (0.0.0.0)
    (0.0.0.0)
    (0.0.0.0)
```

```
        (0.0.0.0)
        (0.0.0.0)
        (0.0.0.0)
    Reply to request 0 (16 ms).    Received packet has options
     Total option bytes= 28, padded length=28
     Record route:
        (14.14.14.4)
        (12.12.12.1)
        (20.1.1.2)
        (12.12.12.2)
        (14.14.14.1)
        (4.4.4.4)
        <*>
     End of list
    Success rate is 100 percent (1/1), round-trip min/avg/max = 16/16/16 ms
```

以上输出表明在 R4 上访问 20.1.1.2 时走 R4→R1→R2→R1→R4 的路径。

② 跟踪路由器 R4 上访问 30.1.1.3 的路径。

```
R4#ping
Protocol [ip]:
Target IP address: 30.1.1.3
Repeat count [5]: 1
Datagram size [100]:
Timeout in seconds [2]:
Extended commands [n]: y
Source address or interface: 4.4.4.4
Type of service [0]:
Set DF bit in IP header? [no]:
Validate reply data? [no]:
Data pattern [0xABCD]:
Loose, Strict, Record, Timestamp, Verbose[none]: r
Number of hops [ 9 ]: 4
Loose, Strict, Record, Timestamp, Verbose[RV]:
Sweep range of sizes [n]:
Type escape sequence to abort.
Sending 1, 100-byte ICMP Echos to 30.1.1.3, timeout is 2 seconds:
Packet sent with a source address of 4.4.4.4
Packet has IP options:    Total option bytes= 19, padded length=20
 Record route: <*>
    (0.0.0.0)
    (0.0.0.0)
    (0.0.0.0)
    (0.0.0.0)
 Reply to request 0 (16 ms).    Received packet has options
  Total option bytes= 20, padded length=20
  Record route:
     (34.34.34.4)
     (30.1.1.3)
     (34.34.34.3)
     (4.4.4.4)
```

```
<*>
End of list
Success rate is 100 percent (1/1), round-trip min/avg/max = 16/16/16 ms
```

以上输出表明在路由器 R4 上访问 30.1.1.3 时走 R4→R3→R4 路径。

本 章 小 结

BGP 是一种用在 AS 之间的动态路由协议，是当前唯一广泛使用的 EGP 版本。与 OSPF、RIP 和 EIGRP 等 IGP 协议不同，其着眼点在于通过丰富的属性对路由实现灵活的过滤和最佳选择。本章介绍了 BGP 特征、属性、消息类型及格式、BGP 邻居状态、BGP 路由决策和 BGP 路由抑制等内容，并用实验演示和验证了 IBGP 和 EBGP 基本配置、BGP 验证、路由抑制和 EBGP 多跳、BGP 地址聚合、路由反射器配置、联邦配置、团体配置以及用 BGP 属性控制选路等。

第8章 分支连接

宽带技术的飞速发展为企业总部、分支机构、供应商和远程工作人员之间实现安全、可靠并且经济的网络连接提供了可能。宽带通信系统指可以在 Internet 和其他网络上高速地提供数据、语音和视频等服务的先进通信系统。信息传输可通过数字用户线路（DSL）技术、光缆技术、同轴电缆技术、无线技术以及卫星技术来实现。随着企业分支机构和远程工作者的人数不断增加，企业也越来越需要将分支机构、小型办公室和家庭办公室（SOHO）以及其他远程工作人员连接在一起，这样不仅让企业能够节省管理成本，统一部署管理策略，它还会影响社会的社群结构，对环境保护产生积极作用，同时在设计企业网络架构时，必须考虑连接性、安全性、可用性等方面。PPPoE 技术可以在以太网上传输 PPP 帧，可以满足 ISP 侧重的 PPP 身份验证、计费和链路管理功能以及用户对以太网连接的易用性和偏爱性。IPSec VPN 可以在公共网络上创建专用隧道，进而可以提高 Internet 上数据传输的安全性。GRE 可以在 IP 隧道内封装各种协议数据包。

8.1 分支连接概述

8.1.1 公共 WAN 基础设施远程连接

远程工作者要连接到企业站点，首先要通过宽带网络连接到 Internet，国内的 ISP 提供多种方式的宽带接入。用户在选择时，应考虑每种宽带类型的优点和缺点，其中可能包括成本、速度、安全性以及实施或安装的难易度。

① DSL（Digital Subscriber Line，数字用户线）：是一种能够通过普通电话线提供宽带数据业务的技术，分为非对称 DSL（Asymmetric Digital Subscriber Line，ADSL）和对称 DSL（Symmetric Digital Subscriber Line，SDSL）。ADSL 为用户提供的下行带宽比上行带宽要宽，而 SDSL 提供的上行带宽和下行带宽相同。DSL 的不同变体支持不同带宽，传输速率取决于本地环路的实际长度以及环路布线的类型和状况，比如要让 ADSL 服务满足要求，环路距离必须短于 5.46 km。

② 电缆：由有线电视服务提供商提供的一种连接方式，电缆系统使用在网络之间传输射频（Radio Frequency，RF）信号的同轴电缆。现代电缆系统为客户提供先进的电信服务，包括高速 Internet 接入、数字有线电视以及住宅电话服务。电缆调制解调器（Cable Modem）将 Internet 信号与该电缆承载的其他信号分离，并提供与计算机或 LAN 的以太网连接。由于采用共享结构，随着用户的增多，个人用户的接入速率会有所下降，安全保密性也欠佳。

③ FTTX：以光纤为传输媒介，为家庭和企业等终端用户提供接入电信局端的服务。FTTX 包括 FTTH、FTTO、FTTB 和 FTTC。

- FTTH：Fiber To The Home，光纤到户；
- FTTO：Fiber To The Office，光纤到办公室；
- FTTB：Fiber To The Building，光纤到大楼；
- FTTC：Fiber To The Curb，光纤到路边。

④ 无线：采用无线技术使用免授权的无线频谱收发数据，包括 WiFi、WiMax 和卫星等。

⑤ 蜂窝网络：移动客户端使用无线电波通过附近运营商的基站进行通信。随着 3G/4G 技术的广泛使用，蜂窝网络速度不断提高。蜂窝宽带接入包含的标准有 cdma2000、WCDMA、TD-CDMA、TD-LTE 和 FDD-LTE 等。

⑥ 虚拟专用网络：当远程工作人员或远程办公室员工使用宽带服务通过互联网接入公司网络时，会带来一定的安全风险。使用虚拟专用网络（Virtual Private Network，VPN）技术可以解决网络访问的安全问题。VPN 是公共网络（如互联网）之上多个专用网络之间的加密连接技术，是企业网络在互联网上的延伸。

8.1.2 专用 WAN 基础设施远程连接

① 租用线路：租用线路（Leased Line）是从 ISP 租用的在客户站点和 ISP 的广域网之间或者在客户的两个站点之间的永久的通信电路。因为提供专用永久性服务，通常被称为专线，而用租用线路构建的广域网通常被称为专用 WAN。租用线路一般按月支付费用，通常比较昂贵。

② 综合业务数字网络：综合业务数字网络（Integrated Services Digital Network，ISDN）是一种电路交换技术，能够用公共交换电话网络（Public Switched Telephone Network，PSTN）本地环路传输数字信号，从而实现更高容量的交换连接。ISDN 接口有基本速率接口（Basic Rate Interface，BRI）和主速率接口（Primary Rate Interface，PRI）2 种。其中目前使用较多的是 PRI，在北美可以提供 23B 信道+1D 信道，总比特率可达 1.544 Mbps；在欧洲、澳大利亚和世界其他地区，可以提供 30B 信道+1D 信道，总比特率可达 2.048 Mbps。借助于 PRI ISDN 可以实现 VoIP、视频会议和无延时、无抖动的高带宽连接。

③ 以太网 WAN：运营商通过光纤布线提供以太网 WAN 服务。以太网 WAN 服务曾有过许多名称，包括城域以太网（MetroE）、MPLS（Multi-Protocol Label Switching，多协议标签交换）以太网（EoMPLS）和虚拟专用 LAN 服务（Virtual Private LAN Service，VPLS）。目前以太网 WAN 已经得到普及和应用，而传统的帧中继（Frame Relay）和异步传输模式（Asynchronous Transfer Mode，ATM）WAN 链路基本很少使用了。

④ 甚小口径终端：甚小口径终端（Very Small Aperture Terminal，VSAT）是一种利用卫星通信创建专用 WAN 的解决方案。VSAT 是一个小型卫星天线，类似于家庭互联网和电视使用的天线。VSAT 创建专用 WAN，同时提供到远程位置的连接。

⑤ 多协议标签交换：MPLS 是一种运营商技术，目前用的最多的是 MPLS VPN 技术。MPLS VPN 是指采用 MPLS 技术在骨干宽带 IP 网络上构建企业 IP 专网，实现跨地域、安全、高速、可靠的数据、语音、图像多业务通信，并结合 QoS 服务、流量工程等相关技术，将公众网可靠的性能、良好的扩展性和丰富的功能与专用网的安全、灵活和高效结合在一起。

8.2 PPPoE 概述

8.2.1 PPPoE 简介

运营商希望通过同一台接入设备来连接远程的多个主机，同时接入设备能够提供访问控制和计费功能。在众多的接入技术中，把多个主机连接到接入设备的最经济的方法就是以太网，而 PPP 可以提供良好的访问控制和计费功能，于是产生了在以太网上传输 PPP 数据包的技术，即 PPPoE（PPP over Ethernet）。PPPoE 通过以太网连接创建 PPP 隧道，利用以太网将大量主机组成网络，通过远端接入设备连入 Internet，并运用 PPP 协议对接入的每台主机进行控制，具有适用范围广、安全性高、计费方便的特点。PPPoE 技术解决了用户上网收费等实际应用问题，得到了宽带接入运营商的认可并被广泛应用。

8.2.2 PPPoE 数据包类型

PPPoE 通过下面 5 种类型的数据包来建立和终结 PPPoE 会话。

① PPPoE 发现初始（PPPoE Active Discovery Initiation，PADI）数据包：用户主机发向 PPPoE 服务器的探测数据包，目的 MAC 地址为广播地址。

② PPPoE 发现提供（PPPoE Active Discovery Offer，PADO）数据包：PPPoE 服务器收到 PADI 数据包之后回应的数据包，目的 MAC 地址为客户端主机的 MAC 地址。

③ PPPoE 发现请求（PPPoE Active Discovery Request，PADR）数据包：用户主机收到 PPPoE 服务器回应的 PADO 数据包后，以单播方式发起的请求数据包，目的地址为此用户选定的 PPPoE 服务器的 MAC 地址。

④ PPPoE 会话确认（PPPoE Active Discovery Session Confirmation，PADS）数据包：PPPoE 服务器分配一个唯一的会话进程 ID，并通过 PADS 数据包发送给主机。

⑤ PPPoE 发现终止（PPPoE Active Discovery Terminate，PADT）数据包：当用户或者服务器需要终止 PPPoE 会话时，可以发送 PADT 数据包。

8.2.3 PPPoE 会话建立过程

PPPoE 会话建立过程可分为 3 个阶段，即发现（Discovery）阶段、会话（Session）阶段和终止（Terminate）阶段，PPPoE 会话建立过程如图 8-1 所示。

（1）发现（Discovery）阶段

① PPPoE Client 广播发送一个 PADI 数据包，在此数据包中包含 PPPoE Client 想要得到的服务类型信息。

② 所有的 PPPoE Server 收到 PADI 数据包之后，将其中请求的服务与自己能够提供的服务进行比较，如果可以提供，则单播回复一个 PADO 数据包。

③ 根据网络的拓扑结构，PPPoE Client 可能收到多个 PPPoE Server 发送的 PADO 数据包，PPPoE Client 选择最先收到的 PADO 数据包对应的 PPPoE Server 作为自己的 PPPoE Server，并单播发送一个 PADR 数据包。

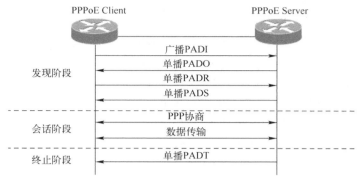

```
PPPoED  Active Discovery Terminate (PADT)
PPPoED  Active Discovery Initiation (PADI)
PPPoED  Active Discovery Offer (PADO) AC-Name='RTB00e0fcba0c51'
PPPoED  Active Discovery Request (PADR) AC-Name='RTB00e0fcba0c51'
PPPoED  Active Discovery Session-confirmation (PADS) AC-Name='RTB00e0fcba0c51'
```

图 8-1　PPPoE 会话建立过程

④ PPPoE Server 产生一个唯一的会话 ID（Session ID），标识和 PPPoE Client 的这个会话，通过发送一个 PADS 数据包，把会话 ID 发送给 PPPoE Client，会话建立成功后便进入 PPPoE 会话阶段。

完成上述 4 个步骤后，PPPoE Server 和 PPPoE Client 通信双方都会知道 PPPoE 的会话 ID 以及对方以太网 MAC 地址，它们共同确定了唯一的 PPPoE 会话。

（2）会话（Session）阶段

PPPoE 会话过程中的 PPP 协商和普通的 PPP 协商方式一致。PPPoE 会话的 PPP 协商成功后，就可以承载 PPP 数据包。在 PPPoE 会话阶段，所有的以太网数据包都是单播发送的。

（3）终止（Terminate）阶段

进入 PPPoE 会话阶段后，PPPoE Client 和 PPPoE Server 都可以通过发送 PADT 数据包的方式来结束 PPPoE 连接。PADT 数据包可以在会话建立以后的任意时刻单播发送。在发送或接收到 PADT 后，就不允许再使用该会话发送 PPP 数据流量了。

在 PPPoE 发现阶段以太网帧的类型字段值为 0x8863，而在 PPPoE 会话阶段以太网帧的类型字段值为 0x8864。PPPoE 数据包格式如图 8-2 所示，各字段的含义如下。

0	7	8	15	16	23	24	31
版本	类型	代码		会话ID			
长度				载荷			

```
⊟ PPP-over-Ethernet Session
    0001 .... = Version: 1
    .... 0001 = Type: 1
    Code: Session Data (0x00)
    Session ID: 0x0001
    Payload Length: 62
⊟ Point-to-Point Protocol
    Protocol: IP (0x0021)
```

图 8-2　PPPoE 数据包格式

① 版本（4 比特）：PPPoE 协议中规定该字段的值为 0x01。
② 类型（4 比特）：PPPoE 协议中规定该字段的值为 0x01。
③ 代码（8 比特）：对于 PPPoE 的不同阶段该字段的内容也是不一样的，PADI 数据包该字段值为 0x09，PADO 数据包该字段值为 0x07，PADR 数据包该字段值为 0x19，PADS 数据包该字段值为 0x65，PADT 数据包该字段值为 0xA7。
④ 会话 ID（16 比特）：当 PPPoE Server 还未分配唯一的会话 ID 给用户主机时，该字段的内容填充必须为 0x0000，一旦主机获取了会话 ID 后，那么在后续的所有报文中该字段必须填充这个唯一的会话 ID 值。
⑤ 长度（16 比特）：用来表示 PPPoE 数据包中载荷的长度。
⑥ 载荷：也称数据域，在 PPPoE 的不同阶段该字段的数据内容会有很大不同。在 PPPoE 发现阶段，该字段会填充一些标记；而在 PPPoE 的会话阶段，该字段则携带的是 PPP 的数据。

8.3 隧道技术概述

8.3.1 GRE 简介

GRE（Generic Routing Encapsulation，通用路由封装）最早是由 Cisco 提出的，而目前它已经成为了一种标准，被定义在 RFC 1701、RFC 1702、RFC 2784 以及 RFC 2890 中。其中 RFC 2890 是基于 RFC 2784 的增强版，最新版本的 Cisco IOS 使用 RFC 2890。GRE 是一种封装协议，它定义了如何用一种网络协议去封装另一种网络协议的方法。GRE 属于 VPN 的第三层隧道（Tunnel）协议，所谓隧道是指包括数据封装、传输和解封装在内的全过程。GRE 只提供数据包的封装，它并没有加密功能来防止网络侦听和攻击，所以在实际环境中它常和 IPSec 一起使用，由 IPSec 提供用户数据的安全性和完整性。例如，GRE 可以封装组播数据（如 OSPF、EIGRP、视频和 VoIP 等）并在 GRE 隧道中传输，而 IPSec 目前只能对单播数据进行加密保护。对于组播数据需要在 IPSec 隧道中传输的情况，可以先建立 GRE 隧道，对组播数据进行 GRE 封装，再对封装后的数据进行 IPSec 加密，从而实现组播数据在 IPSec 隧道中的加密传输。GRE 的主要应用就是在 IP 网络中承载 IP 以及非 IP 数据，GRE 特征如下。
① GRE 是一种无状态协议，不提供流量控制。
② GRE 至少增加 24 字节的开销，包括一个 20 字节 IPv4 头部和无任何附加选项的 4 字节的 GRE 头部。
③ GRE 具备多协议性，可以将 IP 以及非 IP 数据封装在隧道内。
④ GRE 允许组播流量和动态路由协议数据包穿越隧道。
⑤ GRE 的安全特性相对较弱。
当 GRE 用 IPv4 作为封装协议时，IPv4 协议号为 47。GRE 数据包头没有统一的格式，每

个厂商具体实现的时候会有所差别，本章讨论的是 RFC 2890 推荐使用的 GRE 头部格式。GRE 封装和包头格式如图 8-3 所示，各字段含义如下。

```
Internet Protocol, Src: 202.96.134.1 (202.96.134.1), Dst: 61.0.0.4 (61.0.0.4)
    Version: 4
    Header length: 20 bytes
  ⊕ Differentiated Services Field: 0x00 (DSCP 0x00: Default; ECN: 0x00)
    Total Length: 128
    Identification: 0x000a (10)
  ⊕ Flags: 0x00
    Fragment offset: 0
    Time to live: 253
    Protocol: GRE (47)
  ⊕ Header checksum: 0x2fdf [correct]
    Source: 202.96.134.1 (202.96.134.1)
    Destination: 61.0.0.4 (61.0.0.4)
⊟ Generic Routing Encapsulation (IP)
  ⊕ Flags and version: 0x2000
    Protocol Type: IP (0x0800)
    GRE Key: 0x0001e240
  ⊕ Internet Protocol, Src: 172.16.14.1 (172.16.14.1), Dst: 172.16.14.4 (172.16.14.4)
```

图 8-3　GRE 封装和包头格式

① C（第 0 位）：校检和存在位，当设置为 1 时存在校验和字段，同时该字段包含 2 字节的有效信息。

②（第 1 位）：在 RFC 2890 中无意义，设置为 0。

③ K（第 2 位）：密钥存在位，当设置为 1 时，在 GRE 包头中包含 4 字节的密钥字段。

④ S（第 3 位）：序列号存在位，当设置为 1 时，在 GRE 包头中包含 4 字节的序列号字段。

⑤ 版本（第 13～15 位）：版本号必须为 0。

⑥ 协议类型（2 字节）：指封装在 GRE 中原始有效载荷中的数据包的协议类型。例如，如果原始有效载荷是 IP 数据包，该字段被设置为 0x0800。

⑦ 校验和：用来保证 GRE 包头和有效载荷的完整性。

⑧ 密钥：是一个用来验证已封装 GRE 数据包的数字，范围为 0～4294967295，为 GRE 提供一种较弱的安全性。隧道两端只接收密钥相同的 GRE 数据包。密钥需要在隧道两端手动配置。

⑨ 序列号：网络两端可以使用序列号跟踪接收到数据包的顺序，并且可以选择性地丢弃乱序到达的数据包。

8.3.2 IPSec VPN 简介

采用 GRE 技术的一个重要问题是数据包在 Internet 上传输是不安全的。IPSec（Internet Protocol Security）VPN 使用先进的加密技术和隧道来实现在 Internet 上建立安全的端到端私有网络。IPSec VPN 的基础是数据机密性、数据完整性、身份验证和防重放攻击。

① 数据机密性：一个常见的安全性考虑是防止窃听者截取数据。数据机密性旨在防止消息的内容被未经身份验证或未经授权的来源拦截。VPN 利用封装和加密机制来实现数据机密性。常用的加密算法包括 DES、3DES 和 AES。

② 数据完整性：数据完整性确保数据在源主机和目的主机之间传送时不被篡改。VPN 通常使用哈希算法来确保数据完整性。哈希算法类似于校验和，但更可靠，它可以确保没有人更改过数据的内容。常用的验证算法包括 MD5 和 SHA-1。

③ 身份验证：身份验证确保信息来源的真实性，并传送到真实目的地。常用的方法包括预共享密码和数字证书等。

④ 防重放攻击：IPSec 接收方可检测并拒绝接收过时或重复的数据包。

8.3.3 AH 和 ESP

IPSec 协议不是一个单独的协议，它是 IETF IPSec 工作组为了在 IP 层提供通信安全而制定的一整套协议标准，包括安全协议，如 AH 和 ESP、IKE（Internet Key Exchange）和用于验证及加密的一些算法等。RFC 2401 定义了 IPSec 的基本结构。要深入了解 IPSec 安全协议，必须先理解 IPSec 的两种工作模式。

（1）隧道（Tunnel）模式

该模式中，原始 IP 数据包被封装在新的 IP 数据包中，并在两者之间插入一个 IPSec 包头（AH 或 ESP），IPSec 隧道模式封装如图 8-4 所示。

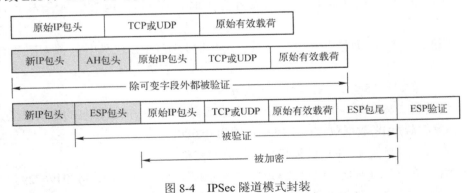

图 8-4 IPSec 隧道模式封装

（2）传输（Transport）模式

该模式中，在 IP 数据包包头和高层协议包头之间插入一个 IPSec 包头（AH 或 ESP）。新的 IP 数据包包头和原始的 IP 数据包包头相同，只是 IP 协议字段被改为 50(ESP)或 51(AH)，IPSec 传输模式如图 8-5 所示。

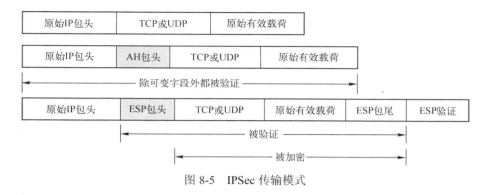

图 8-5　IPSec 传输模式

（3）IPSec 安全协议

IPSec 安全协议包括 2 种：AH 和 ESP。

① AH（Authentication Header，验证包头）。

不要求或不允许有机密性时使用，可以提供数据完整性和身份验证。AH 不提供数据包的数据机密性（加密）。AH 协议单独使用时提供的保护较脆弱。AH 包头格式如图 8-6 所示。

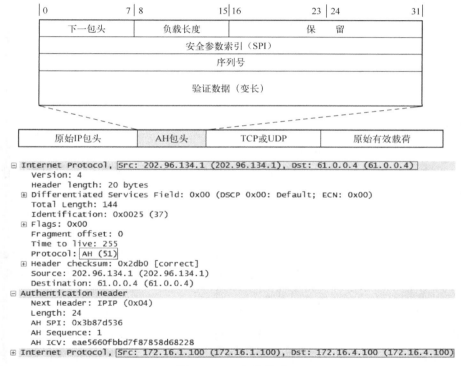

图 8-6　AH 包头格式

② ESP（Encapsulating Security Payload，封装安全有效负载）。

ESP 提供数据机密性、完整性和身份验证。IP 数据包加密可以隐藏数据及源主机和目的主机的身份。ESP 可验证内部 IP 数据包和 ESP 包头，从而提供数据来源验证和数据完整性检查，因此使用较多。ESP 包头格式如图 8-7 所示，在数据包捕获过程中，只能看到前两个字段，其他字段均被加密。

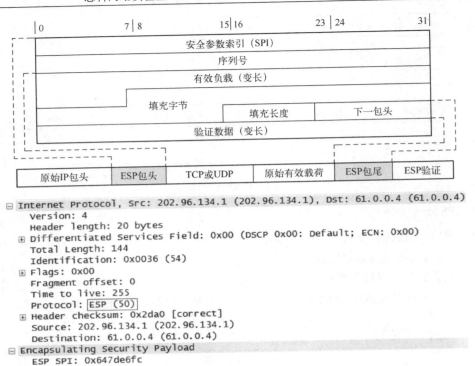

图 8-7　ESP 包头格式

8.3.4　安全关联和 IKE

安全关联（Security Association，SA），是 IPSec 的基本部件，是通信对等体间对某些要素的约定，例如，使用哪种协议、封装模式、加密算法、预共享密钥以及密钥的生存周期等。SA 分为两种：IKE（Internet Key Exchange）SA 和 IPSec SA。SA 是单向的，在两个对等体之间的双向通信，最少需要两个 SA 来分别对两个方向的数据流进行安全保护。同时，如果两个对等体同时使用 AH 和 ESP 来进行安全通信，则每个对等体都会针对每一种协议来构建一个独立的 SA。SA 由一个三元组来唯一标识，这个三元组包括 SPI（Security Parameters Index，安全参数索引）、目的 IP 地址、安全协议号（AH 或 ESP）。通过 IKE 协商建立的 SA 具有生存周期，生存周期有以下 2 种定义方式。

① 基于时间的生存周期：定义了一个 SA 从建立到失效的时间。
② 基于流量的生存周期：定义了一个 SA 允许处理的最大流量。

生存周期到达指定的时间或指定的流量，SA 就会失效。在 SA 失效前，IKE 将为 IPSec 协商建立新的 SA，这样，在旧的 SA 失效前新的 SA 就已经准备好。在新的 SA 开始协商而没有协商好之前，继续使用旧的 SA 保护通信。在新的 SA 协商好之后，则立即采用新的 SA 保护通信。

IKE 使 IPSec 可以在不安全的网络上安全地验证身份、分发密钥、建立 IPSec SA，该协议建立在由 ISAKMP（Internet Security Association and Key Management Protocol，Internet 安全关联和密钥管理协议）定义的框架上。IKE 协商采用 UDP 数据包格式，默认端口是 500。

IKE 分以下两个阶段为 IPSec 进行密钥协商并建立 SA。

第一阶段：让 IKE 对等体验证对方并确定会话密钥，即建立一个 ISAKMP SA。第一阶段有主模式（Main Mode）和积极模式（Aggressive Mode）2 种 IKE 方法。

第二阶段：为 IPSec 协商具体的 SA，建立最终的用于 IP 数据安全传输的 IPSec SA。

第一阶段采用主模式，第二阶段采用快速模式进行 IKE 协商。IKE 协商的过程如图 8-8 所示。

```
11 33.303000  202.96.134.1   61.0.0.4       ISAKMP  Identity Protection (Main Mode)
12 33.537000  61.0.0.4       202.96.134.1   ISAKMP  Identity Protection (Main Mode)
13 33.646000  202.96.134.1   61.0.0.4       ISAKMP  Identity Protection (Main Mode)
14 33.865000  61.0.0.4       202.96.134.1   ISAKMP  Identity Protection (Main Mode)
15 34.006000  202.96.134.1   61.0.0.4       ISAKMP  Identity Protection (Main Mode)
16 34.178000  61.0.0.4       202.96.134.1   ISAKMP  Identity Protection (Main Mode)
17 34.303000  202.96.134.1   61.0.0.4       ISAKMP  Quick Mode
18 34.521000  61.0.0.4       202.96.134.1   ISAKMP  Quick Mode
19 34.646000  202.96.134.1   61.0.0.4       ISAKMP  Quick Mode
```

图 8-8　IKE 协商的过程

8.3.5　IPSec 操作步骤

IPSec 的目标是用必要的安全服务保护通信数据，它的操作可分为以下 5 个步骤。

① 定义感兴趣的数据流：应用 ACL 来匹配感兴趣的数据流。数据包处理方式分 3 种：应用 IPSec、绕过 IPSec 以明文发送或丢弃。丢弃是指发现在策略中定义为加密数据，但实际它并未加密，那么丢弃该数据包。

② IKE 阶段 1：该阶段用于协商 IKE 策略集、验证对等体并在对等体之间建立安全通道，包括主模式和积极模式 2 种模式。主模式的主要结果是为对等体之间的后续信息交换建立一个安全通道。在发送端和接收端有如下 3 次双向信息交换。

第一次信息交换：在两个对等体之间协商用于保证 IKE 通信安全的算法和散列，结果是 ISAKMP 被商定。

第二次信息交换：使用 DH 交换来产生共享密钥 SKEYID，并且衍生出以下其他 3 个密钥。
- SKEYID_d：被用于计算后续 IPSec 密钥资源；
- SKEYID_a：被用于后续 IKE 消息的数据完整性验证；
- SKEYID_e：被用于提供后续 IKE 消息的加密。

第三次信息交换：验证对等体身份，包括预共享密钥、RSA 签名和 RSA 加密的 nonces 三种验证方法。

积极模式较主模式而言，信息交换次数和信息较少。在这种模式中不提供身份保护，交换的信息都是以明文传递的。在第一次交换中，几乎所有需要交换的信息都被压缩到所建议的 IKE SA 中一起发给对端，接收方返回所需内容，等待确认，然后由发送端确认最后的协商结果。

③ IKE 阶段 2：该阶段 IPSec 参数被协商，执行以下功能。
- 协商 IPSec 安全性参数和 IPSec 转换集；
- 建立 IPSec 的 SA；
- 定期重协商 IPSec 的 SA，以确保安全性；
- 当使用 PFS（Perfect Forward Secrecy，完美前向保密）时，可执行额外的 DH 交换。

IKE 阶段 2 只有一种模式，即快速模式（Quick Mode）。快速模式协商一个共享的 IPSec 策略，获得共享的、用于 IPSec 安全算法的密钥资源，并建立 IPSec SA。快速模式也用于 IPSec

SA 生命期过期后重新协商一个新的 IPSec SA。该阶段的最终目的是在对等体间建立一个安全的 IPSec 会话。在会话建立之前，对等体要协商所需的加密和验证算法，这些内容被统一到 IPSec 转换集（Transform Set）中。IPSec 转换集在对等体之间交换，如转换集匹配，则 IPSec 会话的流程继续进行；如果转换集不匹配，则终止协商。

④ 数据传输：在完成 IKE 阶段 2 之后，将通过安全通道在主机之间传输数据流。

⑤ IPSec 终止：管理员手工删除会话，或者空闲时间到会话被自动删除。

8.4 配置 PPPoE

8.4.1 实验 1：配置 ADSL

1. 实验目的

通过本实验可以掌握：
① ATM 接口配置。
② VPDN 配置。
③ NAT 配置。
④ DHCP 配置。

2. 实验拓扑

配置 ADSL 连接到 Internet（使用 HWIC-1 ADSL 模块）实验拓扑如图 8-9 所示。

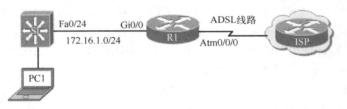

图 8-9　配置 ADSL 连接到 Internet（使用 HWIC-1 ADSL 模块）实验拓扑

3. 实验步骤

（1）实验准备

配置路由器接口 IP 地址、DHCP 服务器，计算机 PC1 的 IP 地址通过 DHCP 方式获得。

```
R1(config)#interface gigabitEthernet 0/0
R1(config-if)#no shutdown
R1(config-if)#ip address 172.16.1.1  255.255.255.0
R1(config)#ip dhcp pool ADSL
R1(dhcp-config)#network 172.16.1.0 255.255.255.0
R1(dhcp-config)#default-router 172.16.1.1
R1(dhcp-config)#dns-server 202.96.134.133
R1(dhcp-config)#domain-name abc.com
```

配置 PC1 的 IP 地址通过 DHCP 自动获得，测试计算机和路由器之间的连通性成功。

（2）在 R1 上配置 ADSL

```
R1(config)#vpdn enable
//由于 ADSL 的 PPPoE 应用是通过虚拟拨号来实现的，所以在路由器中需要使用 VPDN 功能
R1(config)#interface atm0/0/0              //进入 ATM 接口
R1(config-if)#no shutdown                  //开启 ATM 接口
R1(config-if)#no ip address
R1(config-if)#dsl operating-mode auto      //配置 ADSL 的操作模式为自动
R1(config-if)#pvc 8/35
//设置 PVC 的相关参数，即 VCI 和 VPI 的值，如果不清楚请向 ISP 查询
R1(config-if-atm-vc)#pppoe-client dial-pool-number 1
//配置该接口是拨号池 1 的成员
R1(config)#interface dialer 1              //创建虚拟的拨号接口
R1(config-if)#ip address negotiated        //配置接口的 IP 地址动态从 ISP 获得
R1(config-if)#ip mtu 1492
//配置接口下的 MTU，默认为 1500，由于 PPPoE 数据包包头会占用 8 字节，所以减小 8 字节
R1(config-if)#dialer pool 1                //创建拨号池
R1(config)#interface dialer 1              //创建拨号接口
R1(config-if)#encapsulation ppp            //配置 PPP 封装
R1(config-if)#ppp authentication pap chap callin
//配置 PPP 验证方法，此处指明可以支持 pap 和 chap 两种方法验证，取决于电信端的设置，参数
callin 并不是对电信端进行 pap 或 chap 验证，其含义是只对客户端拨入服务器的行为进行单向验证，也就是
只让电信的服务器端验证拨入的客户端，而客户端不需要验证服务端，并且会忽略服务端发来的验证请求
R1(config-if)#ppp pap sent-username test@163.gd password test123
//配置 PPP 的 pap 验证发送的信息
R1(config-if)#ppp chap hostname test@163.gd
//配置 PPP 的 chap 验证发送的用户名
R1(config-if)#ppp chap password test123    //配置 PPP 的 chap 验证发送的密码
//以上 PPP 的验证同时配置了 pap 和 chap 验证，如果已经明确地知道电信端的 PPP 验证方法，则
只需要配置相应的 PPP 验证方法即可，没有必要两种验证方法都配置
```

（3）在 R1 上配置 NAT

```
R1(config)#interface dialer 1
R1(config-if)#ip nat outside               //配置 NAT 外部接口
R1(config-if)#exit
R1(config)#interface gigabitEthernet 0/0
R1(config-if)#ip nat inside                //配置 NAT 内部接口
R1(config-if)#exit
R1(config)#access-list 1 permit 172.16.1.0 0.0.0.255   //NAT 转换的 ACL
R1(config)#ip nat inside source list 1 interface dialer 1 overload  //配置 PAT
R1(config)#ip route 0.0.0.0 0.0.0.0 dialer 1           //配置静态默认路由
```

4. 实验调试

（1）查看 ATM 接口信息

```
R1#show interfaces atm 0/0/0
ATM0/0/0 is up, line protocol is up
```

```
        Hardware is DSLSAR (with Alcatel ADSL Module)      //ADSL 模块信息
        MTU 4470 bytes, sub MTU 4470, BW 768 Kbit, DLY 660 usec,
             reliability 255/255, txload 42/255, rxload 40/255
        Encapsulation ATM, loopback not set                 //接口采用 ATM 封装
        Encapsulation(s): AAL5   AAL2, PVC mode             //ATM 封装格式和虚链路模式

        23 maximum active VCs, 256 VCs per VP, 1 current VCCs
        VC Auto Creation Disabled.
        VC idle disconnect time: 300 seconds                //VC 空闲断开连接时间为 300 秒
        （此处省略部分输出）
```

（2）查看 ATM 接口信息

```
    ① R1#show ip interface ATM 0/0/0
    ATM0/0/0 is up, line protocol is up                 //接口工作正常
        Internet protocol processing disabled
    （此处省略部分输出）
    ② R1#show ip interface dialer 1
    Dialer1 is up, line protocol is up
        Internet address is 113.88.232.216/32           //ISP 分配
        Broadcast address is 255.255.255.255
        Address determined by IPCP                      //IP 地址通过 IPCP 获得
        MTU is 1492 bytes                               //接口 MTU
    （此处省略部分输出）
```

（3）查看 DSL 接口信息

```
    R1#show dsl interface atm 0/0/0
    ATM0/0/0
    Alcatel 20150 chipset information                  //ADSL 模块的芯片信息
                        ATU-R (DS)              ATU-C (US)
    Modem Status:      Showtime (DMTDSL_SHOWTIME)     // Modem 状态
    DSL Mode:          ITU G.992.1 (G.DMT) Annex A    //DSL 操作模式
    （此处省略部分输出）
```

（4）查看 PPPoE 会话信息

```
    R1#show pppoe session
       1 client session
    Uniq ID  PPPoE  RemMAC          Port         Source    VA      State
             SID    LocMAC                                 VA-st
    N/A      8552   0018.82ab.70ba  ATM0/0/0     Di1       Vi1     UP
                    0013.c3b4.0b20  VC:  8/35                      UP
```

以上输出显示了 PPPoE 会话的信息，包括会话的 ID、本地和远端的 MAC 地址、VC 信息以及状态等。

（5）测试网络

在计算机上打开浏览器上网，测试上网是否正常。

8.4.2 实验 2：配置 PPPoE 服务器和客户端

1．实验目的

通过本实验可以掌握：
① PPPoE 客户端配置。
② PPPoE 服务器配置。

2．实验拓扑

配置 PPPoE 服务器和客户端实验拓扑如图 8-10 所示。

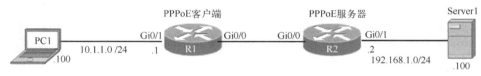

图 8-10 配置 PPPoE 服务器和客户端实验拓扑

3．实验步骤

（1）配置路由器 R1

```
R1(config)#interface gigabitEthernet0/1
R1(config-if)#no shutdown
R1(config-if)#ip address 10.1.1.1  255.255.255.0
R1(config-if)#exit
R1(config)#interface Dialer0                    //创建拨号接口
R1(config-if)#ip address negotiated             //IP 地址采用 PPP 协商方式获得
R1(config-if)#encapsulation ppp                 //配置 PPP 封装
R1(config-if)#dialer pool 1                     //配置拨号池
R1(config-if)#dialer-group 1                    //配置拨号组
R1(config-if)#ppp chap hostname cisco           //采用 chap 验证时发送的用户名
R1(config-if)#ppp chap password   cisco         //采用 chap 验证时发送的密码
R1(config-if)#mtu 1492                          //配置接口下的 MTU
R1(config-if)#ip tcp adjust-mss 1450
//调整 TCP 三次握手期间的 mss 值来防止丢弃 TCP 会话
R1(config-if)#exit
R1(config)#interface gigabitEthernet0/0
R1(config-if)#pppoe enable group global         //开启 PPPoE
R1(config-if)#pppoe-client dial-pool-number 1
//将物理端口与虚拟拨号端口进行关联
R1(config-if)#exit
R1(config)#ip route 0.0.0.0 0.0.0.0 Dialer0     //配置默认路由，拨号接口为出接口
```

（2）配置路由器 R2

```
R2(config)#username cisco password cisco        //PPP chap 验证的用户名和密码
R2(config)#ip local pool cisco 172.16.1.10 172.16.1.20    //创建本地地址池
R2(config)#bba-group pppoe ABC                  //创建 BBA（Broadband Aggregation）组
R2(config-bba-group)#virtual-template 1         //关联一个虚拟模板
```

```
R2(config-bba-group)#exit
R2(config)#interface virtual-template 1              //创建虚拟模板
R2(config-if)#ip address 172.16.1.2 255.255.255.0
R2(config-if)# encapsulation ppp                     //配置 PPP 封装
R2(config-if)#peer default ip address pool cisco
//使用本地地址池为客户端分配 IP 地址

R2(config-if)#ppp authentication chap                //配置 PPP 验证方式为 chap
R2(config-if)#mtu 1492                               //配置接口上的 MTU
R2(config-if)#exit
R2(config)#interface gigabitEthernet0/0
R2(config-if)# pppoe enable group ABC                //开启 PPPoE
R2(config-if)#no shutdown
R2(config-if)#exit
R2(config)#interface gigabitEthernet0/1
R2(config-if)# ip address 192.168.1.2 255.255.255.0
R2(config-if)#no shutdown
R2(config-if)#exit
R2(config-if)#ip route 10.1.1.0 255.255.255.0 172.16.1.10
```

4. 实验调试

（1）查看所有的 PPPoE 会话信息

```
① R1#show pppoe session all
Total PPPoE sessions 1
session id: 1
local MAC address: f872.ead6.f4c8, remote MAC address: f872.ea69.1c78
virtual access interface: Vi2, outgoing interface: Gi0/0
  VLAN Priority: 0
      29 packets sent, 0 received
      836 bytes sent, 0 received
② R2#show pppoe session all
Total PPPoE sessions 1
session id: 1
local MAC address: f872.ea69.1c78, remote MAC address: f872.ead6.f4c8
virtual access interface: Vi1.1, outgoing interface: Gi0/0
      49 packets sent, 49 received
      1122 bytes sent, 1116 received
```

以上①和②输出表明 PPPoE 客户端和服务器显示的 PPPoE 会话的信息，包括会话 ID、本地和远程 MAC 地址、虚拟访问接口、路由器的出接口，以及发送和接收数据包的个数和字节数。

（2）查看路由表

```
① R1#show ip route connected
（此处省略路由代码部分）
Gateway of last resort is 0.0.0.0 to network 0.0.0.0
      10.0.0.0/8 is variably subnetted, 2 subnets, 2 masks
```

```
C       10.1.1.0/24 is directly connected, GigabitEthernet0/1
L       10.1.1.1/32 is directly connected, GigabitEthernet0/1
        172.16.0.0/32 is subnetted, 2 subnets
C       172.16.1.2 is directly connected, Dialer0
//接口 PPP 封装的特性，对方接口的地址会在本地路由表中生成主机路由
C       172.16.1.10 is directly connected, Dialer0
//该条路由是通过 PPP 的 IPCP 协商从 R2 的本地地址池分配的 IP 地址
② R2#show ip route connected
（此处省略路由代码部分）
        172.16.0.0/16 is variably subnetted, 3 subnets, 2 masks
C       172.16.1.0/24 is directly connected, Virtual-Access1.1
L       172.16.1.2/32 is directly connected, Virtual-Access1.1
C       172.16.1.10/32 is directly connected, Virtual-Access1.1
//分配给 R1 拨号接口的地址，由于链路是 PPP 封装，所以本地路由表会出现此主机路由
        192.168.1.0/24 is variably subnetted, 2 subnets, 2 masks
C       192.168.1.0/24 is directly connected, GigabitEthernet0/1
L       192.168.1.2/32 is directly connected, GigabitEthernet0/1
```

（3）查看拨号接口信息

```
R1#show ip interface brief | include Dialer0
Dialer0         172.16.1.10      YES IPCP    up          up
//拨号接口的 IP 地址通过 PPP 的 IPCP 协商获得
```

8.5 配置隧道

8.5.1 实验 3：配置 GRE 隧道

1．实验目的

通过本实验可以掌握：
① GRE 的工作原理和特征。
② Tunnel 接口的配置和特征。
③ GRE 隧道的配置和调试方法。

2．实验拓扑

配置 GRE 隧道实验拓扑如图 8-11 所示。

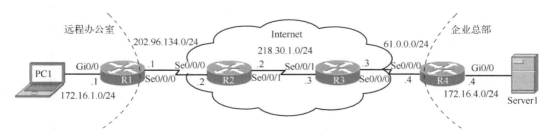

图 8-11　配置 GRE 隧道实验拓扑

3. 实验步骤

本实验中，路由器 R2 和 R3 模拟 Internet，R1 和 R4 通过静态路由连接到 Internet 上。路由器 R1 的 Gi0/0 接口模拟远程办公室所在的局域网，路由器 R4 的 Gi0/0 接口模拟企业总部所在的局域网。使用 GRE 隧道将远程办公室与企业总部进行连接，并且在远程办公室和企业总部间运行 EIGRP 路由协议，实现两地网络连通。同时需要在路由器 R1 和 R4 上配置 NAT，使两地的网络也可以访问 Internet。

（1）配置路由器 R1

```
R1(config)#interface tunnel 0
//创建 Tunnel 接口，编号为 0，Tunnel 接口的编号本地有效，不必和对端的相同
R1(config-if)#tunnel source serial0/0/0
//配置 Tunnel 的源接口，路由器将以此接口的地址作为源地址重新封装数据包，也可以直接输入接口的地址
R1(config-if)#tunnel destination 61.0.0.4
//配置 Tunnel 的目的 IP 地址，路由器将以此地址作为目的地址重新封装数据包
R1(config-if)#tunnel mode gre ip
//配置隧道的模式，默认就是 gre ip
R1(config-if)#ip address 172.16.14.1 255.255.255.0
//配置隧道接口下的 IP 地址，创建该隧道后，可以把隧道比作一条专线
R1(config-if)#tunnel key 123456           //配置验证的 key，提供隧道建立的安全性
R1(config-if)#exit
R1(config)#ip route 0.0.0.0 0.0.0.0 serial0/0/0    //默认路由指向 Internet
R1(config)#router eigrp 1
R1(config-router)#network 172.16.1.1 0.0.0.0
R1(config-router)#network 172.16.14.1 0.0.0.0
R1(config-router)#passive-interface gigabitEthernet0/0
R1(config)#interface serial0/0/0
R1(config-if)#ip nat outside
R1(config)#interface gigabitEthernet0/0
R1(config-if)#ip nat inside
R1(config)#access-list 10 permit 172.16.1.0 0.0.0.255
R1(config)#ip nat inside source list 10 interface serial0/0/0 overload
```

（2）配置路由器 R2

```
R2(config)#ip route 61.0.0.0 255.255.255.0 serial0/0/1
```

（3）配置路由器 R3

```
R3(config)#ip route 202.96.134.0 255.255.255.0 serial0/0/1
```

（4）配置路由器 R4

```
R4(config)#interface tunnel 0
R4(config-if)#tunnel source serial0/0/0
R4(config-if)#tunnel destination 202.96.134.1
R4(config-if)#tunnel mode gre ip
R4(config-if)#ip address 172.16.14.4 255.255.255.0
R4(config-if)#tunnel key 123456
```

```
R4(config)#ip route 0.0.0.0 0.0.0.0 serial0/0/0
R4(config)#router eigrp 1
R4(config-router)#network 172.16.4.4 0.0.0.0
R4(config-router)#network 172.16.14.4 0.0.0.0
R4(config-router)#passive-interface gigabitEthernet0/0
R4(config)#interface serial0/0/0
R4(config-if)#ip nat outside
R4(config)#interface gigabitEthernet0/0
R4(config-if)#ip nat inside
R4(config)#access-list 10 permit 172.16.4.0 0.0.0.255
R4(config)#ip nat inside source list 10 interface serial0/0/0 overload
```

4．实验调试

（1）查看隧道接口信息

```
R1#show interfaces tunnel 0
Tunnel0 is up, line protocol is up                   //隧道接口状态
  Hardware is Tunnel                                 //接口硬件是隧道
  Internet address is 172.16.14.1/24                 //隧道接口 IP 地址
  MTU 17912 bytes, BW 100 Kbit/sec, DLY 50000 usec,
     reliability 255/255, txload 1/255, rxload 1/255
  Encapsulation TUNNEL, loopback not set             //隧道封装
  Keepalive not set
  Tunnel linestate evaluation up
  Tunnel source 202.96.134.1 (Serial0/0/0), destination 61.0.0.4    //隧道源和目的地
  Tunnel Subblocks:
     src-track:
        Tunnel0 source tracking subblock associated with Serial0/0/0
        Set of tunnels with source Serial0/0/0, 1 member (includes iterators), on interface <OK>
        //以上四行是隧道源跟踪的情况
  Tunnel protocol/transport GRE/IP          //隧道协议为 GRE，传输协议为 IP
    Key 0x1E240, sequencing disabled        //隧道验证的 Key，123456 转换成 16 进制就是 1E240
    Checksumming of packets disabled
//以上两行表示序列号和校验和位为 0，即 GRE 包头没有相应的字段，Key 位为 1，即 GRE 包
头包含 Key
  Tunnel TTL 255 , Fast tunneling enabled   //隧道 TTL 值，启用快速建立隧道
  Tunnel transport MTU 1472 bytes    //GRE 会额外增加 24 字节开销，再加上 Key 选项的 4 字节，
一共 28 字节，所以 MTU=1500-24-4=1472
  (此处省略部分输出)
```

（2）查看 GRE 隧道的建立情况

```
R1#debug tunnel
Tunnel Interface debugging is on
00:41:59: Tunnel0: GRE/IP encapsulated 202.96.134.1->61.0.0.4 (linktype=7, len=88)
//此行显示 GRE 封装模式、封装的源地址和目的地址以及链路类型
00:42:03: Tunnel0: GRE/IP to classify 61.0.0.4->202.96.134.1 (len=88 type=0x800 ttl=252 tos=0xC0)
00:42:03: Tunnel0: GRE/IP to decaps 61.0.0.4->202.96.134.1 (len=88 ttl=252)
00:42:03: Tunnel0: GRE decapsulated IP packet (linktype=7, len=60)
//以上三行显示 GRE 封装模式、协议类型、TTL 值、链路类型以及解封装过程
```

(3) 查看 IP 数据包发送和接收信息

```
R1#debug ip packet detail
R1#ping 172.16.4.4 source 172.16.1.1 repeat 1
01:10:40: IP: tableid=0, s=172.16.1.1 (local), d=172.16.4.4 (Tunnel0), routed via FIB
01:10:40: IP: s=172.16.1.1 (local), d=172.16.4.4 (Tunnel0), len 100, sending
01:10:40:            ICMP type=8, code=0            //ICMP 的类型和代码，表示 echo 的 request 数据包
//以上三行显示了源地址为 172.16.1.1，目的地址为 172.16.4.4 的数据包匹配路由表的情况，确定出接口为 Tunnel0
01:10:40: IP: s=202.96.134.1 (Tunnel0), d=61.0.0.4 (Serial0/0), len 128, sending, proto=47
//用 Tunnel0 接口指定的源地址和目的地址对数据包进行 GRE 封装并发送，新的 IP 包的协议字段为 47
01:10:40: IP: tableid=0, s=61.0.0.4 (Serial0/0), d=202.96.134.1 (Serial0/0), routed via RIB
01:10:40: IP: s=61.0.0.4 (Serial0/0), d=202.96.134.1 (Serial0/0), len 128, rcvd 3, proto=47
//以上显示了收到 R4 发送的 GRE 封装的数据包
01:10:40: IP: tableid=0, s=172.16.4.4 (Tunnel0), d=172.16.1.1 (FastEthernet1/0), routed via RIB
01:10:40: IP: s=172.16.4.4 (Tunnel0), d=172.16.1.1, len 100, rcvd 4
01:10:40:            ICMP type=0, code=0            //ICMP 的类型和代码，表示 echo 的 reply 数据包
//以上显示了本地解 GRE 封装后数据包的情况
```

(4) 查看路由表

```
① R1#show ip route eigrp
     172.16.0.0/24 is subnetted, 3 subnets
D       172.16.4.0 [90/26882560] via 172.16.14.4, 00:03:37, Tunnel0
② R4#show ip route eigrp
     172.16.0.0/24 is subnetted, 3 subnets
D       172.16.1.0 [90/26882560] via 172.16.14.1, 00:04:07, Tunnel0
```

以上①和②输出表明两端互相学到内部网络的路由，路由的下一跳为隧道另一端的地址，出接口为隧道接口。

(5) 执行 ping 命令

```
① R1#ping 172.16.4.4 source gigabitEthernet0/0
Type escape sequence to abort.
Sending 5, 100-byte ICMP Echos to 172.16.4.4, timeout is 2 seconds:
Packet sent with a source address of 172.16.1.1
!!!!!
Success rate is 100 percent (5/5), round-trip min/avg/max = 52/52/52 ms
```

以上输出表明远程办公室已经可以和企业总部通信了。

```
② R1#ping 61.0.0.3 source gigabitEthernet0/0
Type escape sequence to abort.
Sending 5, 100-byte ICMP Echos to 61.0.0.3, timeout is 2 seconds:
Packet sent with a source address of 172.16.1.1
!!!!!
Success rate is 100 percent (5/5), round-trip min/avg/max = 28/28/32 ms
```

以上测试表明远程办公室和 Internet 的通信成功。

8.5.2 实验 4：配置 Site To Site VPN

1．实验目的

通过本实验可以掌握：
① Site to Site VPN 的概念。
② Site to Site VPN 的配置和调试方法。

2．实验拓扑

实验拓扑如图 8-11 所示。

3．实验步骤

当采用 GRE 封装数据包时，数据在 Internet 上传输时是不安全的，本实验要采用 IPSec VPN 解决该问题。Site to Site 是指把一个局域网和另一个局域网连接在一起，有时候也称为 LAN-to-LAN。实验中，路由器 R2 和 R3 模拟 Internet，R1 和 R4 连接到 Internet 上，路由器 R1 的 Gi0/0 接口模拟远程办公室所在的局域网，路由器 R4 的 Gi0/0 接口模拟企业总部所在的局域网。要将远程办公室和企业总部进行连接，实现两地网络安全连通。同时需要在路由器 R1 和 R4 上配置 NAT，使两地的网络也可以访问 Internet。

（1）配置路由器 R1

```
R1(config)#crypto isakmp policy 10          //创建一个 isakmp 策略，编号为 10
R1(config-isakmp)#encryption aes
//配置 isakmp 采用的加密算法，默认是 DES
R1(config-isakmp)#authentication pre-share
//配置 isakmp 采用的身份验证算法，这里采用预共享密钥。如果有 CA 服务器，也可以用 CA 进
行身份验证
R1(config-isakmp)#hash sha              //配置 isakmp 采用的 HASH 算法，默认是 SHA
R1(config-isakmp)#group 5               //配置 isakmp 采用的 DH 组，默认为组 1
R1(config-isakmp)#exit
R1(config)#crypto isakmp key cisco address 61.0.0.4
//配置对等体 61.0.0.4 的预共享密钥为 cisco，双方配置的密钥需要一致
R1(config)#crypto ipsec transform-set TRAN esp-aes esp-sha-hmac
//创建一个 IPSec 转换集，名称本地有效，但是双方路由器转换集参数要一致
R1(cfg-crypto-trans)# mode tunnel       //配置隧道的工作模式，默认就是 tunnel 模式
R1(cfg-crypto-trans)#exit
```

【技术要点】

① ISAKMP 策略可以有多个策略，双方路由器将采用编号最小、参数一致的策略，双方至少要有一个策略是一致的，否则协商失败。ISAKMP 工作端口为 UDP 500。

② DH 组可以选择 1，2 或 5，group1 的密钥长度为 768 比特，group2 的密钥长度为 1024 比特，group5 的密钥长度为 1536 比特。

③ 转换集有 ESP 封装、AH 封装、ESP+AH 封装 3 种方式，加密算法有 DES、3DES 和

AES，HASH 包括 MD5 和 SHA 算法。ESP 封装可以提供机密性、完整性、身份验证功能，而 AH 封装仅提供完整性、身份验证功能。实际中 AH 使用得较少。

```
R1(config)#ip access-list extended VPN
R1(config-ext-nacl)#permit ip 172.16.1.0 0.0.0.255 172.16.4.0 0.0.0.255
//定义 VPN 感兴趣流量，用来指明什么样的流量要通过 VPN 加密传输。注意：这里限定的是从远程办公室发出的到达企业总部的流量，对这部分流量进行加密，其他流量（例如到 Internet）不加密
R1(config)#crypto map MAP 10 ipsec-isakmp
//创建加密图，名为 MAP，10 为该加密图编号，名称和编号都本地有效，如果有多个编号，路由器将从小到大逐一匹配
R1(config-crypto-map)#set peer 61.0.0.4              //配置 VPN 对等体的地址
R1(config-crypto-map)#set transform-set TRAN         //配置转换集
R1(config-crypto-map)#match address VPN              //指明 VPN 感兴趣流量
R1(config-crypto-map)#reverse-route static
//配置反向路由注入，这样在路由器中 VPN 会话建立时将产生一条静态路由，static 关键字指明即使 VPN 会话没有建立起来，反向路由也要创建
R1(config-crypto-map)#exit
R1(config)#interface serial0/0/0
R1(config-if)#crypto map MAP                         //在接口下应用创建的加密图
R1(config-if)#ip nat outside
R1(config)#interface gigabitEthernet0/0
R1(config-if)#ip nat inside
R1(config)#access-list 100 deny ip 172.16.1.0 0.0.0.255 172.16.4.0 0.0.0.255
//在执行 NAT 时，排除 VPN 感兴趣流量
R1(config)#access-list 100 permit ip 172.16.1.0 0.0.0.255 any
R1(config)#ip nat inside source list 100 interface serial0/0/0 overload
R1(config)#ip route 0.0.0.0 0.0.0.0 serial 0/0/0
```

（2）配置路由器 R2

```
R2(config)#ip route 61.0.0.0 255.255.255.0 serial0/0/1
```

（3）配置路由器 R3

```
R3(config)#ip route 202.96.134.0 255.255.255.0 serial0/0/1
```

（4）配置路由器 R4

```
R4(config)#crypto isakmp policy 10
R4(config-isakmp)#encryption aes
R4(config-isakmp)#authentication pre-share
R4(config-isakmp)#hash sha
R4(config-isakmp)#group 5
R4(config-isakmp)#exit
R4(config)#crypto isakmp key cisco address 202.96.134.1
R4(config)#crypto ipsec transform-set TRAN esp-aes esp-sha-hmac
R4(cfg-crypto-trans)# mode tunnel
R4(cfg-crypto-trans)#exit
R4(config)#ip access-list extended VPN
R4(config-ext-nacl)#permit ip 172.16.4.0 0.0.0.255 172.16.1.0 0.0.0.255
R4(config)#crypto map MAP 10 ipsec-isakmp
```

```
R4(config-crypto-map)#set peer 202.96.134.1
R4(config-crypto-map)#set transform-set TRAN
R4(config-crypto-map)#reverse-route static
R4(config-crypto-map)#match address VPN
R4(config-crypto-map)#exit
R4(config)#interface serial0/0/0
R4(config-if)#crypto map MAP
R4(config-if)#ip nat outside
R4(config)#interface gigabitEthernet0/0
R4(config-if)#ip nat inside
R4(config)#access-list 100 deny ip 172.16.4.0 0.0.0.255 172.16.1.0 0.0.0.255
R4(config)#access-list 100 permit ip 172.16.4.0 0.0.0.255 any
R4(config)#ip nat inside source list 100 interface serial0/0/0 overload
R4(config)#ip route 0.0.0.0 0.0.0.0 serial 0/0/0
```

4．实验调试

（1）查看路由表

```
① R1#show ip route static
        172.16.0.0/24 is subnetted, 3 subnets, 2 masks
S          172.16.4.0 [1/0] via 61.0.0.4
S*      0.0.0.0/0 is directly connected, Serial0/0/0
② R4#show ip route static
S*      0.0.0.0/0 is directly connected, Serial0/0/0
        172.16.0.0/16 is variably subnetted, 3 subnets, 2 masks
S          172.16.1.0/24 [1/0] via 202.96.134.1
```

以上①和②输出表明路由器 R1 和 R2 上已经有静态路由存在了，即使 VPN 隧道还没有建立，该路由是通过反向路由注入添加到路由表中的，下一跳为 VPN 对端的公网 IP 地址。

（2）查看 IKE 第一阶段和第二阶段具体信息

```
R1#debug crypto isakmp
```

从远程办公室 ping 总部的网络，显示如下：

```
R1#ping 172.16.4.4 source gigabitEthernet0/0
Type escape sequence to abort.
Sending 5, 100-byte ICMP Echos to 172.16.4.4, timeout is 2 seconds:
Packet sent with a source address of 172.16.1.1
00:22:36: isakmp: received ke message (1/1)
00:22:36: isakmp:(0:0:N/A:0): SA request profile is (NULL)
00:22:36: isakmp: Created a peer struct for 61.0.0.4, peer port 500
//isakmp 对端地址和 UDP 端口
00:22:36: isakmp: New peer created peer = 0x645AD004 peer_handle = 0x80000002
00:22:36: isakmp: Locking peer struct 0x645AD004, IKE refcount 1 for isakmp_initiator
00:22:36: isakmp: local port 500, remote port 500            //本地和对端端口号
00:22:36: isakmp: set new node 0 to QM_IDLE
00:22:36: insert sa successfully sa = 652E3BF0
00:22:36: isakmp:(0:0:N/A:0):Can not start Aggressive mode, trying Main mode.    //尝试主模式
```

```
00:22:36: isakmp:(0:0:N/A:0):found peer pre-shared key matching 61.0.0.4        //预共享密钥匹配
00:22:36: isakmp:(0:0:N/A:0): constructed NAT-T vendor-07 ID           //NAT 穿越
00:22:36: isakmp:(0:0:N/A:0): constructed NAT-T vendor-03 ID
00:22:36: isakmp:(0:0:N/A:0): constructed NAT-T vendor-02 ID
00:22:36: isakmp:(0:0:N/A:0):Input = IKE_MESG_FROM_IPSEC, IKE_SA_REQ_MM
00:22:36: isakmp:(0:0:N/A:0):Old State = IKE_READY  New State = IKE_I_MM1       //第一个包
00:22:36: isakmp:(0:0:N/A:0): beginning Main Mode exchange
                                        //以下开始主模式交换，显示具体交换的 6 个包的过程
00:22:36: isakmp:(0:0:N/A:0): sending packet to 61.0.0.4 my_port 500 peer_port 500 (I) MM_NO_
STATE
00:22:36: isakmp (0:0): received packet from 61.0.0.4 dport 500 sport 500 Global (I) MM_NO_ STATE
00:22:36: isakmp:(0:0:N/A:0):Input = IKE_MESG_FROM_PEER, IKE_MM_EXCH
00:22:36: isakmp:(0:0:N/A:0):Old State = IKE_I_MM1  New State = IKE_I_MM2
00:22:36: isakmp:(0:0:N/A:0): processing SA payload. message ID = 0
00:22:36: isakmp:(0:0:N/A:0): processing vendor id payload
00:22:36: isakmp:(0:0:N/A:0): vendor ID seems Unity/DPD but major 245 mismatch
00:22:36: isakmp (0:0): vendor ID is NAT-T v7
00:22:36: isakmp:(0:0:N/A:0):found peer pre-shared key matching 61.0.0.4
00:22:36: isakmp:(0:0:N/A:0): local preshared key found
00:22:36: isakmp : Scanning profiles for xauth ...
00:22:36: isakmp:(0:0:N/A:0):Checking isakmp transform 1 against priority 10 policy
00:22:36: isakmp:           encryption DES-CBC
00:22:36: isakmp:           hash MD5
00:22:36: isakmp:           default group 2
Success rate is 80 percent (4/5), round-trip min/avg/max = 72/119/172 ms
00:22:36: isakmp:           auth pre-share
00:22:36: isakmp:           life type in seconds
00:22:36: isakmp:           life duration (VPI) of  0x0 0x1 0x51 0x80
00:22:36: isakmp:(0:0:N/A:0):atts are acceptable. Next payload is 0
00:22:36: isakmp:(0:1:SW:1): processing vendor id payload
00:22:36: isakmp:(0:1:SW:1): vendor ID seems Unity/DPD but major 245 mismatch
00:22:36: isakmp (0:134217729): vendor ID is NAT-T v7
00:22:36: isakmp:(0:1:SW:1):Input = IKE_MESG_INTERNAL, IKE_PROCESS_MAIN_MODE
00:22:36: isakmp:(0:1:SW:1):Old State = IKE_I_MM2  New State = IKE_I_MM2        //第二个包
00:22:36: isakmp:(0:1:SW:1): sending packet to 61.0.0.4 my_port 500 peer_port 500 (I) MM_SA_
SETUP
00:22:36: isakmp:(0:1:SW:1):Input = IKE_MESG_INTERNAL, IKE_PROCESS_COMPLETE
00:22:36: isakmp:(0:1:SW:1):Old State = IKE_I_MM2  New State = IKE_I_MM3        //第三个包
00:22:36: isakmp (0:134217729): received packet from 61.0.0.4 dport 500 sport 500 Global (I) MM_
SA_SETUP
00:22:36: isakmp:(0:1:SW:1):Input = IKE_MESG_FROM_PEER, IKE_MM_EXCH
00:22:36: isakmp:(0:1:SW:1):Old State = IKE_I_MM3  New State = IKE_I_MM4
00:22:36: isakmp:(0:1:SW:1): processing KE payload. message ID = 0
00:22:36: isakmp:(0:1:SW:1): processing NONCE payload. message ID = 0
00:22:36: isakmp:(0:1:SW:1):found peer pre-shared key matching 61.0.0.4
00:22:36: isakmp:(0:1:SW:1):SKEYID state generated
00:22:36: isakmp:(0:1:SW:1): processing vendor id payload
00:22:36: isakmp:(0:1:SW:1): vendor ID is Unity
00:22:36: isakmp:(0:1:SW:1): processing vendor id payload
00:22:36: isakmp:(0:1:SW:1): vendor ID is DPD
```

00:22:36: isakmp:(0:1:SW:1): processing vendor id payload
00:22:36: isakmp:(0:1:SW:1): speaking to another IOS box!
00:22:36: isakmp:(0:1:SW:1):Input = IKE_MESG_INTERNAL, IKE_PROCESS_MAIN_MODE
00:22:36: isakmp:(0:1:SW:1):Old State = IKE_I_MM4   New State = IKE_I_MM4   //第四个包
00:22:36: isakmp:(0:1:SW:1):Send initial contact
00:22:36: isakmp:(0:1:SW:1):SA is doing pre-shared key authentication using id type ID_IPV4_ADDR
00:22:37: isakmp (0:134217729): ID payload
    next-payload : 8
    type   : 1
    address   : 202.96.134.1
    protocol   : 17
    port   : 500
    length   : 12
00:22:37: isakmp:(0:1:SW:1):Total payload length: 12
00:22:37: isakmp:(0:1:SW:1): sending packet to 61.0.0.4 my_port 500 peer_port 500 (I) MM_KEY_EXCH
00:22:37: isakmp:(0:1:SW:1):Input = IKE_MESG_INTERNAL, IKE_PROCESS_COMPLETE
00:22:37: isakmp:(0:1:SW:1):Old State = IKE_I_MM4   New State = IKE_I_MM5   //第五个包
00:22:37: isakmp (0:134217729): received packet from 61.0.0.4 dport 500 sport 500 Global (I) MM_KEY_EXCH
00:22:37: Isakmp:(0:1:SW:1): processing ID payload. message ID = 0
00:22:37: isakmp (0:134217729): ID payload
    next-payload : 8
    type   : 1
    address   : 61.0.0.4
    protocol   : 17
    port   : 500
    length   : 12
00:22:37: isakmp:(0:1:SW:1)::: peer matches *none* of the profiles
00:22:37: isakmp:(0:1:SW:1): processing HASH payload. message ID = 0
00:22:37: isakmp:(0:1:SW:1):SA authentication status: authenticated
00:22:37: isakmp:(0:1:SW:1):SA has been authenticated with 61.0.0.4
00:22:37: isakmp: Trying to insert a peer 202.96.134.1/61.0.0.4/500/, and inserted successfully 645AD004.
00:22:37: isakmp:(0:1:SW:1):Input = IKE_MESG_FROM_PEER, IKE_MM_EXCH
00:22:37: isakmp:(0:1:SW:1):Old State = IKE_I_MM5   New State = IKE_I_MM6   //第六个包

00:22:37: isakmp:(0:1:SW:1):Input = IKE_MESG_INTERNAL, IKE_PROCESS_MAIN_MODE
00:22:37: isakmp:(0:1:SW:1):Old State = IKE_I_MM6   New State = IKE_I_MM6
00:22:37: isakmp:(0:1:SW:1):Input = IKE_MESG_INTERNAL, IKE_PROCESS_COMPLETE
00:22:37: isakmp:(0:1:SW:1):Old State = IKE_I_MM6   New State = IKE_P1_COMPLETE
//IKE 第一阶段完成
00:22:37: isakmp:(0:1:SW:1):**beginning Quick Mode exchange**, M-ID of -933112422
//开始快速模式交换
00:22:37: isakmp:(0:1:SW:1): sending packet to 61.0.0.4 my_port 500 peer_port 500 (I) QM_IDLE
00:22:37: isakmp:(0:1:SW:1):Node -933112422, Input = IKE_MESG_INTERNAL, IKE_INIT_QM
00:22:37: isakmp:(0:1:SW:1):Old State = IKE_QM_READY   New State = IKE_QM_I_QM1
00:22:37: isakmp:(0:1:SW:1):Input = IKE_MESG_INTERNAL, IKE_PHASE1_COMPLETE
00:22:37: isakmp:(0:1:SW:1):Old State = IKE_P1_COMPLETE   New State = IKE_P1_COMPLETE

```
00:22:37: isakmp (0:134217729): received packet from 61.0.0.4 dport 500 sport 500 Global (I)
QM_IDLE
     00:22:37: isakmp:(0:1:SW:1): processing HASH payload. message ID = -933112422
     00:22:37: isakmp:(0:1:SW:1): processing SA payload. message ID = -933112422
     00:22:37: isakmp:(0:1:SW:1):Checking IPSec proposal 1
     00:22:37: isakmp: transform 1, ESP_DES
     00:22:37: isakmp:     attributes in transform:
     00:22:37: isakmp:        encaps is 1 (Tunnel)
     00:22:37: isakmp:        SA life type in seconds
     00:22:37: isakmp:        SA life duration (basic) of 3600
     00:22:37: isakmp:        SA life type in kilobytes
     00:22:37: isakmp:        SA life duration (VPI) of  0x0 0x46 0x50 0x0
     00:22:37: isakmp:        authenticator is HMAC-MD5
     00:22:37: isakmp:(0:1:SW:1):atts are acceptable.
     00:22:37: isakmp:(0:1:SW:1): processing NONCE payload. message ID = -933112422
     00:22:37: isakmp:(0:1:SW:1): processing ID payload. message ID = -933112422
     00:22:37: isakmp:(0:1:SW:1): processing ID payload. message ID = -933112422
     00:22:37: isakmp: Locking peer struct 0x645AD004, IPSEC refcount 1 for for stuff_ke
     00:22:37: isakmp:(0:1:SW:1): Creating IPSec SAs
     00:22:37:         inbound SA from 61.0.0.4 to 202.96.134.1 (f/i)   0/ 0
            (proxy 172.16.4.0 to 172.16.1.0)
     00:22:37:         has spi 0x7E935377 and conn_id 0 and flags 2
     00:22:37:         lifetime of 3600 seconds
     00:22:37:         lifetime of 4608000 kilobytes
     00:22:37:         has client flags 0x0
     00:22:37:         outbound SA from 202.96.134.1 to 61.0.0.4 (f/i) 0/0
            (proxy 172.16.1.0 to 172.16.4.0)
     00:22:37:         has spi 2099998430 and conn_id 0 and flags A
     00:22:37:         lifetime of 3600 seconds
     00:22:37:         lifetime of 4608000 kilobytes
     00:22:37:         has client flags 0x0
     00:22:37: isakmp:(0:1:SW:1): sending packet to 61.0.0.4 my_port 500 peer_port 500 (I) QM_IDLE
     00:22:37: isakmp:(0:1:SW:1):deleting node -933112422 error FALSE reason "No Error"
     00:22:37: isakmp:(0:1:SW:1):Node -933112422, Input = IKE_MESG_FROM_PEER, IKE_QM_ EXCH
     00:22:37: isakmp:(0:1:SW:1):Old State = IKE_QM_I_QM1 New State = IKE_QM_PHASE2_COMPLETE
//IKE 第二阶段完成
     00:22:37: isakmp: Locking peer struct 0x645AD004, IPSEC refcount 2 for from create_transforms
     00:22:37: isakmp: Unlocking IPSEC struct 0x645AD004 from create_transforms, count 1
     00:23:27: isakmp:(0:1:SW:1):purging node -933112422
     Type escape sequence to abort.
     Sending 5, 100-byte ICMP Echos to 172.16.4.4, timeout is 2 seconds:
     Packet sent with a source address of 172.16.1.1
     !!!!!
     Success rate is 100 percent (5/5), round-trip min/avg/max = 72/72/72 ms
```

以上输出信息非常复杂，本节主要关注 IKE 第一阶段和第二阶段交换过程即可。

（3）查看活动的 IPSec VPN 会话信息

```
R1#show crypto engine connections active
Crypto Engine Connections    //加密引擎的连接
   ID     Type      Algorithm           Encrypt   Decrypt   LastSeqN   IP-Address
```

1001	IKE	SHA+AES	0	0	0 202.96.134.1
2001	IPsec	AES+SHA	0	19	19 202.96.134.1
2002	IPsec	AES+SHA	19	0	0 202.96.134.1

以上输出显示活动的 VPN 会话中的 IKE 和 IPSec 的基本情况，包括会话 ID、会话类型、加密和验证算法、加密和解密数据包数量、最后一个包的序号以及本地加密点的 IP 地址，其中 IPSec 的加密和解密是独立的会话，可以看到加密和解密了各 19 个数据包。

（4）查看 ISAKMP 策略信息

```
R1#show crypto isakmp policy
Global IKE policy                                //全局 IKE 策略
Protection suite of priority 10
        encryption algorithm:    AES - Advanced Encryption Standard (128 bit keys).
        //加密算法
        hash algorithm:          Secure Hash Standard              //HASH 算法
        authentication method:   Pre-Shared Key                    //验证方法
        Diffie-Hellman group:    #5 (1536 bit)                     //DH 组
        lifetime:                86400 seconds, no volume limit    //生存时间
```

（5）查看 IPSec 转换集的信息

```
R1#show crypto ipsec transform-set
Transform set TRAN: { esp-aes esp-sha-hmac  }    //配置的转换集名称以及加密和验证算法
   will negotiate = { Tunnel,  },                //工作模式为隧道模式
Transform set default: { esp-aes esp-sha-hmac  } //系统默认的转换集名称以及加密和验证算法
   will negotiate = { Transport,  },             //工作模式为传输模式
```

（6）查看加密图的信息

```
R1#show crypto map
Crypto Map "MAP" 10 ipsec-isakmp             //名为 MAP 的加密图，编号 10 的配置
        Peer = 61.0.0.4                      //VPN 对端地址
        Extended IP access list VPN          //VPN 感兴趣流量
           access-list VPN permit ip 172.16.1.0 0.0.0.255 172.16.4.0 0.0.0.255
        Current peer: 61.0.0.4               //当前 VPN 会话的对端 IP 地址
        Security association lifetime: 4608000 kilobytes/3600 seconds
                //生存时间，即多长时间或者传输了多少字节后重新建立会话，保证数据的安全
        Responder-Only (Y/N): N              //只作为 VPN 响应端
        PFS (Y/N): N                         //没有开启完美前向保密（Perfect Forward Secrecy）
        Mixed-mode : Disabled                //混合模式禁用
        Transform sets={
                TRAN:  { esp-aes esp-sha-hmac  },
                                //使用的转换集 TRAN，包括加密和验证算法
        }
        Reverse Route Injection Enabled      //启用反向路由注入
        Interfaces using crypto map MAP:     //加密图应用的接口
                Serial0/0/0
```

（7）查看 IPSec 会话的安全关联信息

```
R1#show crypto ipsec sa
```

```
interface: Serial0/0/0
    Crypto map tag: MAP, local addr 202.96.134.1        //加密图的名字及本地加密点的接口地址
   protected vrf: (none)
   local  ident (addr/mask/prot/port): (172.16.1.0/255.255.255.0/0/0)
   remote ident (addr/mask/prot/port): (172.16.4.0/255.255.255.0/0/0)
//以上两行显示触发建立 VPN 连接的感兴趣流量
   current_peer 61.0.0.4 port 500                       //当前 VPN 对端和 isakmp 工作端口
     PERMIT, flags={origin_is_acl,}                     //标记为 ACL 定义的流量开始触发 VPN
    #pkts encaps: 19, #pkts encrypt: 19, #pkts digest: 19
    #pkts decaps: 19, #pkts decrypt: 19, #pkts verify: 19
//以上两行是该接口的加密和解密数据包、验证数据包的数量统计
    #pkts compressed: 0, #pkts decompressed: 0
    #pkts not compressed: 0, #pkts compr. failed: 0
    #pkts not decompressed: 0, #pkts decompress failed: 0
    #send errors 1, #recv errors 0
   local crypto endpt.: 202.96.134.1, remote crypto endpt.: 61.0.0.4
//建立 VPN 连接的本地端点和远程端点
   plaintext mtu 1438, path mtu 1500, ip mtu 1500, ip mtu idb Serial0/0/0
   current outbound spi: 0x18100C04(403704836)          //当前出向 spi 值,与对端入向 spi 值相同
   inbound esp sas:                                     //入方向的 esp 安全关联集合
     spi: 0x9ABED570(2596197744)                        //入向 spi 值
       transform: esp-aes esp-sha-hmac ,                //转换集信息
       in use settings ={Tunnel, }                      //工作模式
       conn id: 2001, flow_id: NETGX:1, sibling_flags 80000046, crypto map: MAP
       //VPN 连接的 ID 及加密图,在重新建立 SA 时,连接 ID 自动加 1
       sa timing: remaining key lifetime (k/sec): (4437839/198)   //VPN 连接剩余的生存时间
       IV size: 16 bytes                                //初始化向量(Initialization Vector,IV)长度
       replay detection support: Y                      //支持重放保护
       Status: ACTIVE                                   //VPN 连接状态
   inbound ah sas:
//入方向的 ah 安全关联,由于没有使用 AH 封装,所以没有 AH 安全关联信息
   inbound pcp sas:

   outbound esp sas:                                    //出方向的 esp 安全关联集合
     spi: 0x18100C04(403704836)
       transform: esp-aes esp-sha-hmac ,
       in use settings ={Tunnel, }
       conn id: 2002, flow_id: NETGX:2, sibling_flags 80000046, crypto map: MAP
       sa timing: remaining key lifetime (k/sec): (4437839/198)
       IV size: 16 bytes
       replay detection support: Y
       Status: ACTIVE
   outbound ah sas:
   outbound pcp sas:
```

(8)查看建立 IPSec VPN 对端的信息

```
R1#show cry isakmp peers
Peer: 61.0.0.4 Port: 500 Local: 202.96.134.1        //IPSec VPN 对端 IP 地址、端口和本端 IP 地址
Phase1 id: 61.0.0.4                                 //IKE 第一阶段 id
```

（9）查看建立 IPSec VPN 的预共享密钥

R1#**show crypto isakmp key**
Keyring   Hostname/Address   Preshared Key
default   61.0.0.4   **cisco**

（10）通过 NAT 访问外网

① R1#**ping 218.30.1.2 source gigabitEthernet0/0**
Type escape sequence to abort.
Sending 5, 100-byte ICMP Echos to 218.30.1.2, timeout is 2 seconds:
Packet sent with a source address of 172.16.1.1
!!!!!
Success rate is 100 percent (5/5), round-trip min/avg/max = 1/1/4 ms
② R4#**ping 218.30.1.2 source gigabitEthernet0/0**
Type escape sequence to abort.
Sending 5, 100-byte ICMP Echos to 218.30.1.2, timeout is 2 seconds:
Packet sent with a source address of 172.16.4.4
!!!!!
Success rate is 100 percent (5/5), round-trip min/avg/max = 1/2/4 ms

以上测试表明远程办公室和总部都可以成功通过 NAT 访问外网。

【技术扩展】

IPSec 提供了端到端的 IP 通信的安全性。如果传输过程中经过 PAT 中间设备，就会带来问题。AH 设计的理念决定了 AH 协议不能穿越 PAT 设备。但是 ESP 协议穿越 PAT 设备时同样会带来问题。NAT 穿越（NAT Traversal，NAT-T）就是为解决这个问题而提出的。NAT-T 将 ESP 协议数据包封装到 UDP（目的端口号为 UDP 4500）包中，即在原 ESP 协议的 IP 包头后添加新的 UDP 包头。NAT-T 在 IKE 第一阶段开始探测网络路径中是否存在 PAT 设备，如果发现存在 PAT 设备，IKE 第二阶段会采用 NAT-T，NAT-T 默认是自动开启的，命令为 crypto ipsec nat-transparency udp-encapsulation，不用手工配置。需要注意的是 IPsec 只有在采用 ESP 隧道模式来封装数据时才能与 NAT-T 共存。

8.5.3 实验 5：配置 Remote VPN

1. 实验目的

通过本实验可以掌握：
① Remote VPN 的概念。
② 在路由器上配置 Remote VPN 的方法。
③ VPN Client 软件的使用方法。

2. 实验拓扑

配置 Remote VPN 实验拓扑如图 8-12 所示。

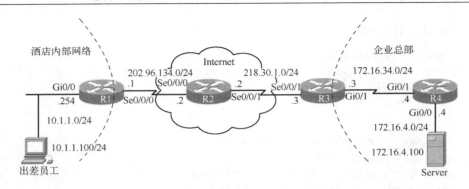

图 8-12 配置 Remote VPN 实验拓扑

3．实验步骤

Remote VPN 对客户端的配置要求较少，很多配置都在 VPN 网关上实现，因此 VPN 网关配置相对较为复杂。这里用路由器 R2 模拟 Internet，路由器 R1 模拟出差员工所在的酒店，酒店通常会采用 NAT 实现上网；路由器 R3 和 R4 模拟企业总部网络，要实现出差员工在酒店内能和企业总部的数据中心网络（172.16.4.0/24）通信，员工的笔记本电脑上要安装 VPN 客户端软件。

（1）配置路由器 R1

```
R1(config)#interface serial0/0/0
R1(config-if)#ip nat outside
R1(config)#interface gigabitEthernet0/0
R1(config-if)#ip nat inside
R1(config)#access-list 10 permit 10.1.1.0 0.0.0.255
R1(config)#ip nat inside source list 10 interface serial0/0/0 overload
R1(config)#ip route 0.0.0.0 0.0.0.0 serial0/0/0
```

（2）配置路由器 R2

由于用一台路由器模拟 Internet，所以，路由器 R2 除了接口地址，不需要其他的配置。

（3）配置路由器 R3

```
R3(config)#aaa new-model                                          //启用 AAA 功能
R3(config)#aaa authentication login CON none
//保护控制台接口，在控制台线性模式下，通过 login authentication CON 调用
R3(config)#aaa authentication login VPNUSERS local                //配置验证方式为本地
R3(config)#aaa authorization network VPN-REMOTE-ACCESS local      //配置网络授权方式为本地
R3(config)#username vpnuser secret cisco        //配置验证 Client 的用户名和密码
R3(config)#crypto isakmp policy 10
R3(config-isakmp)#encr 3des
R3(config-isakmp)#hash sha
R3(config-isakmp)#authentication pre-share
R3(config-isakmp)#group 2                   //如果客户端使用软件客户端，要选择 group 2
R3(config)#crypto ipsec transform-set VPNTRANSFORM esp-3des esp-sha-hmac
R3(config)#ip local pool REMOTE-POOL 172.16.100.1 172.16.100.250
//配置分配给 VPN 客户的 IP 地址池
```

```
R3(config)#ip access-list extended VPN
R3(config-ext-nacl)#permit ip 172.16.0.0 0.0.255.255 any
//定义的 ACL 向客户端指明只有发往该网络的数据包才进行加密，而其他流量（例如，访问酒店
内部或者 Internet 的流量）不要加密，该技术称为隧道分离（Split-Tunneling）。这里对企业总部的各个子网
进行了汇总
R3(config)#crypto isakmp client configuration group VPN-REMOTE-ACCESS
//创建用户组策略，要对该组的属性进行设置。每个连接上来的 VPN Client 都与一个用户组相关
联，如果没有配置特定组，但配置了默认组，用户将和默认组相关联
R3(config-isakmp-group)#key MYVPNKEY           //设置组的密码
R3(config-isakmp-group)#pool REMOTE-POOL        //配置分配给组用户的 IP 地址池
R3(config-isakmp-group)#save-password           //配置客户端允许用户保存组的密码
R3(config-isakmp-group)#acl VPN                 //指明隧道分离所使用的 ACL
R3(config)#crypto map CLIENTMAP isakmp authorization list VPN-REMOTE-ACCESS
//指明 isakmp 的授权方式
R3(config)#crypto dynamic-map DYNMAP 1
R3(config-crypto-map)#set transform-set VPNTRANSFORM
R3(config-crypto-map)#reverse-route
//以上三行创建了一个动态加密图，并指明了加密图的转换集和反向路由注入情况。加密图之所以
是动态的，是因为无法预知客户端的 IP 地址
R3(config)#crypto map CLIENTMAP client configuration address respond
//配置当用户请求 IP 地址时就响应地址请求
R3(config)#crypto map CLIENTMAP 65535 ipsec-isakmp dynamic DYNMAP
//将动态加密图应用于静态加密图中，因为接口下只能应用静态加密图
R3(config)#crypto map CLIENTMAP client authentication list VPNUSERS
//对用户进行验证
R3(config)#crypto isakmp xauth timeout 20       //设置验证超时时间
R3(config)#crypto isakmp keepalive 20
//定义 DPD（Dead Peer Detection）时间
```

【技术要点】

ISAKMP Keepalive 和 DPD 机制主要是用来检测当前 IPSec SA 的可用性，用来实现 IPSec VPN 的高可用性以及避免 IPSec SA 的黑洞。如果发送的 DPD 包对端没有回应就意味着当前的 IPSec SA 已经不可用了，这时 VPN 设备会清除掉 ISAKMP SA 和 IPSec SA，因此不会被动等待 IPSec SA 超时。建立 VPN 连接的两端都要配置 DPD。DPD 包发送机制有以下 2 种。

① 周期性（periodic）发送：周期性地发送 DPD 包，这种机制能够很快地发现问题，但是消耗 CPU 和带宽等资源较多。配置命令为 **crypto isakmp keepalive** *seconds* [*retries*] **periodic**。其中 *retries* 时间是可选配置，例如，DPD 发送时间是 10 秒，*retries* 时间是 2 秒，表示每 10 秒都应该收到邻居一个 DPD 包，但如果到了 10 秒还没收到邻居的 DPD 包，则不会再等 10 秒，而是会在 *retries* 的时间 2 秒后再向对方发送 DPD 包，默认连续发送 5 个 DPD 包，即 10 秒后就认为 VPN 连接失效。

② 按需（on-demand）发送：这是 DPD 包发送的默认机制，如果 VPN 连接正常，既加密和解密没问题，那么就不发送 DPD 包，但是如果对端 VPN 出现问题，这时就需要发送 DPD 包来查询对端的状态。配置命令为 **crypto isakmp keepalive** *seconds [retries]* **on-demand]**。

```
R3(config)#interface serial0/0/1
```

R3(config-if)#**crypto map CLIENTMAP**
R3(config)#**ip route 0.0.0.0 0.0.0.0 serial0/0/1**
R3(config)#**router eigrp 1**
R3(config-router)#**network 172.16.34.3 0.0.0.0**
R3(config-router)#**redistribute static**
//目的是把静态路由发布给 R4，静态路由包含了默认路由以及 VPN 客户连通后反向路由注入产生的主机路由

（4）配置路由器 R4

R4(config)#**router eigrp 1**
R4(config-router)#**network 172.16.4.4 0.0.0.0**
R4(config-router)#**network 172.16.34.4 0.0.0.0**

（5）VPN Client 软件配置

在 PC 上配置 IP 地址为 10.1.1.100/255.255.255.0，网关指向 10.1.1.254，测试能否 ping 通模拟 VPN 网关的 R3（218.30.1.3）。

Cisco 公司提供了 VPN Client 客户端软件，下载后安装即可，安装完毕后需要重启计算机。从【开始】→【Cisco Systems VPN Client】→【VPN Client】菜单中启动 VPN Client 程序，VPN Client 主窗口如图 8-13 所示。

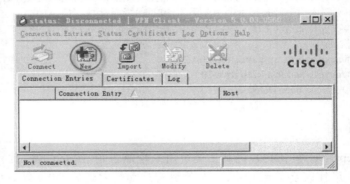

图 8-13　VPN Client 主窗口

单击【New】图标，如图 8-14 所示建立新的 VPN 连接。在【Connection Entry】文本框中输入连接的名字（名字自定，方便自己使用即可），在【Host】文本框中输入 VPN 网关的 IP 地址，在【Name】文本框中输入配置的组名，在【Password】文本框中输入密码（组的密码，这里为 MYVPNKEY，大小写敏感），在【Confirm Password】文本框中再次确认密码，然后单击【Save】按钮保存。

4．实验调试

（1）进行 Remote VPN 连接

在 VPN Client 主窗口双击刚创建的连接，在如图 8-15 所示窗口中输入建立 VPN 连接的用户名和密码（不要和组名、组密码混淆），单击【OK】按钮。

成功连接到 VPN 网关后，窗口会缩小为图标，该图标为一把已经锁上了的小锁，在屏幕

第 8 章 分 支 连 接

右下角的位置，如此行， ![icons] 所示。

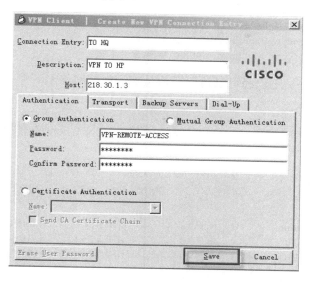

图 8-14 建立新的 VPN 连接

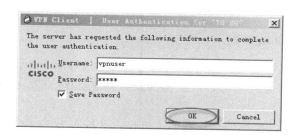

图 8-15 输入建立 VPN 连接的用户名和密码

在客户机上使用 **ipconfig** 命令应该可以看到获取了一个 IP 地址，如下：

```
C:\>ipconfig
Ethernet adapter 本地连接:                          //此连接为真实物理网卡
    Connection-specific DNS Suffix   . :
    IP Address. . . . . . . . . . . . : 10.1.1.100
    Subnet Mask . . . . . . . . . . . : 255.255.255.0
    Default Gateway . . . . . . . . . : 10.1.1.254
Ethernet adapter 本地连接 3:   //此连接为 Cisco Systems VPN 网卡，VPN 连接成功之后会自动启用
    Connection-specific DNS Suffix   . : cisco.com
    IP Address. . . . . . . . . . . . : 172.16.100.1
    Subnet Mask . . . . . . . . . . . : 255.255.0.0
    Default Gateway . . . . . . . . . :

C:\>ping 172.16.4.4
Pinging 172.16.4.4 with 32 bytes of data:
Reply from 172.16.4.4: bytes=32 time=40ms TTL=254
Reply from 172.16.4.4: bytes=32 time=35ms TTL=254
Reply from 172.16.4.4: bytes=32 time=49ms TTL=254
```

```
Reply from 172.16.4.4: bytes=32 time=47ms TTL=254
Ping statistics for 172.16.4.4:
    Packets: Sent = 4, Received = 4, Lost = 0 (0% loss),
Approximate round trip times in milli-seconds:
    Minimum = 35ms, Maximum = 49ms, Average = 42ms
```

以上输出表示出差员工已经可以访问公司总部的内部网络了。

（2）检查路由表

在 VPN 客户端上检查路由表，当 VPN 连通后，VPN Client 软件会增加到达企业总部的路由，如下：

```
C:\>route print
IPv4 Route Table
Active Routes:
Network Destination        Netmask          Gateway        Interface    Metric
        0.0.0.0            0.0.0.0         10.1.1.254      10.1.1.100      20
       10.1.1.0        255.255.255.0       10.1.1.100      10.1.1.100      20
      10.1.1.100       255.255.255.255      127.0.0.1       127.0.0.1      20
      172.16.0.0         255.255.0.0      172.16.100.3    172.16.100.3       1
     172.16.100.1     255.255.255.255       127.0.0.1       127.0.0.1      20
    172.16.255.255    255.255.255.255    172.16.100.3    172.16.100.3      20
（省略）
Default Gateway: 10.1.1.254
```

从 PC 的路由表中可以看出，只有发往公司总部的 172.16.0.0/16 的流量才从 VPN 网关 172.16.100.3 发出，而其他流量都是从正常接口（10.1.1.254）发出的，因为默认网关就是 10.1.1.254。

路由器 R3 上也多了一条指向客户端的主机路由（掩码为 32 位），该路由会被 EIGRP 重分布到路由器 R4。R3 和 R4 的路由表显示如下：

```
R3#show ip route | include D|S
D       172.16.4.0/24 [90/30720] via 172.16.34.4, 00:50:57, GigabitEthernet0/1
S       172.16.100.1/32 [1/0] via 202.96.134.1
S*      0.0.0.0/0 is directly connected, Serial0/0/1
R4#show ip route | include D
D EX    172.16.100.1/32 [170/20514560] via 172.16.34.3, 00:06:38, GigabitEthernet0/1
D*EX 0.0.0.0/0 [170/20514560] via 172.16.34.3, 00:51:07, GigabitEthernet0/1
```

（3）执行 telnet 命令

在 PC 上远程登录路由器 R4，具体如下：

```
C:\>telnet 172.16.4.4
```

在路由器 R4 上执行 who 命令，显示如下：

```
R4#who
    Line        User      Host(s)            Idle       Location
*  0 con 0                idle              00:00:00
 514 vty 0                idle              00:00:04   172.16.100.1
```

| Interface | User | Mode | Idle | Peer Address |

以上输出表明 VPN Client 是以 Server 动态分配的地址为源和 Server 的网段进行通信的，该地址可以在任意地址段，和 VPN Client 当前连接的网络（10.1.1.0/24）没有直接关系。

（4）在客户端查看 VPN 统计信息

双击屏幕右下角小锁，可以打开 VPN 主窗口，选择【Status】→【Statistics】菜单可以看到 VPN 统计信息，如图 8-16 所示。图上的信息很明了，不在这里赘述。本地路由和安全路由详细情况如图 8-17 所示，该图显示了网络的流量，其中本地流量（Local LAN 数量）数据不加密，而 VPN 流量（Secured Routes 流量）数据加密。

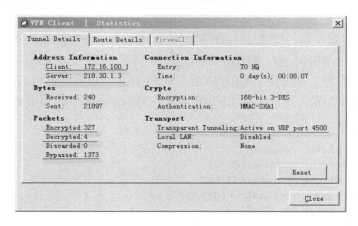

图 8-16　VPN 统计信息

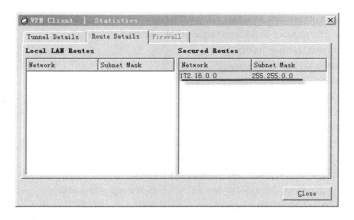

图 8-17　本地路由和安全路由详细情况

8.5.4　实验 6：配置 GRE Over IPSec

1．实验目的

通过本实验可以掌握：

① GRE 的工作原理和特征。
② IPSec 的工作过程和特征。
③ GRE Over IPSec 的工作过程和封装方法。
④ GRE Over IPSec 的应用场合。
⑤ GRE Over IPSec 的配置和调试方法。

2．实验拓扑

实验拓扑如图 8-11 所示。

3．实验步骤

在本章 8.5.1 实验 3 中，完成了 GRE 的配置，将远程办公室和企业总部进行连接，并且在远程办公室和企业总部间运行 EIGRP 路由协议。本实验通过 IPSec 实现路由协议 EIGRP 流量的保护。实现思想：首先建立 GRE 隧道，对 EIGRP 组播数据进行 GRE 封装，再对封装后的数据进行 IPSec 加密，从而实现 EIGRP 流量在 IPSec 隧道中的加密传输。

（1）配置路由器 R1

```
R1(config)#interface tunnel 0
R1(config-if)#tunnel source serial0/0/0
R1(config-if)#tunnel destination 61.0.0.4
R1(config-if)#ip address 172.16.14.1 255.255.255.0
R1(config-if)#tunnel key 123456
R1(config)#ip route 0.0.0.0 0.0.0.0 serial0/0/0
R1(config)#router eigrp 1
R1(config-router)#network 172.16.1.1 0.0.0.0
R1(config-router)#network 172.16.14.1 0.0.0.0
R1(config-router)#passive-interface GigabitEthernet0/0
R1(config)#crypto isakmp policy 10
R1(config-isakmp)#encryption aes
R1(config-isakmp)#authentication pre-share
R1(config-isakmp)#hash sha
R1(config-isakmp)#group 5
R1(config-isakmp)#lifetime 1800
//配置 SA 的生存周期，以秒为单位，默认值为 86400，两端的路由器都要设置相同值，否则 VPN
在正常初始化之后，将会在较短的一个 SA 生存周期到达时中断
R1(config)#crypto isakmp key cisco address 61.0.0.4
R1(config)#crypto ipsec transform-set TRAN esp-aes esp-sha-hmac
R4(cfg-crypto-trans)#mode transport        //工作模式为传输模式，由于 GRE 已经封装了原始数
据包，就不需要 IPSec 再去封装 GRE 添加新的 IP 包头了，这样数据包长度可以节省 20 字节
R1(config)#ip access-list extended VPN
R1(config-ext-nacl)#permit gre host 202.96.134.1 host 61.0.0.4
//IPSec 感兴趣流量为 GRE 流量
R1(config)#crypto map MAP 10 ipsec-isakmp
R1(config-crypto-map)#set peer 61.0.0.4
R1(config-crypto-map)#set transform-set TRAN
R1(config-crypto-map)#set pfs group5             //开启 PFS 功能
R1(config-crypto-map)#match address VPN
```

```
R1(config-crypto-map)#reverse-route static
R1(config)#interface serial0/0/0
R1(config-if)#crypto map MAP
R1(config-if)#ip nat outside
R1(config)#interface GigabitEthernet0/0
R1(config-if)#ip nat inside
R1(config)#access-list 100 deny ip 172.16.1.0 0.0.0.255 172.16.4.0 0.0.0.255
R1(config)#access-list 100 permit ip 172.16.1.0 0.0.0.255 any
R1(config)#ip nat inside source list 100 interface serial0/0/0 overload
```

（2）配置路由器 R2

```
R2(config)#ip route 61.0.0.0 255.255.255.0 serial0/0/1
```

（3）配置路由器 R3

```
R3(config)#ip route 202.96.134.0 255.255.255.0 serial0/0/1
```

（4）配置路由器 R4

```
R4(config)#interface tunnel 0
R4(config-if)#tunnel source serial0/0/0
R4(config-if)#tunnel destination 202.96.134.1
R4(config-if)#ip address 172.16.14.4 255.255.255.0
R4(config-if)#tunnel key 123456
R4(config)#ip route 0.0.0.0 0.0.0.0 serial0/0/0
R4(config)#router eigrp 1
R4(config-router)#network 172.16.4.4 0.0.0.0
R4(config-router)#network 172.16.14.4 0.0.0.0
R4(config-router)#passive-interface GigabitEthernet0/0
R4(config)#crypto isakmp policy 10
R4(config-isakmp)#encryption aes
R4(config-isakmp)#authentication pre-share
R4(config-isakmp)#hash sha
R4(config-isakmp)#group 5
R4(config-isakmp)#lifetime 1800
R4(config)#crypto isakmp key cisco address 202.96.134.1
R4(config)#crypto ipsec transform-set TRAN esp-aes esp-sha-hmac
R4(cfg-crypto-trans)#mode transport
R4(config)#ip access-list extended VPN
R4(config-ext-nacl)#permit gre host 61.0.0.4 host 202.96.134.1
R4(config)#crypto map MAP 10 ipsec-isakmp
R4(config-crypto-map)#set peer 202.96.134.1
R4(config-crypto-map)#set transform-set TRAN
R4(config-crypto-map)#set pfs group5
R4(config-crypto-map)#reverse-route static
R4(config-crypto-map)#match address VPN
R4(config)#interface serial0/0/0
R4(config-if)#crypto map MAP
R4(config-if)#ip nat outside
R4(config)#interface GigabitEthernet0/0
R4(config-if)#ip nat inside
```

```
R4(config)#access-list 100 deny ip 172.16.4.0 0.0.0.255 172.16.1.0 0.0.0.255
R4(config)#access-list 100 permit ip 172.16.4.0 0.0.0.255 any
R4(config)#ip nat inside source list 100 interface serial0/0/0 overload
```

4. 实验调试

（1）查看 IPSec 转换集信息

```
R1#show crypto ipsec transform-set
Transform set TRAN: { esp-aes esp-sha-hmac  }
   will negotiate = { Transport,  },                    //将协商传输模式
```

（2）查看 IPSec 安全关联信息

```
R1#show crypto ipsec sa
PFS (Y/N): Y, DH group: group5           //开启 PFS，DH 交换采用 group5
interface: Serial0/0/0
    Crypto map tag: MAP, local addr 202.96.134.1
   protected vrf: (none)
   local   ident (addr/mask/prot/port): (202.96.134.1/255.255.255.255/47/0)
   remote ident (addr/mask/prot/port): (61.0.0.4/255.255.255.255/47/0)
```
//以上两行显示触发建立 VPN 连接的流量为 GRE 流量，通过 EIGRP 的 Hello 数据包自动触发建立 VPN 连接
```
   current_peer 61.0.0.4 port 500
     PERMIT, flags={origin_is_acl,}
    #pkts encaps: 17, #pkts encrypt: 17, #pkts digest: 17
    #pkts decaps: 16, #pkts decrypt: 16, #pkts verify: 16
    #pkts compressed: 0, #pkts decompressed: 0
    #pkts not compressed: 0, #pkts compr. failed: 0
    #pkts not decompressed: 0, #pkts decompress failed: 0
    #send errors 43, #recv errors 0
     local crypto endpt.: 202.96.134.1, remote crypto endpt.: 61.0.0.4
     path mtu 1500, ip mtu 1500, ip mtu idb Serial0/0/0
     current outbound spi: 0xA26F62F4(2725208820)
     inbound esp sas:
      spi: 0xE5BC0C32(3854306354)
        transform: esp-aes esp-sha-hmac ,
        in use settings ={Transport, }     //工作模式为传输模式
        conn id: 2001, flow_id: NETGX:1, sibling_flags 80000006, crypto map: MAP
        sa timing: remaining key lifetime (k/sec): (4526519/3542)
        IV size: 16 bytes
        replay detection support: Y
        Status: ACTIVE
     inbound ah sas:
     inbound pcp sas:
     outbound esp sas:
      spi: 0xA26F62F4(2725208820)
        transform: esp-aes esp-sha-hmac ,
```

```
                in use settings ={Transport, }
                conn id: 2002, flow_id: NETGX:2, sibling_flags 80000006, crypto map: MAP
                sa timing: remaining key lifetime (k/sec): (4526519/3542)
                IV size: 16 bytes
                replay detection support: Y
                Status: ACTIVE
       outbound ah sas:
       outbound pcp sas:
```

【技术扩展】

使用传统的命令集，受到 IOS 版本的影响，在 GRE Over IPSec 实验中，即使在配置转换集时指定工作模式为传输模式，但是协商的结果也有可能是隧道模式，表示传输模式没有协商成功。为了解决这种问题，这里介绍 IOS 新的配置 GRE Over IPSec 命令集，此处只给出相对于上面实验而言有变化的部分：

```
R1(config)#crypto ipsec profile MYSEC              //配置 IP Sec profile
R1(ipsec-profile)#set transform-set TRAN           //在 IP Sec profile 中调用转换集
R1(config)#interface Tunnel0
R1(config-if)#tunnel protection ipsec profile MYSEC
//将 IPSec profile 关联到 GRE 接口，用于保护 GRE 接口的流量
```

从上面的配置看到，新的命令集不需要定义 MAP，也不需要在物理接口下调用 MAP，更不需要以 GRE 流量作为感兴趣流量，配置相对简单。路由器 R4 的配置步骤相同，这里不再给出。

（3）查看路由表

① R1#**show ip route eigrp**
```
         172.16.0.0/24 is subnetted, 3 subnets
    D    172.16.4.0 [90/26882560] via 172.16.14.4, 00:12:56, Tunnel0
```
② R4#**show ip route eigrp**
```
         172.16.0.0/24 is subnetted, 3 subnets
    D    172.16.1.0 [90/26882560] via 172.16.14.1, 00:13:25, Tunnel0
```

以上输出表明远程办公室和企业总部通过动态路由协议学到了相互的内部网络路由信息。

（4）查看 ISAKMP SA 的详细信息

```
R1#show crypto isakmp sa detail
Codes: C - IKE configuration mode, D - Dead Peer Detection
       K - Keepalives, N - NAT-traversal
       T - cTCP encapsulation, X - IKE Extended Authentication
       psk - Preshared key, rsig - RSA signature
       renc - RSA encryption
IPv4 Crypto isakmp sa
C-id   Local             Remote           I-VRF       Status Encr Hash Auth DH Lifetime Cap.
1005   202.96.134.1      61.0.0.4                     ACTIVE aes  sha  psk  5  00:07:07
       Engine-id:Conn-id =    SW:5
```

8.5.5 实验 7：配置 Redundancy VPN

1. 实验目的

通过本实验可以掌握：
① Redundancy VPN 的工作原理。
② HSRP 的工作原理。
③ Redundancy VPN 的应用场合。
④ Redundancy VPN 的配置和调试方法。

2. 实验拓扑

配置 Redundancy VPN 实验拓扑如图 8-18 所示。

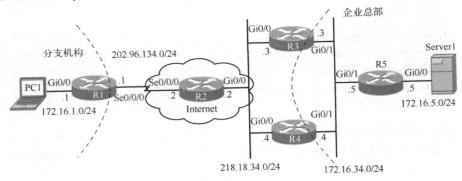

图 8-18　配置 Redundancy VPN 实验拓扑

3. 实验步骤

Redundancy VPN 是实现 VPN 高可用性（High Availability，HA）的方法之一，它利用了 HSRP 技术。本实验中，在路由器 R1 与路由器 R3 和 R4 之间配置 IPsec Site-to-Site VPN，其中 R3 和 R4 通过 HSRP 虚拟成单台 VPN 路由器，虚拟 IP 地址为 218.18.34.254，并以该虚拟 IP 地址为源和路由器 R1 建立 VPN 连接，在任何时候，R3 和 R4 中只在 HSRP 组中处于活动 (Active) 路由器负责与对端路由器 R1 建立 VPN 连接，对于路由器 R1 来讲，它并不知道与它建立 VPN 连接的设备到底是路由器 R3 还是 R4。R3 和 R4 同时还与内部路由器 R5 运行 EIGRP 路由协议，并且把在路由器 R3 和 R4 上反向路由注入的静态路由重分布进 EIGRP 进程中，将去往路由器 R1 的感兴趣流量的路由信息通告给路由器 R5。本实验要保护的是路由器 R1 上 172.16.1.0/24 和 R5 上 172.16.5.0/24 两个网络之间的流量。

（1）配置路由器 R1

```
R1(config)#crypto isakmp policy 10
R1(config-isakmp)#encr aes
R1(config-isakmp)#hash sha
R1(config-isakmp)#authentication pre-share
```

```
R1(config-isakmp)#group 5
R1(config)#crypto isakmp keepalive 10 periodic          //定义 DPD 时间
R1(config)#crypto isakmp key cisco address 218.18.34.254
R1(config)#crypto ipsec transform-set TRAN esp-aes esp-sha-hmac
R1(config)#crypto map MAP 10 ipsec-isakmp
R1(config-crypto-map)#set peer 218.18.34.254           //对端地址是 HSRP 的虚拟地址
R1(config-crypto-map)#set transform-set TRAN
R1(config-crypto-map)#set pfs group5
R1(config-crypto-map)#match address VPN
R1(config-crypto-map)#reverse-route static
R1(config)#ip access-list extended VPN
R1(config-ext-nacl)#permit ip 172.16.1.0 0.0.0.255 172.16.5.0 0.0.0.255
R1(config)#interface Serial0/0/0
R1(config-if)#crypto map MAP
R1(config)#ip route 0.0.0.0 0.0.0.0 Serial0/0/0
```

（2）配置路由器 R2

由于用一台路由器模拟 Internet，所以，路由器 R2 除了接口地址，不需要其他的配置。

（3）配置路由器 R3

```
R3(config)#crypto isakmp policy 10
R3(config-isakmp)#encr aes
R3(config-isakmp)#hash sha
R3(config-isakmp)#authentication pre-share
R3(config-isakmp)#group 5
R3(config)#crypto isakmp key cisco address 202.96.134.1
R3(config)#crypto isakmp keepalive 10 periodic
R3(config)#crypto ipsec transform-set TRAN esp-aes esp-sha-hmac
R3(config)#ip access-list extended VPN
R3(config-ext-nacl)#permit ip 172.16.5.0 0.0.0.255 172.16.1.0 0.0.0.255
R3(config)#crypto map MAP 10 ipsec-isakmp
R3(config-crypto-map)#set peer 202.96.134.1
R3(config-crypto-map)#set transform-set TRAN
R3(config-crypto-map)#set pfs group5
R3(config-crypto-map)#match address VPN
R3(config-crypto-map)#reverse-route static
```
//配置反向路由注入，需要注意的是在 Redundancy VPN 的环境下配置 RRI 时，即使配置了关键字 static，在 HSRP 中也只有处于活动的状态路由器才会产生反向路由注入

```
R3(config)#interface GigabitEthernet0/0
R3(config-if)#ip address 218.18.34.3 255.255.255.0
R3(config-if)#standby 1 ip 218.18.34.254
```
//启用 HSRP 功能，设置虚拟 IP 地址和组号，所有属于同一个 HSRP 组的路由器的虚拟地址必须一致

```
R3(config-if)#standby 1 priority 105
```
//配置 HSRP 的优先级，如果不设置该项，默认优先级为 100，该值越大抢占为活动路由器的优先权越高

```
R3(config-if)#standby 1 preempt
```
//配置抢占能力，即允许该路由器在优先级最高时成为活动路由器，默认为非抢占

```
R3(config-if)#standby 1 authentication md5 key-string cisco
```
//配置验证密码，防止非法设备加入到 HSRP 组中，同一个组的验证密码必须一致

```
R3(config-if)#standby 1 name VPNHA                    //配置 HSRP 的名字
R3(config-if)#standby 1 track GigabitEthernet0/1 30
//配置端口跟踪，如果该接口发生故障，优先级降低 30。应该选取合适的降低的值，使得路由器
R4 在抢占时能成为活动路由器
R3(config-if)#crypto map MAP redundancy VPNHA         //配置冗余加密图
R3(config)#access-list 10 permit 172.16.1.0
R3(config)#route-map REDIS permit 10
R3(config-route-map)#match ip address 10
R3(config)#router eigrp 1
R3(config-router)#redistribute static route-map REDIS
//重分布 RRI 路由到 EIGRP 进程中
R3(config-router)#network 172.16.34.3 0.0.0.0
R3(config)#ip route 202.96.134.0 255.255.255.0 218.18.34.2
```

（4）配置路由器 R4

```
R4(config)#crypto isakmp policy 10
R4(config-isakmp)#encr aes
R4(config-isakmp)#hash sha
R4(config-isakmp)#authentication pre-share
R4(config-isakmp)#group 5
R4(config)#crypto isakmp key cisco address 202.96.134.1
R4(config)#crypto isakmp keepalive 10 periodic
R4(config)#crypto ipsec transform-set TRAN esp-aes esp-sha-hmac
R4(config)#ip access-list extended VPN
R4(config-ext-nacl)#permit ip 172.16.5.0 0.0.0.255 172.16.1.0 0.0.0.255
R4(config)#crypto map MAP 10 ipsec-isakmp
R4(config-crypto-map)#set peer 202.96.134.1
R4(config-crypto-map)#set transform-set TRAN
R4(config-crypto-map)#set pfs group5
R4(config-crypto-map)#match address VPN
R4(config-crypto-map)#reverse-route static
R4(config)#interface GigabitEthernet0/0
R4(config-if)#ip address 218.18.34.4 255.255.255.0
R4(config-if)#standby 1 ip 218.18.34.254
R4(config-if)#standby 1 preempt
R4(config-if)#standby 1 authentication md5 key-string cisco
R4(config-if)#standby 1 name VPNHA
R4(config-if)#crypto map MAP redundancy VPNHA
R4(config)#access-list 10 permit 172.16.1.0
R4(config)#route-map REDIS permit 10
R4(config-route-map)#match ip address 10
R4(config)#router eigrp 1
R4(config-router)#redistribute static route-map REDIS
R4(config-router)#network 172.16.34.4 0.0.0.0
R4(config)#ip route 202.96.134.0 255.255.255.0 218.18.34.2
```

（5）配置路由器 R5

```
R5(config)#router eigrp 1
R5(config-router)#network 172.16.5.5 0.0.0.0
R5(config-router)#network 172.16.34.5 0.0.0.0
```

4．实验调试

（1）测试 VPN 的高可用性

本测试通过从路由器 R1 ping 多个数据包测试 VPN 的高可用性。操作步骤如下。

① 对 IP Sec VPN 感兴趣流在路由器 R1 上执行 ping 命令，即执行 **ping 172.16.5.5 source 172.16.1.1 repeat 2000** 命令。

② 过一会儿，把路由器 R3 的 Gi0/1 接口 **shutdown**。

③ 再过一会儿，把路由器 R3 的 Gi0/1 接口 **no shutdown**。

完成以上三个步骤后，在 R1 显示的 ping 操作的结果如下：

```
R1#ping 172.16.5.5 source 172.16.1.1 repeat 2000
Type escape sequence to abort.
Sending 2000, 100-byte ICMP Echos to 172.16.5.5, timeout is 2 seconds:
Packet sent with a source address of 172.16.1.1
!!!!!!!!!!!!!!!!!!!!!!!!!!!!!!!!!!!!!!!!!!!!!!!!!!!!
!!!!!!!!!!!!!!!!!!!!!!!!!!!!!!!!!!!!!!!!!!!!!!!!!!!!
!!!!!!!!!!!!!!!!!!!!!!!!!!!!!!!!!!!!!!!!!!!!!!!!!!!!
!!!!!!!!!!!!!!!!!!!!!!!!!!!!!!!!!!!!!!!........!!
//上述中断是因为路由器 R3 的 Gi0/1 接口被 shutdown，HSRP 切换，然后重新建立 VPN 连接
!!!!!!!!!!!!!!!!!!!!!!!!!!!!!!!!!!!!!!!!!!!!!!!!!!!!
!!!!!!!!!!!!!!!!!!!!!!!!!!!!!!!!!!!!!!!!!!!!!!!!!!!!
!!!!!!!!!!!!!!!!!!!!!!!!!!!!!!!!!!!!!!!!!!!!!!!!!!!!
!!!!!!!!!!!!!!!!!!!!!!!!!!!!!!!!!!!!!!!!!!!!!!!!!!!!
!!!!!!!!!!!!!!!!!!!!!!!!!!!!!!!!!!!!!!!!!!!!!!!!!!!!
!!!!!!!!!!!!!!!!!!!!!!!!!!!!!!!!..........
......!!!!!!!!!!!!!!!!!!!!!!!!!!!!!!!!!!!!!!!!!!!!
//上述中断是因为路由器 R3 的 Gi0/1 接口被 no shutdown，HSRP 切换，然后重新建立 VPN 连接
!!!!!!!!!!!!!!!!!!!!!!!!!!!!!!!!!!!!!!!!!!!!!!!!!!!!
!!!!!!!!!!!!!!!!!!!!!!!!!!!!!!!!!!!!!!!!!!!!!!!!!!!!
!!!!!!!!!!!!!!!!!!!!!!!!!!!!!!!!!!!!!!!!!!!!!!!!!!!!
```

【技术扩展】

以上测试结果显示，Redundancy VPN 并不完美，因为在 HSRP 主路由器 R3 不可用时，HSRP 备用路由器 R4 在接替主路由器的工作后，需要重新与 R1 建立 VPN 连接，这必定会造成用户的数据丢包。如果在任何时候，HSRP 主路由器和备用路由器同时存在 VPN 连接，那么在备用路由器接替主路由器的工作时就不需要再次重新建立 VPN 连接，就能直接使用备用路由器当前的 VPN 连接，这样就不会丢那么多数据包了，切换时间也会大大缩短，这才是真正的高可用性，实现了主路由器和备用路由器之间的无缝切换，并且对用户来说是完全透明的。VPN 的状态切换（Stateful Failover）就可以实现上述功能。更详细的信息和配置请读者自己查阅相关技术文档。

（2）查看路由表

① R5#**show ip route eigrp**

D EX   172.16.1.0 [170/30720] via 172.16.34.3, 00:00:57, GigabitEthernet0/1

以上输出的路由信息表明路由条目 172.16.1.0 是通过路由器 R3 学到的，因为 R3 与 R1 建立了 VPN 连接，通过反向路由注入的路由条目 172.16.1.0 再经过重分布进入 EIGRP，被路由器 R5 学到。

② R5#**show ip route eigrp**
　　172.16.0.0/24 is subnetted, 4 subnets
D EX   172.16.1.0 [170/30720] via 172.16.34.4, 00:00:02, GigabitEthernet0/1

以上输出的路由信息表明路由条目 172.16.1.0 是通过路由器 R4 学到的，因为路由器 R3 的 Gi0/1 接口被 shutdown，路由器 R4 成为 HSRP 活动路由器，R4 与 R1 建立了 VPN 连接，通过反向路由注入的路由条目 172.16.1.0 再经过重分布进入 EIGRP 网络，被路由器 R5 学到。

（3）查看活动的 IPSec VPN 会话信息

R1#**show crypto engine connections active**
Crypto Engine Connections
　ID　Type　Algorithm　　　　Encrypt　Decrypt　IP-Address
　1004　IKE　　SHA+AES　　　　0　　　　0　　　218.18.34.254
　2007　IPsec　AES+SHA　　　　0　　　　368　　218.18.34.254
　2008　IPsec　AES+SHA　　　　368　　　0　　　218.18.34.254

以上输出表明路由器 R3 是用 HSRP 的虚拟地址建立 VPN 连接的，如果路由器 R4 成为 HSRP 活动路由器，也同样会用 HSRP 的虚拟地址建立 VPN 连接。

（4）查看 HSRP 摘要信息

① R3#**show standby brief**
　　　　　　　　　　P indicates configured to preempt.
　　　　　　　　　　　|
Interface　Grp　Pri P State　Active　　　　Standby　　　　Virtual IP
Fa0/0　　　1　　105 P **Active**　local　　　　218.18.34.4　　　218.18.34.254

以上输出表明路由器 R3 是活动路由器。

② R3#**show standby brief**
　　　　　　　　　　P indicates configured to preempt.
　　　　　　　　　　　|
Interface　Grp　Pri　P　State　　Active　　　　Standby　　Virtual IP
Fa0/0　　　1　　75　P　**Standby**　218.18.34.4　　local　　　218.18.34.254

以上输出表明当路由器 R3 的 Gi0/1 接口被 shutdown 后，HSRP 优先级降为 75，路由器 R3 成为 HSRP Standby 路由器，R4 抢占成为 HSRP 活动路由器。

8.5.6　实验 8：配置 DMVPN

1．实验目的

通过本实验可以掌握：
① mGRE 的工作原理。

② NHRP 的工作原理。
③ DMVPN 的工作原理和应用场合。
④ DMVPN 的配置和调试方法。

2．实验拓扑

配置 DMVPN 实验拓扑如图 8-19 所示。

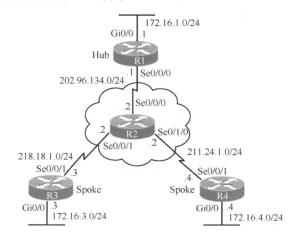

图 8-19　配置 DMVPN 实验拓扑

3．实验步骤

出于对经济的考虑，目前大多数企业的网络都采用星形结构（Hub-and-Spoke）组网。在需要建立 IPSec 隧道并运行动态路由协议的需求下，通过 GRE Over IPSec 可以解决，但是必须知道对等体 IP 地址，在实际案例中，Spoke 端的 IP 地址通常是动态从 ISP 获得的，而且 GRE Over IPSec 只可以建立点到点的隧道，随着 Spoke 的增加，配置会愈加烦琐，且不便于维护和排错。DMVPN（Dynamic Multipoint VPN，动态多点 VPN）可以很好地解决上面的问题，Hub 端不需要更改配置，为 Hub 和 Spoke 提供了全互连（Full Mesh）的 VPN 连通性，更好地实现大型网络的扩展。DMVPN 是 Cisco 私有协议，是通过多点 GRE（mGRE）和下一跳解析协议（Next Hop Resolution Protocol，NHRP）与 IPSec 相结合实现的。在 DMVPN 解决方案中，利用 IPSec 实现加密功能，利用多点 GRE（mGRE）建立隧道，利用 NHRP 解决分支节点的动态地址问题。NHRP 采用客户端/服务器模式工作，Hub 是服务器，Spoke 是客户端。本实验中，在路由器 R1、R3 和 R4 之间配置 DMVPN，其中 R1 为 Hub，模拟总部，R3 和 R4 为 Spoke，模拟分支机构，R2 模拟 Internet。路由器 R1、R3 和 R4 之间运行 EIGRP 路由协议，解决内网间路由连通性问题。R1 和 R3 以及 R1 和 R4 之间建立 Hub-to-Spoke 永久隧道，R3 和 R4 之间按需建立 Spoke-to-Spoke 临时隧道。

（1）配置路由器 R1

```
R1(config)#crypto isakmp policy 10
R1(config-isakmp)# encryption 3des
R1(config-isakmp)#hash md5
```

```
R1(config-isakmp)#authentication pre-share
R1(config-isakmp)#group 2
R1(config)#crypto isakmp key cisco address 0.0.0.0
//由于 Spoke 端的 IP 地址未知，有可能通过 ISP 动态获得，所以采用动态方式，Spoke 端 IP 为 0.0.0.0
R1(config)#crypto ipsec transform-set TRAN esp-3des esp-sha-hmac
R1(cfg-crypto-trans)#mode transport            //工作模式为传输模式
R1(config)#crypto ipsec profile PRO            //配置 ipsec profile
R1(ipsec-profile)#set transform-set TRAN       //在 ipsec profile 中调用转换集
R1(config)#interface Tunnel0                   //创建用于 DMVPN 的 mGRE 隧道接口
R1(config-if)#ip address 192.168.134.1 255.255.255.0
R1(config-if)#ip mtu 1440                      //配置隧道接口可以发送 IP 数据包的 MTU，以字节为
单位，Cisco 推荐为 1440 或 1436
R1(config-if)#no ip next-hop-self eigrp 1
    //由于是 Hub-and-Spoke 拓扑结构，在 Spoke 之间通信时，默认是通过 Hub 来转发的，配置此命
令，Spoke 之间的路由下一跳直接指向相应的 Spoke 的 Tunnel 接口的 IP 地址
R1(config-if)#ip nhrp authentication cisco123
    //配置 NHRP 验证密码，同一个 mGRE 中所有点（包括 Hub 和所有 Spoke）的密码必须相同
R1(config-if)#ip nhrp map multicast dynamic
    //启用 Hub 路由器上自动添加 Spoke 路由器到 NHRP 映射中，用来实现动态 VPN 连接
R1(config-if)#ip nhrp network-id 100
    //配置 NHRP 网络标识号，标识 DMVPN 所连接的 NBMA 网络，在接口下激活 NHRP，范围为 1～
4294967295，同一个 mGRE 中所有点的标识号必须相同
R1(config-if)#ip nhrp holdtime 300
    //配置 NHRP 维持时间，表示改变 NHRP NBMA 地址可以在多长时间内被通告才算有效，单位是
秒，默认为 7200 秒，范围 1～65535，NHRP 的 Spoke 端将以维持时间的 1/3 为周期向 Hub 端注册
R1(config-if)#no ip split-horizon eigrp 1      //关闭 EIGRP 水平分割
R1(config-if)#tunnel source Serial0/0/0        //指定隧道源
R1(config-if)#tunnel mode gre multipoint       //配置接口采用 mGRE 封装模式，这种模式不需
要配置建立隧道的目的 IP 地址
R1(config-if)#tunnel key 123456
    //配置隧道验证 key，范围为 0～4294967295，同一个 mGRE 中所有点的 key 必须相同
R1(config-if)#tunnel protection ipsec profile PRO
    //将 ipsec profile 关联到 mGRE 接口，用于保护 mGRE 接口的流量
R1(config)#ip route 0.0.0.0 0.0.0.0 202.96.134.2
R1(config)#router eigrp 1
R1(config-router)#network 172.16.1.1 0.0.0.0
R1(config-router)#network 192.168.134.1 0.0.0.0
R1(config-router)#passive-interface gigabitEthernet0/0
```

（2）配置路由器 R2

由于用路由器 R2 模拟 Internet，所以除了接口地址，不需要其他的配置。

（3）配置路由器 R3

```
R3(config)#crypto isakmp policy 10
R3(config-isakmp)#encryption 3des
R3(config-isakmp)#hash md5
R3(config-isakmp)#authentication pre-share
R3(config-isakmp)#group 2
R3(config)#crypto isakmp key cisco address 0.0.0.0
R3(config)#crypto ipsec transform-set TRAN esp-3des esp-sha-hmac
```

```
R3(cfg-crypto-trans)#mode transport
R3(config)#crypto ipsec profile PRO
R3(ipsec-profile)#set transform-set TRAN
R3(config)#interface Tunnel0
R3(config-if)#ip address 192.168.134.3 255.255.255.0
R3(config-if)#ip mtu 1440
R3(config-if)#ip nhrp authentication cisco123
R3(config-if)#ip nhrp map multicast 202.96.134.1
//开启能向 Hub 发送组播信息的功能,从而可以在 Spoke 和 Hub 之间激活动态路由选择协议
R3(config-if)#ip nhrp map 192.168.134.1 202.96.134.1
//配置 Hub 的 GRE 接口地址和公网 IP 地址静态映射
R3(config-if)#ip nhrp network-id 100
R3(config-if)#ip nhrp holdtime 300
R3(config-if)#ip nhrp nhs 192.168.134.1
//将 Hub 路由器指定为 NHRP 服务器,注意是虚拟 IP 地址
R3(config-if)#tunnel source Serial0/0/1
R3(config-if)#tunnel mode gre multipoint
R3(config-if)#tunnel key 123456
R3(config-if)#tunnel protection ipsec profile PRO
R3(config)#ip route 0.0.0.0 0.0.0.0 218.18.1.2
R3(config)#router eigrp 1
R3(config-router)#network 172.16.3.3 0.0.0.0
R3(config-router)#network 192.168.134.3 0.0.0.0
R3(config-router)#passive-interface gigabitEthernet0/0
```

(4)配置路由器 R4

```
R4(config)#crypto isakmp policy 10
R4(config-isakmp)#encryption 3des
R4(config-isakmp)#hash md5
R4(config-isakmp)#authentication pre-share
R4(config-isakmp)#group 2
R4(config)#crypto isakmp key cisco address 0.0.0.0
R4(config)#crypto ipsec transform-set TRAN esp-3des esp-sha-hmac
R4(cfg-crypto-trans)#mode transport
R4(config)#crypto ipsec profile PRO
R4(ipsec-profile)#set transform-set TRAN
R4(config)#interface Tunnel0
R4(config-if)#ip address 192.168.134.4 255.255.255.0
R4(config-if)#ip mtu 1440
R4(config-if)#ip nhrp authentication cisco123
R4(config-if)#ip nhrp map multicast 202.96.134.1
R4(config-if)#ip nhrp map 192.168.134.1 202.96.134.1
R4(config-if)#ip nhrp network-id 100
R4(config-if)#ip nhrp holdtime 300
R4(config-if)#ip nhrp nhs 192.168.134.1
R4(config-if)#tunnel source Serial0/0/1
R4(config-if)#tunnel mode gre multipoint
R4(config-if)#tunnel key 123456
R4(config-if)#tunnel protection ipsec profile PRO
R4(config)#ip route 0.0.0.0 0.0.0.0 211.24.1.2
```

```
R4(config)#router eigrp 1
R4(config-router)#network 172.16.4.4 0.0.0.0
R4(config-router)#network 192.168.134.4 0.0.0.0
R4(config-router)#passive-interface gigabitEthernet0/0
```

4．实验调试

（1）查看 NHRP 的缓存信息

① R1#**show ip nhrp**　//查看 NHRP 的缓存信息，可以通过执行 clear ip nhrp 命令清除 NHRP 的缓存内容

```
192.168.134.3/32 via 192.168.134.3
    Tunnel0 created 00:34:04, expire 00:04:15
    Type: dynamic, Flags: unique registered
    NBMA address: 218.18.1.3
192.168.134.4/32 via 192.168.134.4
    Tunnel0 created 00:33:30, expire 00:04:48
    Type: dynamic, Flags: unique registered used
    NBMA address: 211.24.1.4
```

以上输出显示 Hub 路由器 R1 上存在 Spoke 路由器 R3 和 R4 的 mGRE 接口地址和对应的公网地址（NBMA address）的映射，并且显示为动态映射，即 Spoke 自动注册完成。

② R3#**show ip nhrp**
```
192.168.134.1/32 via 192.168.134.1
    Tunnel0 created 00:36:22, never expire
    Type: static, Flags: used
    NBMA address: 202.96.134.1
```

以上输出显示 Spoke 路由器 R3 上存在到 Hub 路由器 R1 的 mGRE 接口地址和对应的公网地址的映射，并且显示为静态映射，即手工配置，所以该条目不过期。

（2）查看 NHRP 的信息统计情况

```
R1#show ip nhrp traffic
Tunnel0: Max-send limit:100Pkts/10Sec, Usage:0%          //最大发送限制
  Sent: Total 76
          0 Resolution Request   0 Resolution Reply   0 Registration Request
         76 Registration Reply   0 Purge Request      0 Purge Reply
          0 Error Indication     0 Traffic Indication
  Rcvd: Total 76
          0 Resolution Request   0 Resolution Reply   76 Registration Request
          0 Registration Reply   0 Purge Request       0 Purge Reply
          0 Error Indication     0 Traffic Indication
```

以上输出表明接口 Tunnel0 的 NHRP 的信息统计情况，其中，收到 NHRP 注册请求数据包 76 个，发送 NHRP 注册应答数据包 76 个。

（3）查看 NHRP 服务器的详细信息

```
R3#show ip nhrp nhs detail
Legend: E=Expecting replies, R=Responding
```

Tunnel0:192.168.134.1 RE   req-sent **27**   req-failed 1   repl-recv **27** (00:00:17 ago)

以上输出包括发送请求、请求失败以及收到应答的数据包数量。最近一次向 Hub 注册发生在 17 秒以前，因为维持时间配置为 300 秒，所以当该值到达 100（300/3）秒后清零并重新向 Hub 端发送注册请求。

（4）查看 IKE 对等体的信息

① R1#**show crypto isakmp peers**
Peer: 211.24.1.4 Port: **500** Local: 202.96.134.1
　Phase1 id: 211.24.1.4
Peer: 218.18.1.3 Port: **500** Local: 202.96.134.1
　Phase1 id: 218.18.1.3
② R3#**show crypto isakmp peers**
Peer: 202.96.134.1 Port: **500** Local: 218.18.1.3
　Phase1 id: 202.96.134.1
③ R4#**show crypto isakmp peers**
Peer: 202.96.134.1 Port: **500** Local: 211.24.1.4
　Phase1 id: 202.96.134.1

以上①、②和③输出信息说明路由器 R1 和 R3 以及路由器 R1 和 R4 已经使用公网地址建立 IKE 的对等体关系，也说明 Spoke 和 Hub 端自动建立了永久的 VPN 连接。

（5）查看 ISAKMP SA 的信息

R1#**show crypto isakmp sa**
IPv4 Crypto ISAKMP SA
dst　　　　　src　　　　　state　　　　conn-id status
202.96.134.1　211.24.1.4　**QM_IDLE**　　1002 **ACTIVE**
202.96.134.1　218.18.1.3　**QM_IDLE**　　1001 **ACTIVE**

（6）查看活动的 IPSec VPN 会话信息

show crypto engine connections active
① R1#**show crypto engine connections active**

ID	Type	Algorithm	Encrypt	Decrypt	IP-Address
1001	IKE	MD5+3DES	0	0	202.96.134.1
1002	IKE	MD5+3DES	0	0	202.96.134.1
2003	IPsec	3DES+SHA	0	351	202.96.134.1
2004	IPsec	3DES+SHA	351	0	202.96.134.1
2005	IPsec	3DES+SHA	0	336	202.96.134.1
2006	IPsec	3DES+SHA	353	0	202.96.134.1

② R3#**show crypto engine connections active**

ID	Type	Algorithm	Encrypt	Decrypt	IP-Address
1001	IKE	MD5+3DES	0	0	218.18.1.3
2001	IPsec	3DES+SHA	0	357	218.18.1.3
2002	IPsec	3DES+SHA	340	0	218.18.1.3

③ R4#**show crypto engine connections active**

ID	Type	Algorithm	Encrypt	Decrypt	IP-Address
1001	IKE	MD5+3DES	0	0	211.24.1.4
2003	IPsec	3DES+SHA	0	364	211.24.1.4

| | 2004 | IPsec | 3DES+SHA | | 365 | 0 | 211.24.1.4 |

以上①、②和③输出显示了活动 IPSec VPN 会话的基本信息，包括会话的 ID、类型、实现数据私密性和完整性的算法、加密和解密的数据包的数量等。

（7）查看 IPSec 会话的安全关联信息

```
R1#show crypto ipsec sa
    interface: Tunnel0
      Crypto map tag: Tunnel0-head-0, local addr 202.96.134.1
    protected vrf: (none)
    local   ident (addr/mask/prot/port): (202.96.134.1/255.255.255.255/47/0)
    remote ident (addr/mask/prot/port): (218.18.1.3/255.255.255.255/47/0)
    current_peer 218.18.1.3 port 500
      PERMIT, flags={origin_is_acl,}
     #pkts encaps: 532, #pkts encrypt: 532, #pkts digest: 532
     #pkts decaps: 531, #pkts decrypt: 531, #pkts verify: 531
     #pkts compressed: 0, #pkts decompressed: 0
     #pkts not compressed: 0, #pkts compr. failed: 0
     #pkts not decompressed: 0, #pkts decompress failed: 0
     #send errors 0, #recv errors 0
      local crypto endpt.: 202.96.134.1, remote crypto endpt.: 218.18.1.3
      path mtu 1500, ip mtu 1500, ip mtu idb Serial0/0/0
      current outbound spi: 0x5AE827AF(1525163951)
      inbound esp sas:
       spi: 0xB86C490C(3094104332)
         transform: esp-3des esp-sha-hmac ,
         in use settings ={Transport, }
         conn id: 2005, flow_id: NETGX:5, sibling_flags 80000006, crypto map: Tunnel0-head-0
         sa timing: remaining key lifetime (k/sec): (4501009/1297)
         IV size: 8 bytes
         replay detection support: Y
         Status: ACTIVE
      inbound ah sas:
      inbound pcp sas:
      outbound esp sas:
       spi: 0x5AE827AF(1525163951)
         transform: esp-3des esp-sha-hmac ,
         in use settings ={Transport, }
         conn id: 2006, flow_id: NETGX:6, sibling_flags 80000006, crypto map: Tunnel0-head-0
         sa timing: remaining key lifetime (k/sec): (4501006/1297)
         IV size: 8 bytes
         replay detection support: Y
         Status: ACTIVE
      outbound ah sas:
      outbound pcp sas:
    protected vrf: (none)
    local   ident (addr/mask/prot/port): (202.96.134.1/255.255.255.255/47/0)
    remote ident (addr/mask/prot/port): (211.24.1.4/255.255.255.255/47/0)
    current_peer 211.24.1.4 port 500
      PERMIT, flags={origin_is_acl,}
```

```
        #pkts encaps: 530, #pkts encrypt: 530, #pkts digest: 530
        #pkts decaps: 532, #pkts decrypt: 532, #pkts verify: 532
        #pkts compressed: 0, #pkts decompressed: 0
        #pkts not compressed: 0, #pkts compr. failed: 0
        #pkts not decompressed: 0, #pkts decompress failed: 0
        #send errors 1, #recv errors 0
          local crypto endpt.: 202.96.134.1, remote crypto endpt.: 211.24.1.4
          path mtu 1500, ip mtu 1500, ip mtu idb Serial0/0/0
          current outbound spi: 0xA9035A2C(2835569196)
          inbound esp sas:
            spi: 0xE3FF17EC(3825145836)
              transform: esp-3des esp-sha-hmac ,
              in use settings ={Transport, }
              conn id: 2003, flow_id: NETGX:3, sibling_flags 80000006, crypto map: Tunnel0-head-0
              sa timing: remaining key lifetime (k/sec): (4439552/1295)
              IV size: 8 bytes
              replay detection support: Y
              Status: ACTIVE
          inbound ah sas:
          inbound pcp sas:
          outbound esp sas:
            spi: 0xA9035A2C(2835569196)
              transform: esp-3des esp-sha-hmac ,
              in use settings ={Transport, }
              conn id: 2004, flow_id: NETGX:4, sibling_flags 80000006, crypto map: Tunnel0-head-0
              sa timing: remaining key lifetime (k/sec): (4439552/1295)
              IV size: 8 bytes
              replay detection support: Y
              Status: ACTIVE
          outbound ah sas:
          outbound pcp sas:
```

以上输出为路由器 R1 的 IPSec SA 信息，请注意重点标注的信息，具体的描述请参考前面的实验。

（8）查看 EIGRP 路由和邻居信息

```
    ① R1#show ip route eigrp
        172.16.0.0/24 is subnetted, 3 subnets
    D      172.16.4.0 [90/26882560] via 192.168.134.4, 00:49:57, Tunnel0
    D      172.16.3.0 [90/26882560] via 192.168.134.3, 00:49:55, Tunnel0
    ② R3#show ip route eigrp
        172.16.0.0/24 is subnetted, 3 subnets
    D      172.16.4.0 [90/28162560] via 192.168.134.4, 00:50:14, Tunnel0
    D      172.16.1.0 [90/26882560] via 192.168.134.1, 00:50:14, Tunnel0
    ③ R4#show ip route eigrp
        172.16.0.0/24 is subnetted, 3 subnets
172.16.0.0/24 is subnetted, 3 subnets
    D      172.16.1.0 [90/26882560] via 192.168.134.1, 00:50:37, Tunnel0
    D      172.16.3.0 [90/28162560] via 192.168.134.3, 00:50:35, Tunnel0
    ④ R3#show ip eigrp neighbors
```

```
IP-EIGRP neighbors for process 1
IP-EIGRP neighbors for process 1
H   Address              Interface    Hold    Uptime      SRTT    RTO     Q     Seq
                                              (sec)       (ms)            Cnt   Num
0   192.168.134.1        Tu0          13      00:51:29    40      1392    0     7
⑤ R4#show ip eigrp neighbors
IP-EIGRP neighbors for process 1
H   Address              Interface    Hold    Uptime      SRTT    RTO     Q     Seq
                                              (sec)       (ms)            Cnt   Num
0   192.168.134.1        Tu0          10      00:50:59    85      2088    0     6
```

由于在路由器 R1 上配置了 **no ip next-hop-self eigrp 1** 命令，让 Hub 路由器 R1 不再将从一个 Spoke 收到的路由信息发给另一个 Spoke 时将下一跳地址改为自己，而是保持原来的下一跳地址不变，即 Spoke 到 Spoke 的下一跳地址还是源 Spoke 的而不是 Hub 路由器的，从而实现 Spoke 与 Spoke 之间的通信不需要通过 Hub 路由器中转。所以 R3 和 R4 的 EIGRP 邻居虽然都是 R1，但是 R3 和 R4 学到的内网的路由条目的下一跳地址都是对端的 IP 地址，而不是 R1 的 IP 地址。

（9）实现 Spoke 到 Spoke 的通信

```
① R4#ping 172.16.3.3 source 172.16.4.4
Type escape sequence to abort.
Sending 5, 100-byte ICMP Echos to 172.16.3.3, timeout is 2 seconds:
Packet sent with a source address of 172.16.4.4
!!!!!
Success rate is 100 percent (5/5), round-trip min/avg/max = 240/339/484 ms
② R3#show crypto engine connections active
Crypto Engine Connections
  ID    Type    Algorithm       Encrypt     Decrypt     IP-Address
  1001  IKE     MD5+3DES        0           0           218.18.1.3
  1002  IKE     MD5+3DES        0           0           218.18.1.3
  1003  IKE     MD5+3DES        0           0           218.18.1.3
  2001  IPsec   3DES+SHA        0           785         218.18.1.3
  2002  IPsec   3DES+SHA        767         0           218.18.1.3
  2007  IPsec   3DES+SHA        0           5           218.18.1.3
  2008  IPsec   3DES+SHA        5           0           218.18.1.3
③ R4#show crypto engine connections active
Crypto Engine Connections
  ID    Type    Algorithm       Encrypt     Decrypt     IP-Address
  1001  IKE     MD5+3DES        0           0           211.24.1.4
  1002  IKE     MD5+3DES        0           0           211.24.1.4
  1003  IKE     MD5+3DES        0           0           211.24.1.4
  2003  IPsec   3DES+SHA        0           779         211.24.1.4
  2004  IPsec   3DES+SHA        779         0           211.24.1.4
  2007  IPsec   3DES+SHA        0           5           211.24.1.4
  2008  IPsec   3DES+SHA        5           0           211.24.1.4
④ R4#show ip nhrp detail
192.168.134.1/32 via 192.168.134.1
     Tunnel0 created 00:57:55, never expire
     Type: static, Flags: used
```

```
            NBMA address: 202.96.134.1
        192.168.134.3/32 via 192.168.134.3
            Tunnel0 created 00:03:21, expire 00:01:38
            Type: dynamic, Flags: router
            NBMA address: 218.18.1.3
        192.168.134.4/32 via 192.168.134.4
            Tunnel0 created 00:03:21, expire 00:01:38
            Type: dynamic, Flags: router unique local
            NBMA address: 211.24.1.4
            (no-socket)
        Requester: 192.168.134.3 Request ID: 2
```

以上①、②、③和④输出信息表明了在 Spoke 到 Spoke 的临时 VPN 连接建立后，R3 和 R4 的 ISAKMP SA、NHRP 映射以及活动 VPN 连接的变化情况。

【提示】

如果上面实验内网之间运行的是 OSPF 路由协议，要注意 mGRE 接口的 OSPF 网络类型，因为 mGRE 接口默认 OSPF 网络类型为 "point-to-point"，现在要和两个路由器建立 OSPF 邻居关系就出现了问题，因此必须改变 mGRE 接口的 OSPF 网络类型为 "point-to-multipoint"。

本 章 小 结

随着企业规模不断扩大以及管理成本和环保的需要，企业总部、分支机构、小型办公室和家庭办公室（SOHO）以及其他远程工作人员需要连接在一起。因此在设计网络架构时，必须考虑连接性、成本、安全性和可用性等方面的问题。本章介绍了 ADSL 技术、GRE 技术以及 IPSec VPN 技术，并用实验演示和验证了 ADSL 连接到 Internet 的实现方法、GRE 隧道配置、Site To Site VPN 配置、GRE Over IPSec 配置、Redundancy VPN 配置和 DMVPN 配置等内容。

第 9 章 IPv6

　　IPv4 的设计思想成功地造就了目前的国际互联网，其核心价值体现在简单、灵活和开放性上。但随着新应用的不断涌现以及接入互联网用户的增加，传统的 IPv4 协议已经难以支持互联网的进一步扩张和新业务的开展，比如端到端的应用、实时应用和服务质量保证等。无论是 NAT，还是 CIDR 等技术都是缓解 IP 地址短缺的手段，而 IPv6 才是解决地址短缺的最终方法，IPv6 能够解决 IPv4 存在的许多问题，如地址短缺和服务质量等。IPv6 是由 IETF 设计的下一代互联网协议，在 RFC 2460 中定义，目的是取代现有的互联网协议 IPv4。

9.1 IPv6 概述

9.1.1 IPv6 特点

　　面对 IPv4 地址的枯竭、越来越庞大的 Internet 路由表和缺乏端到端 QoS 保证等缺点，IPv6（Internet Protocol Version 6）的实施是必然的趋势。IPv6 对 IPv4 进行了大量的改进，其主要特征如下：

　　① 128 比特的地址方案（3.4×10^{38} 个地址）提供足够大的地址空间，充足的地址空间将极大地满足网络智能设备（如个人数字助理、移动电话、家庭网络接入设备、智能游戏终端、安保监控设备和 IPTV 等）对 IP 地址增长的需求。

　　② 多等级编址层次有助于路由聚合，提高了路由选择的效率。

　　③ 无须网络地址转换（NAT），实现端到端的通信更加便捷。

　　④ IPv6 地址自动配置功能支持即插即用，使得在 Internet 上大规模部署新设备，如 IOT (Internet of Things) 设备成为可能。IPv6 支持有状态和无状态两种地址自动配置方式。

　　⑤ IPv6 中没有广播地址，它的功能被组播地址所代替，ARP 广播被本地链路组播代替。

　　⑥ IPv6 对数据包头进行了简化，不需要处理校验和，因此减少处理器开销并节省网络带宽，也有助于提高网络设备性能和转发效率。

　　⑦ IPv6 中流标签字段使得设备无须查看传输层信息就可以实现流量区分，因此可以提供更好的 QoS 保障。

　　⑧ IPv6 协议内置移动性和安全性。移动性让设备在不中断网络连接的情况下在网络中移动。IPv6 将 IPSec 作为标准配置，使得所有终端的通信安全都能得到保证，实现端到端的安全通信。

　　⑨ 在 IPv6 中引入了扩展包头的概念，用扩展包头代替了 IPv4 包头中存在的可变长度的选项，进一步提高了路由性能和效率。

9.1.2 IPv6 地址与基本包头格式

IPv4 地址采用点分十进制格式表示，而 IPv6 地址采用冒号分十六进制格式表示。IPv6 由网络前缀和接口 ID 两部分组成。IPv6 使用"IPv6 地址 / 前缀长度"的格式表示 IPv6 地址的网络部分，前缀长度范围为 0～128。典型 IPv6 前缀长度为/64。以下是一个 IPv6 地址的例子：

2020:00D3:0000:0000:02BB:00FE:0000:2019 是一个完整的 IPv6 地址。从这个例子可以看到手工配置和管理 IPv6 地址的难度，也看到了自动配置 IPv6 地址、网关和 DNS 的必要性。

如下规则可以简化 IPv6 地址的表示：

① IPv6 地址中每个 16 位分组中的前导零位可以被去除。

② 可以将冒号十六进制格式中相邻的连续零位合并，用双冒号"::"表示，但是"::"在一个 IPv6 地址中只能出现一次。通过上述两条规则，上述的 IPv6 地址可以简化为 2020:D3::2BB:FE:0:2019。

IPv6 数据包基本包头长度固定为 40 字节，其格式如图 9-1 所示，各字段的含义如下。

0	7|8	15|16	23|24	31
版本	流量类型	流标签		
有效载荷长度		下一包头	跳数限制	
源IPv6地址				
目的IPv6地址				

图 9-1 IPv6 数据包基本包头格式

① 版本（4 比特）：对于 IPv6，该字段的值为 6。

② 流量类型（8 比特）：该字段以 DSCP（Differentiated Services Code Point，区分服务编码点）标记一个 IPv6 数据包，以此指明数据包应当如何处理，提供 QoS 服务。

③ 流标签（20 比特）：在 IPv6 协议中，该字段是新增加的，用来标记 IPv6 数据的一个流，让路由器或者交换机基于流而不是数据包来处理数据，该字段也可用于 QoS。

④ 有效载荷长度（16 比特）：该字段标识有效载荷的长度，所谓有效载荷指的是紧跟 IPv6 包头的数据包其他部分。

⑤ 下一包头（8 比特）：该字段定义紧跟 IPv6 基本包头的信息类型，信息类型可能是高层协议，如 TCP 或 UDP，也可能是一个新增的可扩展包头。

⑥ 跳数限制（8 比特）：该字段定义了 IPv6 数据包所经过的最大跳数。

⑦ 源 IPv6 地址（128 比特）：该字段标识发送方的 IPv6 源地址。

⑧ 目的 IPv6 地址（128 比特）：该字段标识 IPv6 数据包的目的地址。

图 9-2 是通过 Wireshark 软件抓取的 IPv6 数据包包头捕获信息，该数据包是两个节点用 ping 命令测试连通性的时候捕获的。其中，版本字段值为 0x06，流量类型字段值为 0x00，流标签字段值为 0x00000，有效载荷长度字段值为 1460（0x05b4），下一包头字段值为 58

（0x3a），跳数限制字段值为 63（0x3f），源 IPv6 地址字段值为 2012::1，目的 IPv6 地址字段值为 2014:4444::4。

```
Internet Protocol Version 6, Src: 2012::1, Dst: 2014:4444::4
    0110 .... = Version: 6
    .... 0000 0000 .... .... .... .... .... = Traffic class: 0x00 (DSCP: CS0, ECN: Not-ECT)
    .... .... .... 0000 0000 0000 0000 0000 = Flowlabel: 0x00000000
    Payload length: 1460
    Next header: ICMPv6 (58)
    Hop limit: 63
    Source: 2012::1
    Destination: 2014:4444::4
    [Source GeoIP: Unknown]
    [Destination GeoIP: Unknown]
```

图 9-2　IPv6 数据包包头捕获信息

IPv6 地址可以通过手工静态配置、EUI-64 格式自动配置、无状态地址自动配置和有状态自动配置方式获得。其中 EUI-64 格式自动配置功能是在接口的 MAC 地址中间插入固定的 FFFE 来生成 64 比特的 IPv6 地址的接口标识符（接口 ID），其工作过程如下：

① 在 48 比特的 MAC 地址的 OUI（前 24 比特）和序列号（后 24 比特）之间插入一个固定数值 FFFE，如 MAC 地址为 0050:3EE4:4C89，那么插入固定数值后的结果是 0050:3EFF:FEE4:4C89。

② 将上述结果的第一字节的第 7 比特（从左到右数）反转，因为在 MAC 地址中，第 7 位为 1 表示本地唯一，为 0 表示全球唯一，而在 EUI-64 格式中，第 7 位为 1 表示全球唯一，为 0 表示本地唯一。上面例子第 7 位反转后的结果为 0250:3EFF:FEE4:4C89。

③ 加上 IPv6 前缀构成一个完整的 IPv6 地址，如 2020:1212::250:3EFF:FEE4:4C89。

9.1.3　IPv6 扩展包头

IPv6 扩展包头实现了 IPv4 包头中选项字段的功能，并进行了扩展，每一个扩展包头都包含一个下一包头（Next Header）字段，用于指明下一个扩展包头的类型。IPv6 扩展包头如图 9-3 所示。目前 IPv6 定义的扩展包头有逐跳选项包头、目的地选项包头、路由选择包头、分段包头、AH 包头、ESP 包头和上层包头，具体描述如下。

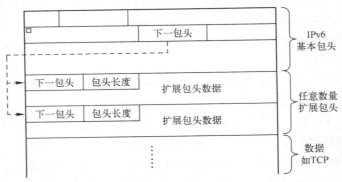

图 9-3　IPv6 扩展包头

① 逐跳（Hop-by-Hop）选项包头：对应的下一包头值为 0，指在数据包传输过程中，每

个路由器都必须检查和处理，如组播侦听者（MLD）和资源预留协议（RSVP）等。其中，MLD 用于支持组播的 IPv6 路由器和网络上的组播组成员之间交换成员状态信息。

② 目的地（Destination）选项包头：对应的下一包头值为 60，指最终的目的节点和路由选择包头指定的节点都对其进行处理。如果存在路由选择扩展包头，则每一个指定的中间节点都要处理这些选项；如果没有路由选择扩展包头，则只有最终目的节点需要处理这些选项。

③ 路由选择（Routing）包头：对应的下一包头值为 43，IPv6 的源节点可以利用路由选择扩展包头指定数据包从源到目的地需要经过的中间节点的列表。

④ 分段（Fragment）包头：对应的下一包头值为 44，当 IPv6 数据包长度大于链路 MTU 时，源节点负责对数据包进行分段，并在分段扩展包头中提供数据包重组信息。高层应该尽量避免发送需要分段的数据包。

【技术扩展】

由于 IPv6 包头中没有分段字段，IPv6 的分段功能只能由源节点和目的节点实现，这样就简化了包头。路径 MTU 发现（Path MTU Discovery，PMD）协议用来发现从源到达目的地的最佳的 MTU，RFC 1981 对其进行描述，其工作过程如下：正在执行路径 MTU 发现的源节点向目的节点发送允许的最大 MTU，如果一条中间链路无法处理该长度的数据包，则向源节点发送一个 ICMPv6 类型 2 的数据包超长错误消息，然后源节点将发送 MTU 值较小的数据包。这个过程将一直重复，直到不再收到 ICMPv6 类型 2 的消息为止，然后源节点就可以使用最新的 MTU 作为路径 MTU。路径 MTU 发现每 5 分钟执行一次，以便查看整个链路的 MTU 变化情况。可以使用 show ipv6 mtu 命令查看缓存中的每个目的地的 MTU 值。

⑤ AH 包头（Authentication Header）：对应的下一包头值为 51，提供身份验证、数据完整性检查和防重放保护。

⑥ ESP（Encapsulating Security Payload）包头：对应的下一包头值为 50，提供身份验证、数据机密性、数据完整性检查和防重放保护。

⑦ 上层（Upper-layer）包头：通常用于传输数据，如 TCP 对应的下一包头值为 6，UDP 对应的下一包头值为 17，OSPF 对应的下一包头值为 89，EIGRP 对应的下一包头值为 88，ICMPv6 对应的下一包头值为 58。

带扩展包头的 IPv6 数据包实例如图 9-4 所示。

如果使用扩展包头，在 RFC 1883 中规定了它们应该出现的顺序。仅逐跳选项扩展包头使用严格的定义，它必须直接跟在 IPv6 的基本包头的后面，因为这样可以更容易地被传输节点找到。建议扩展包头的顺序如下。

① IPv6 基本包头。
② 逐跳选项包头。
③ 目的地选项包头（只有在路由选择包头中指定的中间节点才必须处理）。
④ 路由选择包头。
⑤ 分段包头。
⑥ AH 包头。
⑦ ESP 包头。

⑧ 目的地选项包头（只有最后的目的节点必须处理）。
⑨ 上层包头。

| IPv6基本包头
next header=51 | AH扩展包头
Next header=89 | OSPFv3数据包 |

```
Internet Protocol Version 6, Src: fe80::cda5:22ff:fed4:353b (fe80::cda5:22ff:fed4:353b), Dst: ff02::5 (ff02::5)
  0110 .... = Version: 6
  .... 1110 0000 .... .... .... .... .... = Traffic class: 0x000000e0
  .... .... .... 0000 0000 0000 0000 0000 = Flowlabel: 0x00000000
  Payload length: 52
  Next header: AH (0x33)
  Hop limit: 1
  Source: fe80::cda5:22ff:fed4:353b (fe80::cda5:22ff:fed4:353b)
  Destination: ff02::5 (ff02::5)
  Authentication Header
    Next Header: OSPF IGP (0x59)
    Length: 24
    AH SPI: 0x000003e8
    AH Sequence: 11
    AH ICV: 07fb27a34012f736ddd94445
  Open Shortest Path First
```

图 9-4　带扩展包头的 IPv6 数据包实例

9.1.4　IPv6 地址类型

IPv6 地址有三种类型：单播、任意播和组播，在每种类型地址中又包括一种或者多种地址，如单播 IPv6 地址包括链路本地地址、站点本地地址、可聚合全球地址、环回地址和未指定地址；任意播 IPv6 地址包括链路本地地址、站点本地地址和可聚合全球地址；多播 IPv6 地址包括指定地址和请求节点地址。下面介绍几个常用的 IPv6 地址。

（1）链路本地（Link Local）地址

当在一个节点或者接口下启用 IPv6 协议栈时，节点的接口自动配置一个链路本地地址，该地址前缀为 FE80::/10，然后通过 EUI-64 方式扩展构成。链路本地地址主要用于节点自动地址配置、邻居发现、路由器发现以及路由更新等。

（2）可聚合全球单播地址

IANA（The Internet Assigned Numbers Authority，互联网数字分配机构）分配 IPv6 地址空间中的一个 IPv6 地址前缀作为可聚合全球单播地址，通常由 48 比特的全局前缀、16 比特子网 ID 和 64 比特的接口 ID 组成。IPv6 可聚合全球单播地如图 9-5 所示。当前 IANA 分配的可聚合全球单播地址是以二进制 001 开头的，前 16 比特地址范围为 2000~3FFF，即 2000::/3，占整个 IPv6 地址空间的 12.5%。

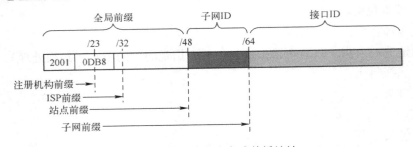

图 9-5　IPv6 可聚合全球单播地址

（3）IPv4 兼容地址

IPv4 兼容的 IPv6 地址是由过渡机制使用的特殊单播 IPv6 地址，目的是在主机和路由器上自动创建 IPv4 隧道以在 IPv4 网络上传送 IPv6 数据包。虽然这种过渡方式实现比较简单，但是实际上已被摒弃。

（4）环回地址

单播地址 0:0:0:0:0:0:0:1 被称为环回地址。节点用它来向自身发送 IPv6 包，它不能分配给任何物理接口。

（5）不确定地址

单播地址 0:0:0:0:0:0:0:0 被称为不确定地址。它不能分配给任何节点，用于特殊用途，如默认路由。

（6）组播地址

组播地址用来标识一组接口，发送给组播地址的数据流同时传输到多个组成员。一个接口可以加入多个组播组。IPv6 组播地址由前缀 FF00::/8 定义，其结构如图 9-6 所示。IPv6 的组播地址都是以 FF 开头的。

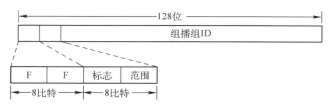

图 9-6 IPv6 组播地址结构

① 标志（4 比特）：表示在组播地址中设置的标志。从 RFC 2373 起，定义的唯一标志是 Transient（T）标志，T 标志使用标志字段的低位比特。当设置为 0 时，表示该组播地址是由 IANA 永久分配的；当设置为 1 时，表示该组播地址是临时使用的。

② 范围（4 比特）：表示组播数据准备在 IPv6 网络中发送的范围。以下是 RFC 2373 中定义该字段的值及对应的发送范围：1 表示节点本地，2 表示链路本地，5 表示站点本地，8 表示组织本地，E 表示全局范围。当 IPv6 数据包在以太网链路上传输时，二层数据帧头的类型字段值为 0x86DD，而在 PPP 链路上传输时，IPv6CP 中协议字段的值为 0x8057。在以太网中，IPv6 组播地址和对应的链路层地址映射通过如下方式构造：前 16 比特固定为 0x33:33，再加上 IPv6 组播地址的后 32 比特，如表示本地所有节点的组播地址 FF02::1 在以太网中对应的链路层地址为 33:33:00:00:00:01。

（7）请求节点（Solicited-node）地址

对于节点或路由器的接口上配置的每个单播和任意播地址，都自动启动一个对应的请求节点地址。请求节点地址受限于本地链路。请求节点组播地址由前缀 FF02::1:FF00:0/104 加上单播 IPv6 地址的最后 24 比特构成。请求节点地址可用于重复地址检测（Duplicate Address Detection，DAD）和邻居地址解析等。

（8）任意播（AnyCast）地址

任意播地址是分配给多个接口的全球单播地址，发送到该接口的数据包被路由到路径最优的目标接口。目前，任意播地址不能用作源地址，只能作为目的地址，且仅分配给路由器。任意播的出现不仅缩短了服务响应的时间，而且也可以减轻网络承载流量的负担。

9.1.5　IPv6 邻居发现协议（NDP）

邻居发现协议（Neighbor Discovery Protocol，NDP）是 IPv6 的一个关键协议，它替代在 IPv4 中使用的 ARP、ICMP、路由器发现和 ICMP 重定向等协议。当然，它还提供了其他功能，如前缀发现、邻居不可达检测、重复地址检测、地址自动配置等，NDP 通过以上功能实现 IPv6 的即插即用的重要特性。

NDP 定义的消息使用 ICMPv6 来承载，在 RFC 2461 中详细说明了 5 个新 ICMPv6 消息，包括路由器请求、路由器通告、邻居请求、邻居通告和重定向消息。

① 路由器请求（Router Solicitation，RS）：节点（包括主机或者路由器）启动后，通过 RS 向路由器发出请求，期望路由器立即发送 RA 响应，ICMPv6 类型为 133。

② 路由器通告（Router Advertisement，RA）：路由器周期性地发送 RA 或者以 RA 响应 RS，发送的 RA 包括链路前缀、链路 MTU、跳数限制以及一些标志位信息。ICMPv6 类型为 134。

③ 邻居请求（Neighbor Solicitation，NS）：通过 NS 可以确定邻居的链路层地址、邻居是否可达，完成重复地址检测等。ICMPv6 类型为 135。

④ 邻居通告（Neighbor Advertisement，NA）：NA 被用于响应 NS，同时节点在链路层地址变化时也可以主动发送 NA，以通知相邻节点自己的链路层地址发生改变。ICMPv6 类型为 136。

⑤ 重定向（Redirect）：路由器通过重定向消息通知到目的地有更好的下一跳路由器。ICMPv6 类型为 137。

9.1.6　IPv6 过渡技术

IPv6 技术相比 IPv4 技术而言具有许多优势，然而大面积部署 IPv6 需要一个过程，在此期间 IPv6 会与 IPv4 共存。为了确保过渡的平稳性，人们已制定了出许多策略，包括双栈技术、隧道技术以及协议转换技术等。

1. IPv6/IPv4 双栈技术

双栈技术是 IPv4 向 IPv6 过渡的一种有效的技术。网络中的节点同时支持 IPv4 和 IPv6 协议栈，源节点根据目的节点地址的不同选用不同的协议栈，而网络设备根据数据包的协议类型选择不同的协议栈进行处理和转发。

2. 隧道技术

隧道（Tunnel）是指将一种协议的数据包封装到另外一种协议中进行传输的技术。隧道技术只要求隧道两端的设备同时支持 IPv4 和 IPv6 协议栈。IPv4 隧道技术利用现有的 IPv4 网络为互相独立的 IPv6 网络提供连通性，IPv6 数据包被封装在 IPv4 数据包中穿越 IPv4 网络，实现 IPv6 数据包的透明传输。这种技术的优点是只要求网络的边界设备实现 IPv4/IPv6 双栈和隧道功能，其他节点不需要支持双协议栈，可以最大限度保护现有的 IPv4 网络投资。但是隧道技术不能实现 IPv4 主机与 IPv6 主机的直接通信。隧道可以手工配置，也可自动配置，采用哪种方式取决于对扩展性和管理开销等方面的要求，用于 IPv6 穿越 IPv4 网络的隧道技术主要有：

（1）IPv6 手工隧道

IPv6 手工隧道的源和目的地址需要手工配置，并且为隧道接口配置 IPv6 地址，为被 IPv4 网络分隔的 IPv6 网络提供稳定的点到点连接。如果一个边界设备要与多个设备建立手工隧道，就需要在设备上配置多个隧道。手工隧道的工作模式为 ipv6ip，对应 IPv4 协议字段的值为 41，可以通过命令 **debug ip packet detail** 查看。

（2）GRE 隧道

GRE 隧道和手工隧道非常相似，GRE 隧道也可为被 IPv4 网络分隔的 IPv6 网络提供稳定的点到点连接。需要手工配置隧道源和目的地址以及隧道接口 IPv6 地址。在 Cisco 路由器上，隧道默认的工作模式就是 gre ip，其对应 IPv4 协议字段的值为 47。

（3）6to4 隧道

6to4 隧道是一种自动隧道，也用于将孤立的 IPv6 网络通过 IPv4 网络连接起来，但是它可以建立多点连接。边界设备使用内嵌在 IPv6 地址中的 IPv4 地址自动建立隧道。6to4 隧道使用专用的 IPv6 地址范围 2002::/16，而一个 6to4 前缀可以表示为 2002:IPv4 地址::/48 格式。例如，边界设备的 IPv4 地址为 192.168.99.1（十六进制为 c0a86301），则其 IPv6 地址前缀为 2002:c0a8:6301::/48。6to4 隧道的源 IPv4 地址手工指定，隧道的目的 IPv4 地址根据通过隧道转发的数据包来决定。如果 IPv6 数据包的下一跳目的地址是 6to4 地址，则从数据包的目的地址中提取出 IPv4 地址作为隧道的目的地址。6to4 隧道最大的缺点是只能使用静态路由或 BGP，这是因为其他路由协议都使用链路本地地址来建立邻居关系和交换路由信息，而链路本地地址不符合 6to4 地址的编址要求，因此不能建立 6to4 隧道。

（4）ISATAP 隧道

ISATAP（Intra-Site Automatic Tunnel Addressing Protocol，站点内自动隧道寻址协议）是另外一种 IPv6 自动隧道技术，也用于将孤立的 IPv6 网络通过 IPv4 网络连接起来。与 6to4 地址类似，ISATAP 地址中也内嵌了 IPv4 地址，这使得边界设备很容易地获得建立隧道的目的地址，从而自动创建隧道。但是这两种自动隧道的地址格式不同。6to4 使用 IPv4 地址作为网络 ID，而 ISATAP 使用 IPv4 地址作为接口 ID。ISATAP 地址的接口 ID 由 0000: 5EFE 加 IPv4 地址（十六进制）构成，其中 0000:5EFE 是一个专用的 OUI，用于标识 IPv6 的 ISATAP 地址。例如，边界设备的 IPv4 地址为 192.168.99.1，则 64 比特的接口 ID 为 0000: 5EFE: c0a8:6301。

3. IPv4/IPv6 网络地址转换-协议转换技术

NAT-PT（Network Address Translation-Protocol Translation，网络地址转换-协议转换）是一种 IPv4 网络和 IPv6 网络之间直接通信的过渡方式，也就是说，原 IPv4 网络不需要进行升级改造，所有包括地址和协议在内的转换工作都由 NAT-PT 网络设备来完成。NAT-PT 设备要向 IPv6 网络发布一个 /96 的路由前缀，凡是具有该前缀的 IPv6 包都被送往 NAT-PT 设备。NAT-PT 设备为了支持 NAT-PT 功能，还具有从 IPv6 向 IPv4 网络转发数据包时使用的 IPv4 地址池。此外，通常在 NAT-PT 设备中实现 DNS-ALG（DNS-应用层网关），以帮助提供名称到地址的映射，在 IPv6 网络访问 IPv4 网络的过程中发挥作用。NAT-PT 分为静态 NAT-PT 和动态 NAT-PT。

9.2 配置 IPv6 地址

9.2.1 实验 1：手工配置 IPv6 单播地址

1. 实验目的

通过本实验可以掌握：
① 配置 IPv6 单播地址的方法。
② 手工配置 IPv6 单播地址的方法。
③ 配置采用 EUI-64 方式获得 IPv6 地址的方法。
④ 手工配置 IPv6 链路本地地址的方法。
⑤ 接口下 ND 参数的含义和调整方法。
⑥ 主机上无状态地址自动配置 IPv6 地址的方法。

2. 实验拓扑

手工配置 IPv6 单播地址实验拓扑如图 9-7 所示。

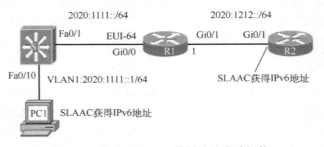

图 9-7　手工配置 IPv6 单播地址实验拓扑

3. 实验步骤

路由器接口的 IPv6 地址可以通过手工静态配置、EUI-64 方式和无状态地址自动配置等方式获得。本实验中，路由器 R1 的 Gi0/0 接口采用 EUI-64 方式配置 IPv6 地址，R1 的 Gi0/1 接口和交换机 S1 的 VLAN1 接口通过手工静态配置 IPv6 地址，R2 的 Gi0/1 接口和 PC1 的

IPv6 地址通过无状态地址自动配置（Stateless Address Auto Configuration，SLAAC）方式获得，这是在网络中没有 DHCPv6 服务器时允许节点自行配置 IPv6 地址的方式。

（1）配置路由器 R1

```
R1(config)#ipv6 unicast-routing             //启用 IPv6 路由功能
R1(config)#interface gigabitEthernet0/0
R1(config-if)#ipv6 enable                   //在接口下启用 IPv6 路由功能会自动生成链路本地地
```
址。如果已经启动 IPv6 路由功能，就不必配置该命令
```
R1(config-if)#ipv6 address 2020:1111::/64 eui-64
```
//接口通过 EUI-64 方式获得 IPv6 地址
```
R1(config-if)#ipv6 mtu 1500                 //配置接口下发送 IPv6 数据包的 MTU
R1(config-if)#ipv6 nd dad attempts 1
```
//确定地址唯一之前，发送 NS 的个数，默认为 1，如果为 0 表示禁止 DAD 功能
```
R1(config-if)#ipv6 nd ra lifetime 1800
```
//配置 RA 的生存期，范围为 0～9000 秒，默认为 30 分钟（1800 秒）
```
R1(config-if)#ipv6 nd ra interval 200
```
//配置发送 RA 的时间间隔，范围为 3～1800，默认为 200 秒，命令 **ipv6 nd ra suppress** 抑制接口发送 RA
```
R1(config-if)#ipv6 nd prefix 2020:1111::/64 2592000 604800
```
//修改前缀通告的参数，后面两个参数分别是有效生存期和首选生存期，其中有效生存期指用无状态地址自动配置获得 IPv6 地址保持有效状态的时间，单位为秒，默认为 2592000 秒，即 30 天，超过该时间，IPv6 地址被认为无效；首选生存期必须小于或等于有效生存期，单位为秒，默认为 604800 秒，即 7 天，该时间到期后，IPv6 地址不能主动去建立新的连接，但可以在有效生存期没过期之前与其他主机建立连接
```
R1(config-if)#ipv6 nd router-preference medium
```
//配置默认路由器的优先级，默认为 medium
```
R1(config-if)#ipv6 nd reachable-time 1800000
```
//配置远端邻居可达的时间，范围为 0～3600000 毫秒。如果在该时间内没收到邻居的消息，就会从邻居表中删除该邻居表项
```
R1(config-if)#ipv6 nd   ns-interval 1000    //配置 NS 重传间隔
R1(config)#interface GigabitEthernet0/1
R1(config-if)#ipv6 address 2020:1212::1/64
R1(config-if)#ipv6 address fe80::1 link-local
```
//配置链路本地地址，一个接口只能配置一个链路本地地址，如果不配置，通过 EUI-64 方式自动生成

（2）配置路由器 R2

```
R2(config)#ipv6 unicast-routing
R2(config)#interface GigabitEthernet0/1
R2(config-if)#ipv6 address autoconfig
```
//使用 SLAAC 方式配置路由器接口 IPv6 地址、前缀长度和默认网关

（3）配置交换机 S1

```
S1(config)#interface vlan 1
S1(config-if)#ipv6 address 2020:1111::1/64
```

【技术要点】

如果在交换机上启用 IPv6 路由功能，需要修改 SDM（Switch Database Management，交换机数据库管理）模板，要做如下配置：

```
S1(config)#sdm prefer dual-ipv4-and-ipv6 routing
Changes to the running SDM preferences have been stored, but cannot take effect
until the next reload                              //重新启动交换机后生效
Use 'show sdm prefer' to see what SDM preference is currently active.
S1#show sdm prefer
  The current template is "desktop default" template.      //当前支持默认的 IPv4 路由功能
  The selected template optimizes the resources in
  the switch to support this level of features for
  8 routed interfaces and 1024 VLANs.
    number of unicast mac addresses:               6K
    number of IPv4 IGMP groups + multicast routes: 1K
    number of IPv4 unicast routes:                 8K
      number of directly-connected IPv4 hosts:     6K
      number of indirect IPv4 routes:              2K
    number of IPv4 policy based routing aces:      0
    number of IPv4/MAC qos aces:                   512
    number of IPv4/MAC security aces:              1K
  On next reload, template will be "desktop IPv4 and IPv6 routing" template.
//重启后，才能支持 IPv4 和 IPv6 路由功能
S1#reload
S1(config)#ipv6 unicast-routing                    //重启后才能支持 IPv6 路由功能
S1#show sdm prefer
  The current template is "desktop IPv4 and IPv6 routing" template.
  The selected template optimizes the resources in the switch to support this level of features for 8 routed
interfaces and 1024 VLANs.
          （此处省略部分输出）
    number of IPv6 policy based routing aces:      0.25K
    number of IPv6 qos aces:                       0.625k
    number of IPv6 security aces:                  0.5K
```

4. 实验调试

（1）查看接口的 IPv6 配置信息

```
R1#show ipv6 interface gigabitEthernet0/0
GigabitEthernet0/0 is up, line protocol is up
  IPv6 is enabled, link-local address is FE80::FA72:EAFF:FED6:F4C8
//本接口启用 IPv6 路由功能，链路本地地址默认以 FE80::/10 为前缀，通过 EUI-64 格式自动配置，
而串行接口和环回接口会借用第一个以太网接口的 MAC 地址来生成链路本地地址的接口 ID 部分，而且有可
能路由器多个接口的链路本地地址相同，所以在 ping 链路本地地址时，需要指定接口。也可以通过如下命令
ipv6 address fe80::1 link-local 手工配置接口的链路本地地址。一个接口只能有一个链路本地地址
  No Virtual link-local address(es):
  Global unicast address(es):
    2020:1111::FA72:EAFF:FEC8:4F98, subnet is 2020:1111::/64 [EUI]
//全球单播地址及子网，一个接口下可以配置多个 IPv6 单播地址，该单播地址通过 EUI 方式
配置
  Joined group address(es)://接口启用 IPv6 功能后会自动加入一些组播组
    FF02::1                    //表示本地链路上的所有节点
    FF02::2                    //表示本地链路上的所有路由器
    FF02::1:FFC8:4F98          //与本接口链路本地地址和全球单播地址对应的请求节点组播地址
```

```
    MTU is 1500 bytes            //接口 MTU 大小
    ICMP error messages limited to one every 100 milliseconds  //ICMPv6 错误消息发送速率限制
    ICMP redirects are enabled                    //接口启用 ICMPv6 重定向功能
    ICMP unreachables are sent                    //接口可以发送 ICMP 不可达消息
    ND DAD is enabled, number of DAD attempts: 1  //启用重复地址检测，尝试次数为 1
    ND reachable time is 1800000 milliseconds (using 1800000)  //认为邻居的可达时间
    ND advertised reachable time is 1800000 milliseconds       //邻居通告的可达时间
    ND advertised retransmit interval is 1000 milliseconds     //NS 通告重传间隔
    ND router advertisements are sent every 200 seconds        //RA 发送间隔
    ND router advertisements live for 1800 seconds             //RA 的生存期
    ND advertised default router preference is Medium          //默认路由器优先级
    Hosts use stateless autoconfig for addresses.              //启用无状态地址自动配置方式
```

（2）查看计算机 PC1 上通过无状态地址自动配置方式获得的 IPv6 地址

① 启用 TCP/IPv6 协议栈。

此处以 Windows 7 为例说明计算机 PC1 启用 TCP/IPv6 协议栈的步骤：选择桌面上的【网络】图标→右键单击【属性】选项→单击左侧【更改适配器设置】→选择需要启用 TCP/IPv6 协议栈的网卡→右键单击【属性】，在复选框选项选中【Internet 协议版本 6（TCP/IPv6）】→单击【属性】按钮→在【Internet 协议版本 6（TCP/IPv6）】页面中单击【自动获取 IPv6 地址】单选框→单击【确定】按钮。

② 查看计算机 PC1 获得 IPv6 地址。

```
    C:\>ipconfig /all
        以太网适配器 本地连接：
        （此处省略部分输出）
        IPv6 地址 . . . . . . . . . . . : 2020:1111::c10f:8ab2:cb65:bafd（首选）
        //IPv6 地址，通过收到的前缀 2020:1111/64 加上本地网卡 MAC 地址 EUI-64 扩展生成，如果路
由器接口有多个 IPv6 地址，此处就会以相应的前缀自动生成多个 IPv6 地址
        临时 IPv6 地址. . . . . . . . . : 2020:1111::20b9:869c:be61:bb2b（首选）
        //临时 IPv6 地址是 Windows 通过收到的前缀 2020:1111 加上随机接口 ID 自动生成的，可以在
CMD 下以管理员身份执行 netsh interface ipv6 set privacy state=disable 命令，然后将网卡重启，就不会看到
临时 IPv6 地址了
        本地链接 IPv6 地址. . . . . . . : fe80::c10f:8ab2:cb65:bafd%5（首选）
        //该网卡的链路本地地址，其中%后面跟的 5 是该网卡的接口标识
        IPv4 地址 . . . . . . . . . . . : 10.3.24.1（首选）
        子网掩码 . . . . . . . . . . . : 255.255.255.0
        默认网关. . . . . . . . . . . . : fe80::fa72:eaff:ffd6:f4c8%5
    //IPv6 默认网关，即路由器 R1 以太网接口 Gi0/1 的链路本地地址，即使路由器的接口有多个 IPv6
地址，网关都是该链路本地地址
```

③ 查看借助 RS 和 RA 实现 IPv6 无状态地址自动配置过程。

首先在路由器 R1 上执行调试命令 debug ipv6 nd，在计算机 PC1 上启用 TCP/IPv6 协议栈，路由器 R1 会收到 PC1 发送的 RS，然后马上发送 RA，过程如下：

```
    R1#debug ipv6 nd
    02:34:51: ICMPv6-ND: (GigabitEthernet0/0,FE80::FA72:EAFF:FE69:1C78) Received RS
        //路由器 R1 从 Gi0/0 接口收到 RS
    02:34:51: ICMPv6-ND: (GigabitEthernet0/0) Sending solicited RA
```

```
//从 Gi0/0 接口发送 RA
02:34:51: ICMPv6-ND: (GigabitEthernet0/0,FE80::FA72:EAFF:FED6:F4C8) send RA to FF02::1
02:34:51: ICMPv6-ND: (GigabitEthernet0/0,FE80::FA72:EAFF:FED6:F4C8) Sending RA (1800) to
FF02::1
//以上三行说明 R1 发送以 Gi0/0 接口链路本地地址为源，以组播地址 FF02::1 为目的 RA 数据包
02:34:51: ICMPv6-ND:    MTU = 1500 //MTU 值
02:34:51: ICMPv6-ND:      prefix 2020:1111::/64 [LA] 2592000/604800
//IPv6 地址前缀以及有效生存期和首选生存期。如果接口有多个 IPv6 地址，则发送多个前缀
```

（3）查看 R2 Gi0/1 接口通过 SLAAC 方式获得的 IPv6 地址和路由条目

```
① R2#show ipv6 interface brief | section GigabitEthernet0/1
GigabitEthernet0/1       [up/up]
    FE80::FA72:EAFF:FE69:1C79
    2020:1212::FA72:EAFF:FE69:1C79
//通过收到的前缀 2020:1212/64+接口 MAC 地址，采用 EUI-64 方式生成 IPv6 地址
② R2#show ipv6 route
(此处省略路由代码部分)
ND   ::/0 [2/0]
      via FE80::FA72:EAFF:FED6:F4C8, GigabitEthernet0/1
NDp 2020:1212::/64 [2/0]
      via GigabitEthernet0/1, directly connected
L    2020:1212::FA72:EAFF:FE69:1C79/128    [0/0]
      via GigabitEthernet0/1 receive
```

以上输出表明 R2 的 Gi0/1 接口在通过 SLAAC 方式获得 IPv6 地址时，会在路由表中生成 3 条路由条目，第 1 条是管理距离为 2 的 **ND** 默认路由，第 2 条是 R1 的 Gi0/1 接口发送 RA 前缀的管理距离为 2 的 **NDp** 路由，第 3 条是该接口 IPv6 地址的本地路由。

```
③ R2#show ipv6 routers                        //显示邻居路由器通告 RA 的详细信息
Router FE80::FA72:EAFF:FED6:F4C9 on GigabitEthernet0/1, last update 1 min
   Hops 64, Lifetime 1800 sec, AddrFlag=0, OtherFlag=0, MTU=1500
   HomeAgentFlag=0, Preference=Medium
   Reachable time 0 (unspecified), Retransmit time 0 (unspecified)
   Prefix 2020:1212::/64 onlink autoconfig
      Valid lifetime 2592000, preferred lifetime 604800
```

以上输出是路由器 R1 发送的 RA 的详细信息，包括 M 位=0、O 位=0、MTU 值、优先级、前缀/长度、有效生存期和首选生存期等信息。

（4）查看 DAD 工作过程

已经获取一个新地址的节点会把这个新的地址归类为临时状态的地址。在 DAD 完成之前不能使用此地址。

```
R1#debug ipv6 nd
```

① 在路由器 R1 的 Gi0/1 接口上配置 IPv6 地址 2020:1212::1 后，显示的 DAD 过程如下：

```
02:46:46: IPv6-Addrmgr-ND: DAD request for 2020:1212::1 on GigabitEthernet0/1
//需要对地址 2020:1212::1 做 DAD
02:46:46: ICMPv6-ND: (GigabitEthernet0/1,2020:1212::1) Sending DAD NS [Nonce: 2883.6e53.185b]
```

//从接口 Gi0/1 发送 NS，源地址为全 0，目的地址为 2020:1212::1 的节点请求地址

02:46:46: IPv6-Addrmgr-ND: **DAD: 2020:1212::1 is unique.**
//由于没有收到 NA，由此判断地址唯一

② 在路由器 R2 的 Gi0/1 接口上配置相同的 IPv6 地址，在 R2 执行上调试命令 **debug ipv6 nd**，显示信息如下：

02:50:00: IPv6-Addrmgr-ND: **DAD request** for 2020:1212::1 on GigabitEthernet0/1
//需要对地址 2020:1212::1 做 DAD
02:50:00: ICMPv6-ND: (GigabitEthernet0/1,2020:1212::1) **Sending DAD NS [Nonce: 14d8.45b5.d606]**
//从接口 Gi0/1 发送 NS，源地址为全 0，目的地址为 2020:1212::1 的节点请求地址
02:50:00: ICMPv6-ND: (GigabitEthernet0/1,2020:1212::1) **Received NA** from 2020:1212::1
//收到 NA，表明链路上其他接口配置了相同的 IPv6 地址
02:50:00: **%IPV6_ND-4-DUPLICATE: Duplicate address** 2020:1212::1 on GigabitEthernet0/1
//判断 2020:1212::1 地址重复

（5）链路地址解析过程

① 查看链路地址解析过程。

为了使得以下调试信息更加清晰和简洁，将路由器 R2 的 Gi0/1 接口的 IPv6 地址手工配置为 2020:1212::2，然后在路由器 R1 上 ping 2020:1212::2，链路层地址解析过程显示如下：

```
R1#ping 2020:1212::2
Type escape sequence to abort.
Sending 5, 100-byte ICMP Echos to 2020:1212::2, timeout is 2 seconds:
!!!!
Success rate is 100 percent (5/5), round-trip min/avg/max = 1/1/4 ms
02:54:14: ICMPv6-ND: (GigabitEthernet0/1,2020:1212::2) Resolution request
//地址解析请求
02:54:14: ICMPv6-ND: (GigabitEthernet0/1,2020:1212::2) DELETE -> INCMP:
//在 IPv6 邻居表中，2020:1212::2 表项状态从 DELETE -> INCMP（Incomplete），该状态表明正在进行链路层地址解析
02:54:14: ICMPv6-ND: (GigabitEthernet0/1,2020:1212::2) Sending NS
//发送 NS 到目标地址相关联的节点请求地址
02:54:14: ICMPv6-ND: (GigabitEthernet0/1,2020:1212::2) Queued data for resolution
//解析链路层地址
02:54:14: ICMPv6-ND: (GigabitEthernet0/1,2020:1212::2) Received NA from 2020:1212::2
//收到对方发送的 NA
02:54:14: ICMPv6-ND: (GigabitEthernet0/1,2020:1212::2) LLA f872.ea69.1c79
//获得了对方的 MAC 地址
02:54:14: ICMPv6-ND: (GigabitEthernet0/1,2020:1212::2) INCMP -> REACH
//IPv6 地址 2020:1212::2 对应的链路层地址解析成功，状态变为可达
```

② 查看 IPv6 的邻居表。

```
R1#show ipv6 neighbors
IPv6 Address                    Age    Link-layer Addr    State      Interface
2020:1212::2                    0      f872.ea69.1c79     REACH      Gi0/1
```

FE80::C802:EFF:FEF8:8	0	f872.ea69.1c79	REACH	Gi0/1

以上显示的内容类似 IPv4 的 ARP 表，显示了邻居的 IPv6 地址、链路层地址和状态等信息。可以通过命令 **clear ipv6 neighbors** 清除该表项动态产生的条目，也可以通过下面命令添加静态表项，该表项会一直存在邻居表中：

R1(config)#**ipv6 neighbor 2021:1313::3 GigabitEthernet 0/1 0023.3364.4fca**

【技术要点】

IPv6 邻居节点的状态包括如下几种。

① Incomplete（未完成）状态：邻居请求（NS）已经发送，在等待邻居发送的邻居通告（NA），该状态表示正在解析地址，但邻居链路层地址尚未确定。

② Reachable（可达）状态：已经收到邻居的 NA，邻居可达，该状态表示地址解析成功，获得了邻居链路层地址。

③ Stale（陈旧）状态：从收到上一次可达性确认后链路闲置了 30 秒（默认），表示可达时间（Reachable Time）到达，现在不能确定邻居是否可达。

④ Delay（延迟）状态：对处于 Stale 状态的邻居发送 1 个数据包，邻居的状态切换至 Delay，默认 5 秒内，若有 NA 应答或者来自对方应用层的提示信息，则从 Delay 状态切换为 Reachable 状态；否则由 Delay 状态切换为 Probe 状态。

⑤ Probe（探测）状态：该状态下每隔 1 秒（默认）发送 1 次 NS，连续发送 3 次，有应答则切换至 Reachable 状态，无应答则切换至 Empty 状态，即删除条目。

⑥ Empty（空闲）状态：没有邻居节点的缓存表项。

9.2.2 实验 2：通过有状态 DHCPv6 获得 IPv6 地址

1．实验目的

通过本实验可以掌握：
① 有状态 DHCPv6 的工作原理和工作过程。
② 配置有状态 DHCPv6 服务器和客户端的方法。
③ 配置 DHCPv6 中继代理的方法。

2．实验拓扑

通过有状态 DHCPv6 服务器获得 IPv6 地址实验拓扑如图 9-8 所示。本实验中，路由器 R1 作为 DHCPv6 服务器，R2 作为 DHCPv6 中继代理，R3 作为 DHCPv6 客户端。R3 的 Gi0/1 接口通过 DHCPv6 服务器获得地址 / 前缀、DNS 和域名信息，实现有状态 DHCPv6 配置，整个网络配置 RIPng 保证 IPv6 的连通性。

图 9-8 通过有状态 DHCPv6 服务器获得 IPv6 地址实验拓扑

3. 实验步骤

（1）配置路由器 R1

```
R1(config)#ipv6 unicast-routing                          //启用 IPv6 路由
R1(config)#ipv6 dhcp pool DHCPv6_Stateful                //配置 DHCPv6 地址池
R1(config-dhcpv6)#address prefix 2020:2323::/64 lifetime infinite infinite
//配置 DHCPv6 服务器 IPv6 地址池、有效时间和首选时间（单位为秒）
R1(config-dhcpv6)#dns-server 2020:1212::1111             //配置 DNS 服务器地址
R1(config-dhcpv6)#domain-name cisco.com                  //配置域名
R1(config-dhcpv6)#exit
R1(config)#ipv6 router rip cisco
R1(config)#interface GigabitEthernet0/0
R1(config-if)#ipv6 address 2020:1212::1/64
R1(config-if)#ipv6 rip cisco enable
R1(config-if)#ipv6 dhcp server DHCPv6_Stateful preference 100
```
//在接口下启用 DHCPv6 功能，并将 DHCPv6 地址池绑定在接口下，参数 **preference** 指定服务器优先级，范围为 0~255，默认为 0，还可以通过 **rapid-commit** 启动快速分配过程

```
R1(config-if)#ipv6 nd managed-config-flag
```
//将 ICMPv6 的 RA 中的 Flag 字段中 M（Managed Address Configuration）位置 1，此接口发送的 RA 声明 DHCPv6 客户端不使用 SLAAC，而要从有状态 DHCPv6 服务器获取 IPv6 地址／前缀和所有网络配置参数

（2）配置路由器 R2

```
R2(config)#ipv6 unicast-routing
R2(config)#ipv6 router rip cisco
R2(config)#interface GigabitEthernet0/0
R2(config-if)#ipv6 address 2020:1212::2/64
R2(config-if)#ipv6 rip cisco enable
R2(config)#interface GigabitEthernet0/1
R2(config-if)#ipv6 address 2020:2323::2/64
R2(config-if)#ipv6 rip cisco enable
R2(config-if)#ipv6 dhcp relay destination 2020:1212::1
```
//配置 DHCPv6 中继代理，转发来自 DHCPv6 客户端或 DHCPv6 服务器的 DHCPv6 数据包

（3）配置路由器 R3

```
R3(config)#interface gigabitEthernet0/1
R3(config-if)#ipv6 enable
```
//接口下启用 IPv6 会自动创建链路本地地址，作为发送 DHCPv6 消息的源地址
```
R3(config-if)#ipv6 address dhcp
```
//使路由器该接口等同于 DHCPv6 客户端，使用有状态 DHCPv6 配置 IPv6 地址和其他网络参数，还可以通过 **rapid-commit** 参数启动快速分配过程

4. 实验调试

（1）查看 DHCPv6 地址池及其参数

```
R1#show ipv6 dhcp pool
```

```
    DHCPv6 pool: DHCPv6_Stateful
        Address allocation prefix: 2020:2323::/64 valid 4294967295 preferred 4294967295 (1 in use, 0
conflicts)                                              //地址池中分配出去一个地址及租用时间
    DNS server: 2020:1212::1111                         //DNS 服务器
    Domain name: cisco.com                              //域名
    Active clients: 1                                   //活动的客户端数量
```

（2）使用 DHCPv6 获得可路由的地址

```
R1#show ipv6 interface gigabitEthernet0/0 | begin Hosts
    Hosts use DHCP to obtain routable addresses.        //主机使用 DHCPv6 获得可路由的地址
```

（3）查看 IPv6 路由表

```
R3#show ipv6 route
(此处省略路由代码部分)
ND    ::/0 [2/0]
        via FE80::C802:39FF:FEA8:1C, GigabitEthernet0/1
LC    2020:2323::3D21:3D2F:4DAC:136D/128 [0/0]
        via GigabitEthernet0/1, receive
```

以上输出表明 R3 的 Gi0/1 接口在通过 DHCPv6 获得 IPv6 地址时，会在路由表中生成 2 条路由条目，第 1 条是管理距离为 2 的 **ND** 默认路由，第 2 条是该接口 IPv6 地址的本地直链路由。

（4）查看 DHCPv6 的 IPv6 地址绑定情况

```
R1#show ipv6 dhcp binding
Client: FE80::C803:2FFF:FEAC:1C               //DHCPv6 客户端链路本地地址
    DUID: 00030001CA032FAC0006                // DHCPv6 服务器的 DUID（DHCP Unique IDentifier）
    Username : unassigned
    VRF : default
    IA NA: IA ID 0x00050001, T1 43200, T2 69120
    //身份管理标识符（Identity Association Identifier）和地址/前缀租约的 T1 和 T2 时间
        Address: 2020:2323::3D21:3D2F:4DAC:136D              //分配出去的地址
            preferred lifetime INFINITY, valid lifetime INFINITY,   //租用时间
```

（5）查看 DHCPv6 接口信息

```
① R1#show ipv6 dhcp interface
GigabitEthernet0/0 is in server mode              //接口是 DHCPv6 服务器模式
    Using pool: DHCPv6_Stateful                   //DHCPv6 使用的地址池
    Preference value: 100                         // DHCPv6 服务器优先级
    Hint from client: ignored
    Rapid-Commit: disabled                        //没有启用快速交换过程
② R2#show ipv6 dhcp interface
GigabitEthernet0/1 is in relay mode               //接口是 DHCPv6 中继模式
    Relay destinations:                           // DHCPv6 中继目的地址
        2020:1212::1
③ R3#show ipv6 dhcp interface
GigabitEthernet0/1 is in client mode              //接口是 DHCPv6 客户端模式
```

```
  Prefix State is IDLE                    //前缀状态为 IDLE
  Address State is OPEN                   //地址状态为 OPEN，说明通过 DHCPv6 服务器获得 IPv6 地址
  Renew for address will be sent in 11:34:20    //IPv6 地址 Renew 数据包发送时间
  List of known servers:                  //列出知晓的 DHCPv6 服务器
    Reachable via address: FE80::C802:39FF:FEA8:1C
    //经过路由器 R2 的 Gi0/1 接口的链路本地地址到达 DHCPv6 服务器
    DUID: 00030001CA0143D40006           //DHCPv6 客户端的 DUID
    Preference: 100                      //DHCPv6 服务器的优先级
    Configuration parameters:            //配置参数
      IA NA: IA ID 0x00050001, T1 43200, T2 69120
        Address: 2020:2323::3D21:3D2F:4DAC:136D/128
                preferred lifetime INFINITY, valid lifetime INFINITY
      DNS server: 2020:1212::1111
      Domain name: cisco.com
      Information refresh time: 0
  Prefix Rapid-Commit: disabled           //没启用前缀快速交换过程
  Address Rapid-Commit: disabled          //没启用地址快速交换过程
```

9.3 配置 IPv6 路由

9.3.1 实验 3：配置 IPv6 静态路由

1．实验目的

通过本实验可以掌握：
① 启用 IPv6 路由的方法。
② 配置 IPv6 地址的方法。
③ 配置 IPv6 静态路由和总结路由的方法。
④ 配置 IPv6 默认路由的方法。
⑤ 配置计算机网卡 IPv6 地址的方法。
⑥ 查看 IPv6 接口和路由表的方法。

2．实验拓扑

配置 IPv6 静态路由实验拓扑如图 9-9 所示。

3．实验步骤

（1）配置路由器 R1

```
R1(config)#ipv6 unicast-routing                    //启用 IPv6 单播路由
R1(config)#interface GigabitEthernet0/1
R1(config-if)#ipv6 address 2019:1110::1/64         //配置 IPv6 单播地址
R1(config)#interface GigabitEthernet0/2
R1(config-if)#ipv6 address 2019:1111::1/64
R1(config)#interface Serial0/0/0
```

```
R1(config-if)#ipv6 address 2020:1212::1/64
R1(config)#ipv6 route 2020:2323::/64 Serial0/0/0
//配置带送出接口的 IPv6 静态路由
R1(config)#ipv6 route 2021:3333::/64 Serial0/0/0
```

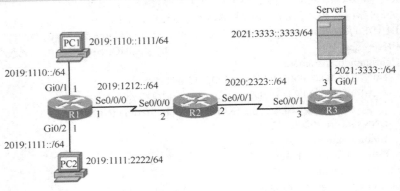

图 9-9　配置 IPv6 静态路由实验拓扑

（2）配置路由器 R2

```
R2(config)#ipv6 unicast-routing
R2(config)#interface Serial0/0/0
R2(config-if)#ipv6 address 2020:1212::2/64
R2(config)#interface Serial0/0/1
R2(config-if)#ipv6 address 2020:2323::2/64
R2(config)#ipv6 route 2019:1110::/31 Serial0/0/0          //配置 IPv6 总结静态路由
R2(config)#ipv6 route 2021:3333::/64 Serial0/0/1
```

【技术要点】

在配置 IPv6 静态路由时，可以使用送出接口方式，也可以使用下一跳地址方式，或者二者结合方式，比如在路由器 R2 上，对于到达前缀 2021:3333::/64 的静态路由可以采用如下 3 种配置之一：

```
R2(config)#ipv6 route 2021:3333::/64 2020:2323::3
//配置带下一跳 IPv6 静态路由
R2(config)#ipv6 route 2021:3333::/64 serial0/0/1 2020:2323::3
//配置带下一跳和全球单播地址结合的 IPv6 静态路由
R2(config)#ipv6 route 2021:3333::/64 serial0/0/1 FE80::FA72:EAFF:FEDB:EA78
//配置带下一跳和链路本地地址结合的 IPv6 静态路由
```

（3）配置路由器 R3

```
R3(config)#ipv6 unicast-routing
R3(config)#interface GigabitEthernet0/1
R3(config-if)#ipv6 address 2021:3333::3/64
R3(config)#interface Serial0/0/1
R3(config-if)#ipv6 address 2020:2323::3/64
R3(config)#ipv6 route ::/0 serial0/0/1 100
//配置 IPv6 默认静态路由，管理距离为 100
```

第 9 章 IPv6

（4）配置计算机 PC1、PC2 和 Server1 的 IPv6 地址

在 Windows 7 环境下配置 IPv6 地址：在【控制面板】→【网络和共享中心】→【更改适配器设置】→【网络连接】页面，选中接入 IPv6 网络的网卡，单击右键，在菜单中单击【属性】，在本地连接属性页面【网络】选项卡中选中【Internet 协议版本 6（TCP/IPv6）】，然后单击【属性】按钮，接下来在【Internet 协议版本 6（TCP/IPv6）属性】页面中，单击【使用以下 IPv6 地址（S）】单选框，填写【IPv6 地址】、【子网前缀长度】和【默认网关】文本框，本节仅以填写 PC1 的 IPv6 地址为例。配置计算机 IPv6 地址、前缀长度和网关地址如图 9-10 所示。需要注意的是网关填写的是 R1 的 Gi0/1 接口的链路本地地址，当然，此处填写 R1 的 Gi0/1 接口的全球单播地址也可以，一般都填写链路本地地址。

图 9-10　配置计算机 IPv6 地址、前缀长度和网关地址

4．实验调试

（1）查看接口 IPv6 地址和状态

```
R1#show ipv6 interface brief
GigabitEthernet0/1       [up/up]
    FE80::FA72:EAFF:FEC8:4F99              //链路本地地址
    2019:1110::1                           //全球单播地址
GigabitEthernet0/2       [up/up]
    FE80::FA72:EAFF:FEC8:4F9A
    2019:1111::1
Serial0/0/0              [up/up]
    FE80::FA72:EAFF:FEC8:4F98
    2020:1212::1
```

（2）查看 IPv6 路由表

以下输出均省略 IPv6 路由代码部分。

```
① R1#show ipv6 route
IPv6 Routing Table - default - 9 entries
C   2019:1110::/64 [0/0]
       via GigabitEthernet0/1, directly connected     //直连 IPv6 路由
L   2019:1110::1/128 [0/0]                            //本地 IPv6 路由，也就是接口的 IPv6 地址
       via GigabitEthernet0/1, receive
```

```
    C    2019:1111::/64 [0/0]
             via GigabitEthernet0/2, directly connected
    L    2019:1111::1/128 [0/0]
             via GigabitEthernet0/2, receive
    C    2020:1212::/64 [0/0]
             via Serial0/0/0, directly connected
    L    2020:1212::1/128 [0/0]
             via Serial0/0/0, receive
    S    2020:2323::/64 [1/0]
             via Serial0/0/0, directly connected
    S    2021:3333::/64 [1/0]
             via Serial0/0/0, directly connected
//以上两条为静态 IPv6 路由，虽然显示为直连，但是该路由条目管理距离为 1，度量值为 0
    L    FF00::/8 [0/0]                //该路由表示所有 IPv6 组播路由
             via Null0, receive        //指向 Null0 接口的路由主要是为了防止路由环路
② R2#show ipv6 route static
    S    2019:1110::/31 [1/0]          //IPv6 总结静态路由
             via Serial0/0/0, directly connected
    S    2018:3333::/64 [1/0]
             via Serial0/0/1, directly connected
③ R3#show ipv6 route static
    S    ::/0 [100/0]                  //管理距离为 100，IPv6 静态默认路由代码 S 后面没有*
             via Serial0/0/1, directly connected    //送出接口
```

（3）使用 ping 命令测试网络连通性

① R1#**ping 2021:3333::3333 source 2019:1111::1**
Type escape sequence to abort.
Sending 5, 100-byte ICMP Echos to 2021:3333::3333, timeout is 2 seconds:
Packet sent with a source address of 2019:1111::1
!!!!!
Success rate is 100 percent (5/5), round-trip min/avg/max = 1/2/4 ms
② R1#**ping 2021:3333::3333 source 2019:1110::1**
Type escape sequence to abort.
Sending 5, 100-byte ICMP Echos to 2021:3333::3333, timeout is 2 seconds:
Packet sent with a source address of 2019:1110::1
!!!!!
Success rate is 100 percent (5/5), round-trip min/avg/max = 1/2/4 ms

9.3.2　实验 4：配置 RIPng

1. 实验目的

通过本实验可以掌握：
① 启用 IPv6 路由的方法。
② 向 RIPng 网络注入默认路由的方法。
③ RIPng 配置和调试的方法。

2. 实验拓扑

配置 RIPng 实验拓扑如图 9-11 所示。

图 9-11　配置 RIPng 实验拓扑

3. 实验步骤

（1）配置路由器 R1

```
R1(config)#ipv6 unicast-routing
R1(config)#ipv6 router rip cisco              //启动 IPv6 RIPng 进程
R1(config-rtr)#split-horizon                  //启用水平分割
R1(config-rtr)#poison-reverse                 //启用毒化反转
R1(config)#interface Loopback0
R1(config-if)#ipv6 address 2019:1111::1/64
R1(config)#interface Serial0/0/0
R1(config-if)#ipv6 address 2020:12::1/64
R1(config-if)#ipv6 rip cisco enable
//在接口下启用 RIPng，进程名字为 cisco，进程名字只有本地含义，如果没有该进程，该命令将自动创建 RIPng 进程
R1(config-if)#ipv6 rip cisco default-information originate
//向 IPv6 RIPng 区域注入一条默认路由（::/0），该命令只从该接口发送默认的 IPv6 路由
R1(config)#ipv6 route ::/0 Loopback0          //配置默认路由
```

（2）配置路由器 R2

```
R2(config)#ipv6 unicast-routing
R2(config)#ipv6 router rip cisco
R2(config-rtr)#split-horizon
R2(config-rtr)#poison-reverse
R2(config)#interface Serial0/0/0
R2(config-if)#ipv6 address 2020:12::2/64
R2(config-if)#ipv6 rip cisco enable
R2(config)#interface Serial0/0/1
R2(config-if)#ipv6 address 2020:23::2/64
R2(config-if)#ipv6 rip cisco enable
```

（3）配置路由器 R3

```
R3(config)#ipv6 unicast-routing
R3(config)#ipv6 router rip cisco
R3(config-rtr)#split-horizon
R3(config-rtr)#poison-reverse
R3(config)#interface Serial0/0/0
R3(config-if)#ipv6 address 2020:34::3/64
```

```
R3(config-if)#ipv6 rip cisco enable
R3(config)#interface Serial0/0/1
R3(config-if)#ipv6 address 2020:23::3/64
R3(config-if)#ipv6 rip cisco enable
```

(4)配置路由器 R4

```
R4(config)#ipv6 unicast-routing
R4(config)#ipv6 router rip cisco
R4(config-rtr)#split-horizon
R4(config-rtr)#poison-reverse
R4(config)#interface Loopback0
R4(config-if)#ipv6 address 2021:4444::4/64
R4(config-if)#ipv6 rip cisco enable
R4(config)#interface Serial0/0/0
R4(config-if)#ipv6 address 2020:34::4/64
R4(config-if)#ipv6 rip cisco enable
```

【技术要点】

RIPng（RIP next generation）是 RIP 的 IPv6 版本，基于 IPv4 的 RIPv2，与其有很多相似的特征，比如更新周期为 30 秒，管理距离为 120，采用跳数作为度量值，最大跳数为 15 等，但是数据包源端口和目的端口都使用 UDP 521 端口进行操作，路由更新采用组播地址 FF02::9。与 IPv4 RIP 一样，RIPng 很少用于现代网络，但是有助于理解 IPv6 网络路由知识。

4. 实验调试

(1) 查看 RIPng 的路由表

以下输出全部省略路由代码部分。

```
① R1#show ipv6 route rip
R    2020:23::/64 [120/2]
      via FE80::FA72:EAFF:FE69:1C78, Serial0/0/0
R    2020:34::/64 [120/3]
      via FE80::FA72:EAFF:FE69:1C78, Serial0/0/0
R    2021:4444::/64 [120/4]
      via FE80::FA72:EAFF:FE69:1C78, Serial0/0/0
② R2#show ipv6 route rip
R    ::/0 [120/2]
     FE80::FA72:EAFF:FED6:F4C8, Serial0/0/0
R    2020:34::/64 [120/2]
      via FE80::FA72:EAFF:FE69:18B8, Serial0/0/1
R    2021:4444::/64 [120/3]
      via FE80::FA72:EAFF:FE69:18B8, Serial0/0/1
```

以上输出表明 R1 确实向 IPv6 RIPng 网络注入了 1 条 IPv6 的默认路由，同时路由器 R2 收到 3 条 RIPng 路由条目。在 RIPng 中，每台路由器在自己的入方向增加路由的度量值，而始发路由的路由器会在发送路由更新时将度量值加一跳。以上所有 IPv6 RIPng 路由条目的下一跳地址均为邻居路由器接口的链路本地地址。可以通过 **show ipv6 rip next-hops** 命令查看 RIPng 的下一跳地址，如下所示：

```
R2#show ipv6 rip next-hops
  RIP process "cisco", Next Hops
    FE80::FA72:EAFF:FED6:F4C8/Serial0/0/0 [2 paths]
    FE80::FA72:EAFF:FE69:18B8/Serial0/0/1 [3 paths]
③ R3#show ipv6 route rip
R   ::/0 [120/3]
       via FE80::FA72:EAFF:FE69:1C78, Serial0/0/1
R   2020:12::/64 [120/2]
       via FE80::FA72:EAFF:FE69:1C78, Serial0/0/1
R   2021:4444::/64 [120/2]
       via FE80::FA72:EAFF:FEC8:4F98, Serial0/0/0
④ R4#show ipv6 route rip
R   ::/0 [120/4]
       via FE80::FA72:EAFF:FE69:18B8, Serial0/0/0
R   2020:12::/64 [120/3]
       via FE80::FA72:EAFF:FE69:18B8, Serial0/0/0
R   2020:23::/64 [120/2]
       via FE80::FA72:EAFF:FE69:18B8, Serial0/0/0
```

（2）显示和 IPv6 路由协议相关的信息

```
R3#show ipv6 protocols
IPv6 Routing Protocol is "connected"          //直连路由
IPv6 Routing Protocol is "application"        //应用路由
IPv6 Routing Protocol is "ND"                 //邻居发现路由
IPv6 Routing Protocol is "rip cisco"          //RIPng 路由
  Interfaces:
    Serial0/0/1
    Serial0/0/0
  Redistribution:
    None
```

以上输出表明启动的 IPv6 RIPng 进程为 cisco，同时在 Se0/0/1 和 Sl0/0/0 接口下启用 RIPng。

（3）查看 RIPng 的数据库

```
R2#show ipv6 rip database
  RIP process "cisco", local RIB
  2020:12::/64, metric 2
      Serial0/0/0/FE80::223:33FF:FE64:4FC8, expires in 174 secs
  2020:23::/64, metric 2
      Serial0/0/1/FE80::223:4FF:FEE5:B220, expires in 173 secs
  2020:34::/64, metric 2, installed
      Serial0/0/1/FE80::223:4FF:FEE5:B220, expires in 173 secs
  2021:4444::/64, metric 3, installed
      Serial0/0/1/FE80::223:4FF:FEE5:B220, expires in 173 secs

  ::/0, metric 2, installed
```

Serial0/0/0/FE80::223:33FF:FE64:4FC8, expires in 174 secs

以上输出显示了 R2 的 RIPng 的数据库,其中 3 条 RIPng 路由条目被放入路由表中。

(4) 查看 RIPng 的更新过程

```
R2#debug ipv6 rip
RIP Routing Protocol debugging is on
R2#clear ipv6 route *
03:41:10: RIPng [default VRF]: a message has been received.
03:41:10: RIPng [Se0/0/1, default VRF]: response received from FE80::FA72:EAFF:FE69:18B8 for process "cisco".
03:41:10:        src=FE80::FA72:EAFF:FE69:18B8 (Serial0/0/1)    //发送更新源地址(接口)
03:41:10:        dst=FF02::9                                     //RIPng 组播更新地址
03:41:10:        sport=521, dport=521, length=92                 //RIPng 数据包的端口
03:41:10:        command=2, version=1, mbz=0, #rte=4
03:41:10:        tag=0, metric=1, prefix=2020:23::/64            //路由前缀
03:41:10:        tag=0, metric=1, prefix=2020:34::/64
03:41:10:        tag=0, metric=2, prefix=2021:4444::/64
03:41:10:        tag=0, metric=16, prefix=::/0
03:41:23: RIPng [Se0/0/1, default VRF]: process "cisco" is sending a multicast update.
//发送组播更新信息
03:41:23:        src=FE80::FA72:EAFF:FE69:1C78
03:41:23:        dst=FF02::9 (Serial0/0/1)
03:41:23:        sport=521, dport=521, length=72
03:41:23:        command=2, version=1, mbz=0, #rte=3
03:41:23:        tag=0, metric=1, prefix=2020:12::/64
03:41:23:        tag=0, metric=1, prefix=2020:23::/64
03:41:23:        tag=0, metric=2, prefix=::/0
03:41:23: RIPng [Se0/0/0, default VRF]: process "cisco" is sending a multicast update.
03:41:23:        src=FE80::FA72:EAFF:FE69:1C78
03:41:23:        dst=FF02::9 (Serial0/0/0)
03:41:23:        sport=521, dport=521, length=92
03:41:23:        command=2, version=1, mbz=0, #rte=4
03:41:23:        tag=0, metric=1, prefix=2020:12::/64
03:41:23:        tag=0, metric=1, prefix=2020:23::/64
03:41:23:        tag=0, metric=2, prefix=2020:34::/64
03:41:23:        tag=0, metric=3, prefix=2021:4444::/64
03:41:25: RIPng [default VRF]: a message has been received.
03:41:25: RIPng [Se0/0/0, default VRF]: response received from FE80::FA72:EAFF:FED6:F4C8 for process "cisco".
03:41:25:        src=FE80::FA72:EAFF:FED6:F4C8 (Serial0/0/0)
03:41:25:        dst=FF02::9
03:41:25:        sport=521, dport=521, length=52
03:41:25:        command=2, version=1, mbz=0, #rte=2
03:41:25:        tag=0, metric=1, prefix=2020:12::/64
03:41:25:        tag=0, metric=1, prefix=::/0
```

以上输出显示了路由器 R2 从接口 Se0/0/0 和 Se0/0/1 发送和接收 RIPng 路由更新信息的过程。

9.3.3 实验 5：配置 OSPFv3

1. 实验目的

通过本实验可以掌握：
① 启用 IPv6 路由的方法。
② 启用运行 OSPFv3 路由协议接口的方法。
③ 配置 OSPFv3 计时器和参考带宽的方法。
④ 向 OSPFv3 网络注入默认路由的方法。
⑤ 配置 OSPFv3 验证的方法。
⑥ 配置 OSPFv3 区域间路由汇总的方法。
⑦ OSPFv3 DR 选举的方法。
⑧ OSPFv3 OE1 和 OE2 路由的区别。
⑨ OSPFv3 链路状态数据库的特征和含义。
⑩ OSPFv3 LSA 的类型和特征。
⑪ OSPFv3 验证和调试的方法。

2. 实验拓扑

配置 OSPFv3 实验拓扑如图 9-12 所示。

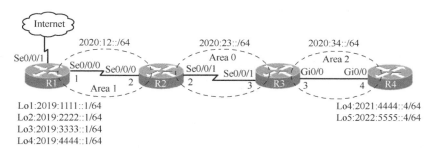

图 9-12　配置 OSPFv3 实验拓扑

3. 实验步骤

在路由器 R1 上向 OSPF 区域注入一条默认路由。在路由器 R4 上重分布直连路由，并用 route-map 对不同的直连路由配置不同的 OSPFv3 外部路由类型和种子度量值。

（1）配置路由器 R1

```
R1(config)#ipv6 unicast-routing
R1(config)#ipv6 router ospf 1                          //启动 OSPFv3 路由进程
R1(config-rtr)#router-id 1.1.1.1                       //定义路由器 ID，IPv4 格式，必须显式配置
R1(config-rtr)#auto-cost reference-bandwidth 1000      //修改计算 Cost 值的参考带宽
R1(config-rtr)#default-information originate metric 30 metric-type 2
//向 OSPFv3 网络注入一条默认路由，种子度量值为 30，类型为 OE2
R1(config-rtr)#passive-interface loopback1             //配置被动接口
```

```
R1(config-rtr)#passive-interface loopback2
R1(config-rtr)#passive-interface loopback3
R1(config-rtr)#passive-interface loopback4
R1(config)#interface range loopback 1 -4
R1(config-if-range)#ipv6 ospf 1 area 1
//接口下启用 OSPFv3 并声明接口所在区域
R1(config-if-range)#ipv6 ospf network point-to-point
//配置环回接口 OSPFv3 网络类型
R1(config)#interface Serial0/0/0
R1(config-if)#ipv6 address 2020:12::1/64
R1(config-if)#ipv6 ospf 1 area 1
R1(config-if)#ipv6 ospf hello-interval 5         //修改 OSPFv3 Hello 间隔
R1(config-if)#ipv6 ospf dead-interval 20         //修改 OSPFv3 Dead 时间
R1(config)#ipv6 route ::/0 serial0/0/1           //配置默认路由
```

（2）配置路由器 R2

```
R2(config)#ipv6 unicast-routing
R2(config)#ipv6 router ospf 1
R2(config-rtr)#router-id 2.2.2.2
R2(config-rtr)#auto-cost reference-bandwidth 1000
R2(config-rtr)#area 0 encryption ipsec spi 1212 esp des 1234567890ABCDEF md5 1234567890
1234567890123456789012
//开启区域 0 的 OSPFv3 数据包验证和加密，指定 IPSec 的 SPI 值以及加密、验证算法
R2(config-rtr)#area 1 range 2019::/16            //配置 OSPFv3 区域间路由汇总
R2(config)#interface Serial0/0/0
R2(config-if)#ipv6 address 2020:12::2/64
R2(config-if)#ipv6 ospf 1 area 1
R2(config-if)#ipv6 ospf hello-interval 5
R2(config-if)#ipv6 ospf dead-interval 20
R2(config)#interface Serial0/0/1
R2(config-if)#ipv6 address 2020:23::2/64
R2(config-if)# ipv6 ospf 1 area 0
```

（3）配置路由器 R3

```
R3(config)#ipv6 unicast-routing
R3(config)#ipv6 router ospf 1
R3(config-rtr)#router-id 3.3.3.3
R3(config-rtr)#auto-cost reference-bandwidth 1000
R3(config-rtr)#area 0 encryption ipsec spi 1212 esp des 1234567890ABCDEF md5 1234567890
1234567890123456789012
R3(config)#interface GigabitEthernet0/0
R3(config-if)#ipv6 address 2020:34::3/64
R3(config-if)#ipv6 ospf 1 area 2
R3(config-if)#ipv6 ospf priority 20              //修改接口优先级，控制 DR 选举，默认为 1
R3(config)#interface Serial0/0/1
R3(config-if)#ipv6 address 2020:23::3/64
R3(config-if)#ipv6 ospf 1 area 0
```

（4）配置路由器 R4

```
R4(config)#ipv6 unicast-routing
R4(config)#interface Loopback4
R4(config-if)#ipv6 address 2021:4444::4/64
R4(config)#interface Loopback5
R4(config-if)#ipv6 address 2022:5555::4/64
R4(config)#ipv6 access-list L1                              //配置 IPv6 ACL
R4(config-ipv6-acl)#permit ipv6 2021:4444::/64 any
R4(config)#ipv6 access-list L2
R4(config-ipv6-acl)#permit ipv6 2022:5555::/64 any
R4(config)#route-map REDIS permit 10
R4(config-route-map)#match ipv6 address L1
R4(config-route-map)#set metric 100
R4(config-route-map)#set metric-type type-1
R4(config)#route-map REDIS permit 20
R4(config-route-map)#match ipv6 address L2
R4(config-route-map)#set metric 200
R4(config)#route-map REDIS permit 30
R4(config)#ipv6 router ospf 1
R4(config-rtr)#router-id 4.4.4.4
R4(config-rtr)#auto-cost reference-bandwidth 1000
R4(config-rtr)#redistribute connected route-map REDIS
//重分布直连路由，并调用 route-map 定义的策略
R4(config)#interface GigabitEthernet0/0
R4(config-if)#ipv6 address 2020:34::4/64
R4(config-if)#ipv6 ospf 1 area 2
```

【技术要点】

OSPFv3 对 OSPFv2 做出了一些必要的改进，OSPFv2 和 OSPFv3 主要差异如表 9-1 所示。

表 9-1 OSPFv2 和 OSPFv3 主要差异

比 较 项	OSPFv2	OSPFv3
通告	IPv4 网络	IPv6 前缀
运行	基于网络	基于链路
源地址	接口 IPv4 地	接口 IPv6 链路本地地址
目的地址	● 邻居接口单播 IPv4 地址 ● 组播 224.0.0.5 或 224.0.0.6 地址	● 邻居 IPv6 链路本地地址 ● 组播 FF02::5 或 FF02::6 地址
通告网络	路由模式下采用 network 命令或接口下采用 ip ospf *process-id* area *area-id* 命令	接口下采用 ipv6 ospf *process-id* area *area-id* 命令
IP 单播路由	IPv4 单播路由，路由器默认启用	IPv6 单播路由，采用 ipv6 unicast-routing 命令启用
同一链路上运行多个实例	不支持	支持，通过 Instance ID 字段来实现
唯一标识邻居	取决于网络类型	通过 Router ID 标识
验证	简单口令或 MD5	使用 IPv6 提供的安全机制来保证自身数据包的安全性

比 较 项	OSPFv2	OSPFv3
包头	● 版本为 2 ● 包头长度为 24 字节 ● 含有验证字段	● 版本为 3 ● 包头长度为 16 字节 ● 去掉了认证字段，增加了 Instance ID 字段 ● OSPFv3 数据包包头格式如图 9-13 所示
LSA	有 Options 字段	取消 Options 字段，新增加了链路 LSA（类型 8）和区域内前缀 LSA（类型 9）

4．实验调试

（1）查看 IPv6 路由表

以下输出全部省略路由代码部分。

① R1#**show ipv6 route ospf**
OI 2020:23::/64 [**110**/1294]
        via FE80::FA72:EAFF:FE69:1C78, Serial0/0/0
OI 2020:34::/64 [**110**/1295]
        via FE80::FA72:EAFF:FE69:1C78, Serial0/0/0
OE2 2022:5555::/64 [**110**/20]
        via FE80::FA72:EAFF:FE69:1C78, Serial0/0/0
OE1 2021:4444::/64 [**110**/1395]
        via FE80::FA72:EAFF:FE69:1C78, Serial0/0/0

② R2#**show ipv6 route ospf**
OE2 ::/0 [110/30], tag 1   //该路由 tag 值为 1
        via FE80::223:33FF:FE64:4FC8, Serial0/0/0
O   2019::/16 [110/648]
        via **Null0**, directly connected   //此路由手工汇总后自动产生，用于防止路由环路
O   2019:1111::/64 [110/648]
        via FE80::FA72:EAFF:FED6:F4C8, Serial0/0/0
O   2019:2222::/64 [110/648]
        via FE80::FA72:EAFF:FED6:F4C8, Serial0/0/0
O   2019:3333::/64 [110/648]
        via FE80::FA72:EAFF:FED6:F4C8, Serial0/0/0
O   2019:4444::/64 [110/648]
        via FE80::FA72:EAFF:FED6:F4C8, Serial0/0/0
OI 2020:34::/64 [110/648]
        via FE80::FA72:EAFF:FE69:18B8, Serial0/0/1
OE2 2022:5555::/64 [110/20]
        via FE80::FA72:EAFF:FE69:18B8, Serial0/0/1
OE1 2021:4444::/64 [110/748]
        via FE80::FA72:EAFF:FE69:18B8, Serial0/0/1

③ R3#**show ipv6 route ospf**
OE1 ::/0 [110/1324], tag 1
        via FE80::FA72:EAFF:FE69:1C78, Serial0/0/1
OI 2019::/16 [110/1295]
        via FE80::FA72:EAFF:FE69:1C78, Serial0/0/1
OI 2020:12::/64 [110/1294]
        via FE80::FA72:EAFF:FE69:1C78, Serial0/0/1

```
   OE2 2022:5555::/64 [110/20]
       via FE80::FA72:EAFF:FEC8:4F98, GigabitEthernet0/0
   OE1 2021:4444::/64 [110/101]
       via FE80::FA72:EAFF:FEC8:4F98, GigabitEthernet0/0
④ R4#show ipv6 route ospf
   OE1 ::/0 [110/1325], tag 1
       via FE80::FA72:EAFF:FE69:18B8, GigabitEthernet0/0
   OI  2019::/16 [110/1296]
       via FE80::FA72:EAFF:FE69:18B8, GigabitEthernet0/0
   OI  2020:12::/64 [110/1295]
       via FE80::FA72:EAFF:FE69:18B8, GigabitEthernet0/0
   OI  2020:23::/64 [110/648]
       via FE80::FA72:EAFF:FE69:18B8, GigabitEthernet0/0
```

以上①、②、③和④输出表明 OSPFv3 的外部路由代码为 **OE2** 或 **OE1**，区域间路由代码为 **OI**，区域内路由代码为 **O**，OSPFv3 管理距离为 110；OE2 和 OE1 路由的区别与 OSPFv2 类似；R1 向 OSPFv3 区域注入了一条度量值为 30 的 OE2 默认路由。

（2）查看 IPv6 路由协议相关信息

```
R2#show ipv6 protocols
IPv6 Routing Protocol is "connected"
IPv6 Routing Protocol is "application"
IPv6 Routing Protocol is "ND"
IPv6 Routing Protocol is "ospf 1"
Router ID 2.2.2.2                                    //路由器 ID
  Area border router                                 //该路由器是 ABR
  Number of areas: 2 normal, 0 stub, 0 nssa          //区域的数量和类型
  Interfaces (Area 0):
    Serial0/0/1
  Interfaces (Area 1):
    Serial0/0/0
  Redistribution:
    None
```

以上输出表明路由器 R2 上启动的 OSPFv3 进程 ID 为 1；在 Se0/0/1 和 Se0/0/0 接口下启用 OSPFv3；Se0/0/1 接口属于区域 0，Se0/0/0 接口属于区域 1。

（3）查看 OSPFv3 链路状态数据库

```
R3#show ipv6 ospf database
 OSPFv3 Router with ID (3.3.3.3) (Process ID 1)      //OSPFv3 路由器 ID 以及进程 ID
            Router Link States (Area 0)              //路由器 LSA
ADV Router      Age       Seq#            Fragment ID   Link count  Bits
2.2.2.2         1907      0x8000000E      0             1           B
3.3.3.3         1831      0x8000000D      0             1           B
            Inter Area Prefix Link States (Area 0)   //区域间前缀 LSA
ADV Router      Age       Seq#            Prefix
2.2.2.2         644       0x80000001      2020:12::/64
2.2.2.2         644       0x80000001      2019::/16
3.3.3.3         630       0x80000001      2020:34::/64
            Inter Area Router Link States (Area 0)   //区域间路由器 LSA
```

ADV Router	Age	Seq#	Link ID	Dest RtrID	
2.2.2.2	57	0x80000001	16843009	1.1.1.1	
3.3.3.3	338	0x80000003	67372036	4.4.4.4	

Link (Type-8) Link States (Area 0)　　　　　　　　//链路 LSA

ADV Router	Age	Seq#	Link ID	Interface
2.2.2.2	1420	0x80000008	7	Se0/0/1
3.3.3.3	1333	0x80000005	7	Se0/0/1

Intra Area Prefix Link States (Area 0)　　　　　　//区域内前缀 LSA

ADV Router	Age	Seq#	Link ID	Ref-lstype	Ref-LSID
2.2.2.2	901	0x80000006	0	0x2001	0
3.3.3.3	832	0x80000006	0	0x2001	0

Router Link States (Area 2)　　　　　　　　　　　//路由器 LSA

ADV Router	Age	Seq#	Fragment ID	Link count	Bits
3.3.3.3	340	0x8000000E	0	1	B
4.4.4.4	335	0x8000000B	0	1	E

Net Link States (Area 2) //网络 LSA

ADV Router	Age	Seq#	Link ID	Rtr count
3.3.3.3	340	0x80000007	4	2

Inter Area Prefix Link States (Area 2)　　　　　　//区域间前缀 LSA

ADV Router	Age	Seq#	Prefix
3.3.3.3	860	0x80000001	2020:23::/64
3.3.3.3	860	0x80000001	2019::/16
3.3.3.3	860	0x80000001	2020:12::/64

Inter Area Router Link States (Area 2)　　　　　　//区域间路由器 LSA

ADV Router	Age	Seq#	Link ID	Dest RtrID
3.3.3.3	103	0x80000001	16843009	1.1.1.1

Link (Type-8) Link States (Area 2)

ADV Router	Age	Seq#	Link ID	Interface
3.3.3.3	1625	0x80000005	4	Gi0/0
4.4.4.4	382	0x80000007	4	Gi0/0

Intra Area Prefix Link States (Area 2)　　　　　　//链路 LSA

ADV Router	Age	Seq#	Link ID	Ref-lstype	Ref-LSID
3.3.3.3	386	0x80000007	4096	0x2002	4

Type-5 AS External Link States　　　　　　　　　//外部 LSA

ADV Router	Age	Seq#	Prefix
1.1.1.1	1362	0x80000003	::/0
4.4.4.4	383	0x80000006	2021:4444::/64
4.4.4.4	383	0x80000006	2022:5555::/64

以上输出显示了路由器 R3 的区域 0 和区域 2 的 OSPFv3 链路状态数据库的信息。

【技术要点】

OSPFv3 的 LSA 和 OSPFv2 的 LSA 的对比如表 9-2 所示。

① 在 OSPFv3 中 IPv6 地址不包含在 OSPFv3 数据包包头中，而是作为有效负载的一部分。OSPFv3 数据包包头格式如图 9-13 所示。OSPFv3 数据包包头新加入实例 ID（Instance ID）字段，如果需要在同一链路上隔离通信，可以在同一条链路上运行多个实例。实例 ID 相同才能彼此通信。默认情况下，实例 ID 为 0。同时 OSPFv3 去掉了 OSPFv2 数据包包头中验证字段，所以 OSPFv3 本身不提供验证功能，而是依赖于 IPv6 扩展包头的验证功能来保证数据包

的完整性和安全性，可以基于接口或者区域对 OSPFv3 数据包进行验证和加密，需要注意的是接口验证优先于区域验证。下面是区域 0 进行验证的实例：

> R2(config-rtr)#**area 0 encryption ipsec spi 1212 esp des 1234567890abcdef md5 12345678901234567890123456789012**

基于链路验证和加密的实例如下：

> R2(config-if)#**ipv6 ospf encryption ipsec spi 500 esp des 1234567890abcdef md5 12345678901234567890123456789012**

② OSPFv3 的路由器 LSA 和网络 LSA 不携带 IPv6 地址，而是将该功能放入区域内前缀 LSA 中，因此路由器 LSA 和网络 LSA 只代表路由器的节点信息。

③ OSPFv3 加入了新的链路 LSA 提供路由器链路本地地址，并列出了链路所有 IPv6 的前缀。

表 9-2　OSPFv3 的 LSA 和 OSPFv2 的 LSA 的对比

OSPFv3 LSA		OSPFv2 LSA	
类型代码	名　称	类型代码	名　称
0x2001	路由器 LSA	1	路由器 LSA
0x2002	网络 LSA	2	网络 LSA
0x2003	区域间前缀 LSA	3	网络汇总 LSA
0x2004	区域间路由器 LSA	4	ASBR 汇总 LSA
0x2005	外部 LSA	5	外部 LSA
0x2007	类型 7 LSA	7	NSSA 外部 LSA
0x2008	链路 LSA		
0x2009	区域内前缀 LSA		

0	7	8	15	16	23	24	31
版本=3		类　型		数据包长度			
路由器 ID							
区域 ID							
校验和				实例 ID		0	

OSPFv3 包头

```
□ Open Shortest Path First
  □ OSPF Header
     OSPF Version: 3
     Message Type: DB Description (2)
     Packet Length: 28
     Source OSPF Router: 1.1.1.1 (1.1.1.1)
     Area ID: 0.0.0.0 (Backbone)
     Packet Checksum: 0xc31d [correct]
     Instance ID: 0 (IPv6 unicast AF)
     Reserved: 0
```

图 9-13　OSPFv3 数据包包头格式

（4）查看 OSPFv3 邻居信息

```
R2#show ipv6 ospf neighbor
Neighbor ID     Pri    State        Dead Time    Interface ID    Interface
3.3.3.3          1     FULL/  -      00:00:38         7          Serial0/0/1
1.1.1.1          1     FULL/  -      00:00:17         6          Serial0/0/0
```

以上输出表明路由器 R2 有 2 个 OSPFv3 邻居，并且状态为 FULL。

（5）查看运行 OSPFv3 接口信息

```
R2#show ipv6 ospf interface Serial0/0/0
Serial0/0/0 is up, line protocol is up
    Link Local Address FE80::FA72:EAFF:FE69:1C78, Interface ID 4
    Area 1, Process ID 1, Instance ID 0, Router ID 2.2.2.2
    Network Type POINT_TO_POINT, Cost: 647
//可以通过 ipv6 ospf cost 命令修改接口 Cost 值
    Transmit Delay is 1 sec, State POINT_TO_POINT,
    Timer intervals configured, Hello 5, Dead 20, Wait 20, Retransmit 5
    （此处省略部分输出）
    Neighbor Count is 1, Adjacent neighbor count is 1
        Adjacent with neighbor 1.1.1.1
    Suppress hello for 0 neighbor(s)
```

以上输出是运行 OSPFv3 路由器接口的基本信息，包括实例 ID、路由器 ID、网络类型、计时器的值以及邻居的数量等信息。

（6）查看 OSPFv3 进程信息

```
① R3#show ipv6 ospf 1       //查看 OSPFv3 进程信息，包括 OSPFv3 进程 ID、路由器 ID、路由器类
型、重分布信息、OSPF 区域信息以及 SPF 算法执行次数等
    Routing Process "ospfv3 1" with ID 3.3.3.3
    （此处省略部分输出）
    It is an area border router                      //ABR 路由器
    （此处省略部分输出）
        Area BACKBONE(0)
        Number of interfaces in this area is 1
        DES Encryption MD5 Auth, SPI 1212            //区域采用 MD5 验证及 IPSec 的 SPI 值
        SPF algorithm executed 3 times               //SPF 算法运行次数
    （此处省略部分输出）
② R4#show ipv6 ospf
    Routing Process "ospfv3 1" with ID 4.4.4.4
    （此处省略部分输出）
    It is an autonomous system boundary router       //ASBR 路由器
    Redistributing External Routes from,
        connected with route-map REDIS               //重分布直连路由并应用 route-map 策略
    （此处省略部分输出）
    Number of external LSA 3. Checksum Sum 0x00D2B2  //外部 LSA 条数
    Number of areas in this router is 1. 1 normal 0 stub 0 nssa   //区域类型和数量
    Graceful restart helper support enabled
    Reference bandwidth unit is 1000 mbps            //计算 Cost 值的参考带宽
```

Area 2
　　Number of interfaces in this area is 1
（此处省略部分输出）

（7）查看活动的 IPSec VPN 会话

```
R2#show crypto engine connections active
Crypto Engine Connections
   ID  Type   Algorithm       Encrypt  Decrypt  LastSeqN  IP-Address
 2001  IPsec  DES+MD5            0       70        0      FE80::FA72:EAFF:FE69:1C78
 2002  IPsec  DES+MD5           70        0        0      FE80::FA72:EAFF:FE69:1C78
```

以上输出给出了建立的 IPSec VPN 会话 ID、使用的加密和验证算法、加密和解密的数据包的数量以及参与加密和验证的路由器接口的链路本地地址。

9.3.4　实验 6：配置 IPv6 EIGRP

1．实验目的

通过本实验可以掌握：
① 启用 IPv6 路由的方法。
② IPv6 EIGRP 配置和调试的方法。
③ IPv6 EIGRP 验证配置方法。
④ IPv6 EIGRP 手工汇总配置方法。

2．实验拓扑

配置 IPv6 EIGRP 实验拓扑如图 9-14 所示。

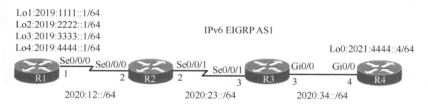

图 9-14　配置 IPv6 EIGRP 实验拓扑

3．实验步骤

本实验中，在路由器 R1 上手工汇总四条环回接口前缀的路由，同时在路由器 R3 和 R4 之间的链路上配置 IPv6 EIGRP 的 MD5 验证。

（1）配置路由器 R1

```
R1(config)#ipv6 unicast-routing
R1(config)#ipv6 router eigrp 1                    //配置 IPv6 EIGRP 进程，进程号为 1
R1(config-rtr)#eigrp router-id 1.1.1.1
//配置 IPv6 EIGRP 路由器 ID，必须显式配置
R1(config-rtr)#passive-interface Loopback1        //配置被动接口
```

```
R1(config-rtr)#passive-interface Loopback2
R1(config-rtr)#passive-interface Loopback3
R1(config-rtr)#passive-interface Loopback4
R1(config-rtr)#no shutdown                           //启动 IPv6 EIGRP 进程，默认已经开启
R1(config)#interface range loopback 1 -4
R1(config-if-range)#ipv6 eigrp 1                     //在接口下激活 IPv6 EIGRP
R1(config)#interface Serial0/0/0
R1(config-if)#ipv6 address 2020:12::1/64
R1(config-if)#ipv6 eigrp 1
R1(config-if)#ipv6 summary-address eigrp 1 2019::/16
//配置 IPv6 EIGRP 路由汇总
```

（2）配置路由器 R2

```
R2(config)#ipv6 unicast-routing
R2(config)#ipv6 router eigrp 1
R2(config-rtr)#eigrp router-id 2.2.2.2
R2(config-rtr)#no shutdown
R2(config)#interface Serial0/0/0
R2(config-if)#ipv6 address 2020:12::2/64
R2(config-if)#ipv6 eigrp 1
R2(config)#interface Serial0/0/1
R2(config-if)#ipv6 address 2020:23::2/64
R2(config-if)#ipv6 eigrp 1
```

（3）配置路由器 R3

```
R3(config)#ipv6 unicast-routing
R3(config)#key chain ccnp                            //配置密钥链
R3(config-keychain)#key 1                            //配置密钥 ID，无须连续，范围为 1～2147483647
R3(config-keychain-key)#key-string cisco123          //配置密钥字符串
R3(config)#ipv6 router eigrp 1
R3(config-rtr)#eigrp router-id 3.3.3.3
R3(config-rtr)#no shutdown
R3(config)#interface GigabitEthernet0/0
R3(config-if)#ipv6 address 2020:34::3/64
R3(config-if)#ipv6 eigrp 1
R3(config-if)#ipv6 authentication mode eigrp 1 md5   //配置验证模式为 MD5
R3(config-if)#ipv6 authentication key-chain eigrp 1 ccnp
//在接口下调用密钥链
R3(config)#interface Serial0/0/1
R3(config-if)#ipv6 address 2020:23::3/64
R3(config-if)#ipv6 eigrp 1
```

（4）配置路由器 R4

```
R4(config)#ipv6 unicast-routing
R4(config)#key chain ccnp
R4(config-keychain)#key 1
R4(config-keychain-key)#key-string cisco123
R4(config)#ipv6 router eigrp 1
R4(config-rtr)#eigrp router-id 4.4.4.4
```

```
R4(config-rtr)#redistribute connected metric 10000 100 255 1 1500
//将直连路由重分布到IPv6 EIGRP 进程中，指定种子度量值
R4(config-rtr)#no shutdown
R4(config)#interface Loopback0
R4(config-if)#ipv6 address 2021:4444::4/64
R4(config)#interface GigabitEthernet0/0
R4(config-if)#ipv6 address 2020:34::4/64
R4(config-if)#ipv6 eigrp 1
R4(config-if)#ipv6 authentication mode eigrp 1 md5
R4(config-if)#ipv6 authentication key-chain eigrp 1 ccnp
```

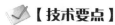【技术要点】

IPv6 的 EIGRP 类似于 IPv4 的 EIGRP，通过交换路由信息来填充 IPv6 路由表。IPv6 的 EIGRP 也使用 DUAL 作为计算引擎，以保证整个路由域中的无环路径和无环备用路径的正常运行。IPv4 的 EIGRP 和 IPv6 的 EIGRP 的主要功能比较如表 9-3 所示。

表 9-3　IPv4 的 EIGRP 和 IPv6 的 EIGRP 的主要功能比较

	IPv4 的 EIGRP	IPv6 的 EIGRP
通告路由	IPv4 网络	IPv6 前缀
距离矢量	是	是
收敛算法	DUAL	DUAL
度量标准	带宽、延迟、可靠性、负载	带宽、延迟、可靠性、负载
传输协议	RTP	RTP
更新方式	增量、部分和限定更新	增量、部分和限定更新
邻居发现维持	Hello 数据包	Hello 数据包
使用组播地址	224.0.0.10	FF02::A
身份验证	MD5、SHA256（命名 EIGRP 支持）	MD5、SHA256（命名 EIGRP 支持）
路由器 ID	与 IP 地址格式相同（可选）	与 IP 地址格式相同（必选）

4．实验调试

（1）查看 IPv6 EIGRP 的路由

以下各命令的输出全部省略 IPv6 路由代码部分。

```
① R1#show ipv6 route eigrp
D    2019::/16 [5/128256]
       via Null0, directly connected   //IPv6 EIGRP 使用 Null0 接口来丢弃与总结网络路由条目匹配
但与所有被总结子网前缀都不匹配的数据包，从而可以有效避免路由环路，该总结路由管理距离为 5
D    2020:23::/64 [90/2681856]
       via FE80::FA72:EAFF:FE69:1C78, Serial0/0/0
D    2020:34::/64 [90/2682112]
       via FE80::FA72:EAFF:FE69:1C78, Serial0/0/0
EX   2021:4444::/64 [170/2707712]
```

```
              via FE80::FA72:EAFF:FE69:1C78, Serial0/0/0
    ② R2#show ipv6 route eigrp
    D    2019::/16 [90/2297856]
              via FE80::FA72:EAFF:FED6:F4C8, Serial0/0/0
    D    2020:34::/64 [90/2172416]
              via FE80::FA72:EAFF:FE69:18B8, Serial0/0/1
    EX   2021:4444::/64 [170/2198016]
              via FE80::FA72:EAFF:FE69:18B8, Serial0/0/1
    ③ R3#show ipv6 route eigrp
    D    2019::/16 [90/2809856]
              via FE80::FA72:EAFF:FE69:1C78, Serial0/0/1
    D    2020:12::/64 [90/2681856]
              via FE80::FA72:EAFF:FE69:1C78, Serial0/0/1
    EX   2021:4444::/64 [170/284160]
              via FE80::FA72:EAFF:FEC8:4F98, GigabitEthernet0/0
    ④ R4#show ipv6 route eigrp
    D    2019::/16 [90/2810112]
              via FE80::FA72:EAFF:FE69:18B8, GigabitEthernet0/0
    D    2020:12::/64 [90/2682112]
              via FE80::FA72:EAFF:FE69:18B8, GigabitEthernet0/0
    D    2020:23::/64 [90/2170112]
              via FE80::FA72:EAFF:FE69:18B8, GigabitEthernet0/0
```

以上①、②、③和④输出说明 IPv6 EIGRP 路由条目的下一跳是其 EIGRP 邻居的链路本地地址，同时 IPv6 EIGRP 也能够区分内部路由和外部路由，内部路由代码为 **D**，管理距离为 90；外部路由代码为 **EX**，管理距离为 170。IPv6 EIGRP 路由度量值的计算方法和 IPv4 EIGRP 的相同。

（2）查看 IPv6 EIGRP 的邻居信息

```
R2#show ipv6 eigrp neighbors
IPv6-EIGRP neighbors for process 1           //进程号为 1 的 IPv6 EIGRP 邻居
H  Address                  Interface    Hold Uptime   SRTT   RTO   Q    Seq
                                         (sec)         (ms)         Cnt  Num
1  Link-local address:      Se0/0/1      14  00:04:38  1037   5000  0    43
   FE80::FA72:EAFF:FE69:18B8
0  Link-local address:      Se0/0/0      14  00:06:21  23     1140  0    26
   FE80::FA72:EAFF:FED6:F4C8
```

以上输出表明路由器 R2 有 2 个 IPv6 EIGRP 邻居，邻居的地址用对方的链路本地地址表示。其他列的含义和 IPv4 EIGRP 的相同。

（3）查看 IPv6 EIGRP 的拓扑表

```
R2#show ipv6 eigrp topology
IPv6-EIGRP Topology Table for AS(1)/ID(2.2.2.2)           //进程号为 1 的 IPv6 EIGRP 拓扑表
Codes: P - Passive, A - Active, U - Update, Q - Query, R - Reply,
       r - reply Status, s - sia Status
P 2019::/16, 1 successors, FD is 2297856
       via FE80::FA72:EAFF:FED6:F4C8 (2297856/128256), Serial0/0/0

P 2020:34::/64, 1 successors, FD is 2170112
```

第 9 章　IPv6

```
                via FE80::FA72:EAFF:FE69:18B8 (2170112/2816), Serial0/0/1
P 2020:12::/64, 1 successors, FD is 2169856
                via Connected, Serial0/0/0
P 2021:4444::/64, 1 successors, FD is 2195712
                via FE80::FA72:EAFF:FE69:18B8 (2195712/281856), Serial0/0/1
P 2020:23::/64, 1 successors, FD is 2169856
                via Connected, Serial0/0/1
```

以上输出显示了路由器 R2 的 IPv6 EIGRP 的拓扑表信息，包括路由前缀的状态、后继路由器、可行距离、所有可行后继路由器及其通告距离、下一跳链路本地地址和出接口等信息。状态代码的含义和 IPv4 EIGRP 的相同。

（4）查看 IPv6 路由协议相关信息

```
R1#show ipv6 protocols
IPv6 Routing Protocol is "connected"
IPv6 Routing Protocol is "application"
IPv6 Routing Protocol is "ND"
IPv6 Routing Protocol is "eigrp 1"              //IPv6 EIGRP 进程 ID 为 1
    EIGRP metric weight K1=1, K2=0, K3=1, K4=0, K5=0
                                                //计算度量值权重因子，默认 K1=K3=0
    Soft SIA disabled
    NSF-aware route hold timer is 240
    Router-ID: 1.1.1.1                          //EIGRP 路由器 ID
    Topology : 0 (base)
        Active Timer: 3 min                     //SIA 计时器
        Distance: internal 90 external 170      //管理距离
        Maximum path: 16                        //支持等价或非等价负载均衡的路由条目
        Maximum hopcount 100                    //IPv6 EIGRP 最大跳数
        Maximum metric variance 1               //variance 值为 1
    Interfaces:                                 //以下五行表示启用 IPv6 EIGRP 的接口
    Serial0/0/0
    Loopback1 (passive)                         //被动接口
    Loopback2 (passive)
    Loopback3 (passive)
    Loopback4 (passive)
    Redistribution:
        None                                    //没有重分布其他 IPv6 路由信息
    Address Summarization:
        2019::/16 for Se0/0/0                   //接口下配置 IPv6 EIGRP 路由手工总结
            Summarizing 4 components with metric 128256  //IPv6 EIGRP 总结路由的初始度量值
```

（5）查看 IPv6 EIGRP 发送和接收数据包统计情况

```
R2#show ipv6 eigrp traffic
IPv6-EIGRP Traffic Statistics for AS 1
    Hellos sent/received: 489/474               //接收和发送的 Hello 数据包的数量
    Updates sent/received: 46/39                //接收和发送的更新数据包的数量
    Queries sent/received: 5/3                  //接收和发送的查询数据包的数量
    Replies sent/received: 2/5                  //接收和发送的应答数据包的数量
```

```
            Acks sent/received: 22/25                          //接收和发送的确认数据包的数量
            SIA-Queries sent/received: 0/0
            SIA-Replies sent/received: 0/0
            Hello Process ID: 230                              //Hello 进程 ID
            PDM Process ID: 203                                //PDM 进程 ID
            Socket Queue: 0/10000/1/0 (current/max/highest/drops)
            Input Queue: 0/10000/1/0 (current/max/highest/drops)
```

（6）查看运行 IPv6 EIGRP 路由协议的接口的信息

```
R2#show ipv6 eigrp interfaces
IPv6-EIGRP interfaces for process 1
                   Xmit Queue    Mean    Pacing Time   Multicast   Pending
Interface   Peers  Un/Reliable   SRTT    Un/Reliable   Flow Timer  Routes
Se0/0/0       1       0/0         23       7/190          286         0
Se0/0/1       1       0/0        1037      7/190         6602         0
```

以上输出信息表明路由器 R2 有 2 个接口运行 IPv6 EIGRP，所有列的含义和 IPv4 EIGRP 的相同。

9.3.5 实验 7：配置 IPv6 集成 IS-IS

1. 实验目的

通过本实验可以掌握：
① 启动 IPv6 集成 IS-IS 路由进程的方法。
② 启用运行 IS-IS 路由协议接口的方法。
③ IPv6 IS-IS 的 L1 和 L2 路由的区别。
④ 配置 IS-IS L1 或 L2 路由器类型的方法。
⑤ 配置 IPv6 IS-IS 接口电路类型的方法。
⑥ 配置 IPv6 IS-IS 区域间路由汇总的方法。
⑦ 向 IPv6 IS-IS 区域注入默认路由的方法。

2. 实验拓扑

配置 IPv6 集成 IS-IS 实验拓扑如图 9-15 所示。

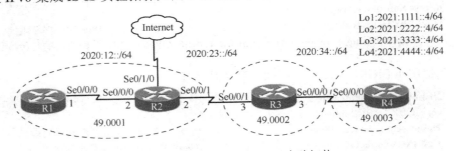

图 9-15　配置 IPv6 集成 IS-IS 实验拓扑

3. 实验步骤

本实验中，在路由器 R4 上手工汇总四条环回接口所在网络的路由，同时在路由器 R2 上向 IPv6 IS-IS 区域注入一条默认路由。

（1）配置路由器 R1

```
R1(config)#ipv6 unicast-routing
R1(config)#router isis cisco                        //启动 IS-IS 路由进程，IS-IS 进程的名字只有本地含义，一台路由器可以启动多个 IS-IS 进程
    R1(config-router)#net 49.0001.1111.1111.1111.00   //配置 NET 地址
    R1(config-router)#is-type level-1                 //将 R1 配置成 L1 路由器
    R1(config)#interface Serial0/0/0
    R1(config-if)#ipv6 address 2020:12::1/64
    R1(config-if)#ipv6 router isis cisco              //接口启用 IPv6 IS-IS
```

（2）配置路由器 R2

```
R2(config)#ipv6 unicast-routing
R2(config)#ipv6 route ::/0 Serial0/1/0
R2(config)#router isis cisco
R2(config-router)#net 49.0001.2222.2222.2222.00
R2(config-router)#address-family ipv6               //进入 IPv6 地址族
R2(config-router-af)#default-information originate
                                                    //向 IPv6 IS-IS 区域注入默认路由
R2(config-router-af)#exit-address-family            //退出 IPv6 地址族
R2(config)#interface Serial0/0/0
R2(config-if)#ipv6 address 2020:12::2/64
R2(config-if)#ipv6 router isis cisco
R2(config)#interface Serial0/0/1
R2(config-if)#ipv6 address 2020:23::2/64
R2(config-if)#ipv6 router isis cisco
```

（3）配置路由器 R3

```
R3(config)#ipv6 unicast-routing
R3(config)#router isis cisco
R3(config-router)#net 49.0002.3333.3333.3333.00
R3(config-router)#is-type level-2-only              //将 R3 配置成 L2 路由器
R3(config)#interface Serial0/0/0
R3(config-if)#ipv6 address 2020:34::3/64
R3(config-if)#ipv6 router isis cisco
R3(config-if)#isis circuit-type level-2-only        //配置接口电路类型
R3(config)#interface Serial0/0/1
R3(config-if)#ipv6 address 2020:23::3/64
R3(config-if)#ipv6 router isis cisco
```

（4）配置路由器 R4

```
R4(config)#ipv6 unicast-routing
R4(config)#router isis cisco
```

```
R4(config-router)#net 49.0003.4444.4444.4444.00
R4(config-router)#is-type level-2-only
R4(config-router)#address-family ipv6
R4(config-router-af)#summary-prefix 2021::/16              //配置区域间路由前缀汇总
R4(config-router-af)#exit-address-family
R4(config)#interface Serial0/0/0
R4(config-if)#ipv6 address 2020:34::4/64
R4(config-if)#ipv6 router isis cisco
R4(config-if)#isis circuit-type level-2-only
R4(config)#interface range loopback 1 -4
R4(config-if-range)#ipv6 router isis cisco
```

4. 实验调试

(1) 查看 IS-IS 数据库

```
show isis database
① R1#show isis database
IS-IS Level-1 Link State Database:
LSPID              LSP Seq Num      LSP Checksum     LSP Holdtime/Rcvd      ATT/P/OL
R1.00-00           * 0x00000004     0xF03C           947/*                  0/0/0
R2.00-00           0x00000004       0x0A1E           1059/1199              1/0/0
② R2#show isis database
Tag cisco:
IS-IS Level-1 Link State Database:
LSPID              LSP Seq Num      LSP Checksum     LSP Holdtime/Rcvd      ATT/P/OL
R1.00-00           0x00000004       0xF03C           921/1199               0/0/0
R2.00-00           * 0x00000004     0x0A1E           1034/*                 1/0/0
IS-IS Level-2 Link State Database:
LSPID              LSP Seq Num      LSP Checksum     LSP Holdtime/Rcvd      ATT/P/OL
R2.00-00           * 0x00000005     0xEB44           1029/*                 0/0/0
R3.00-00           0x00000004       0x2C63           1122/1199              0/0/0
R4.00-00           0x00000005       0x5F66           1131/1198              0/0/0
③ R3#show isis database
Tag cisco:
IS-IS Level-2 Link State Database:
LSPID              LSP Seq Num      LSP Checksum     LSP Holdtime/Rcvd      ATT/P/OL
R2.00-00           0x00000005       0xEB44           939/1199               0/0/0
R3.00-00           * 0x00000004     0x2C63           1034/*                 0/0/0
R4.00-00           0x00000005       0x5F66           1042/1199              0/0/0
④ R4#show isis database
Tag cisco:
IS-IS Level-2 Link State Database:
LSPID              LSP Seq Num      LSP Checksum     LSP Holdtime/Rcvd      ATT/P/OL
R2.00-00           0x00000005       0xEB44           1009/1104              0/0/0
R3.00-00           0x00000004       0x2C63           1009/1199              0/0/0
R4.00-00           * 0x00000005     0x5F66           1018/*                 0/0/0
```

以上输出表明：路由器 R1 为 L1 路由器，只维护 Level-1 的链路状态数据库；路由器 R2 为 L1/L2 路由器，同时为 Level-1 和 Level-2 维护单独的链路状态数据库，也表明所在区域有另一台路由器 R1；路由器 R3 和 R4 为 L2 路由器，只维护 Level-2 的链路状态数据库。

（2）查看和 CLNS 路由协议相关信息

```
R2#show clns protocol
IS-IS Router: cisco
    System Id: 2222.2222.2222.00    IS-Type: level-1-2
    //系统 ID 和 IS-IS 路由器类型
    Manual area address(es):
        49.0001
    Routing for area address(es):
        49.0001
    Interfaces supported by IS-IS:                //启用 IPv6 IS-IS 接口
        Serial0/0/1 - IPv6
        Serial0/0/0 - IPv6
    Redistribute:
        static (on by default)                    //默认重分布静态路由
    Distance for L2 CLNS routes: 110
    RRR level: none
    Generate narrow metrics: level-1-2
    Accept narrow metrics:    level-1-2
    Generate wide metrics:    none
    Accept wide metrics:      none
```

（3）查看 IS-IS 的邻居信息

```
R2#show clns neighbors
System Id      Interface    SNPA      State    Holdtime   Type  Protocol
R1             Se0/0/0      *HDLC*    Up       21         L1    IS-IS
R3             Se0/0/1      *HDLC*    Up       29         L2    IS-IS
```

以上输出表明路由器 R2 有 2 个邻居，路由器 R1 是 **L1** 类型，路由器 R3 是 **L2** 类型。由于 R1 和 R2 以及 R2 和 R3 之间的串行链路采用默认的 HDLC 封装，所以 SNPA 为***HDLC***。如果串行接口采用 PPP 封装，则 SNPA 为***PPP***。

（4）查看 IPv6 路由表

以下输出均省略路由代码部分。

```
① R1#show ipv6 route isis
I1    ::/0 [115/10]
        via FE80::FA72:EAFF:FE69:1C78, Serial0/0/0
I1    2020:23::/64 [115/20]
        via FE80::FA72:EAFF:FE69:1C78, Serial0/0/0
② R2#show ipv6 route isis
I2    2020:34::/64 [115/20]
        via FE80::FA72:EAFF:FE69:18B8, Serial0/0/1
I2    2021::/16 [115/30]
        via FE80::FA72:EAFF:FE69:18B8, Serial0/0/1
③ R3#show ipv6 route isis
I2    ::/0 [115/10]
        via FE80::FA72:EAFF:FE69:1C78, Serial0/0/1
```

```
    I2   2020:12::/64 [115/20]
             via FE80::FA72:EAFF:FE69:1C78, Serial0/0/1
    I2   2021::/16 [115/20]
             via FE80::FA72:EAFF:FEC8:4F98, Serial0/0/0
④ R4#show ipv6 route isis
    I2   ::/0 [115/20]
             via FE80::FA72:EAFF:FE69:18B8, Serial0/0/0
    I2   2020:12::/64 [115/30]
             via FE80::FA72:EAFF:FE69:18B8, Serial0/0/0
    I2   2020:23::/64 [115/20]
             via FE80::FA72:EAFF:FE69:18B8, Serial0/0/0
    IS   2021::/16 [115/10]
             via Null0, directly connected          //此路由是手工汇总后自动产生的,用于防止路由环路
```

以上输出表明：在 IPv6 IS-IS 中，Level-1 的路由代码为 I1， Level-2 的路由代码为 I2，管理距离为 115，接口的 Cost 值默认都为 10；由于 R1 为 L1 路由器，所以只有 I1 的路由条目和一条到最近的 L1/L2 路由器 R2 的默认路由条目；R3 和 R4 都是 L2 路由器，所以只有 I2 的路由条目；R1、R3 和 R4 都收到一条由 R2 注入的默认路由::/0；R2 和 R3 都收到路由器 R4 四个环回接口汇总的路由条目，同时 R4 的路由表自动生成一条指向 Null0 接口的路由代码为 IS 的路由条目，主要是为了避免路由环路。

（5）查看 clns 接口的信息

```
R3#show clns interface Serial0/0/0
Serial0/0/0 is up, line protocol is up
    Checksums enabled, MTU 1500, Encapsulation HDLC
    ERPDUs enabled, min. interval 10 msec.
    CLNS fast switching enabled
    CLNS SSE switching disabled
    DEC compatibility mode OFF for this interface
    Next ESH/ISH in 47 seconds
    Routing Protocol: IS-IS
        Circuit Type: level-2                              //电路类型
        Interface number 0x0, local circuit ID 0x100       //本地电路 ID
        Neighbor System-ID: R4
        //邻居系统 ID，可以通过命令 show isis hostname 查看主机名和系统 ID 的映射关系
        Level-2 Metric: 10, Priority: 64, Circuit ID: R3.00
        //接口 Level-2 的度量值、接口优先级以及电路 ID
        Level-2 IPv6 Metric: 10                            //接口 Cost 值
        Number of active level-2 adjacencies: 1            //该接口活动 L2 邻居的个数
        Next IS-IS Hello in 241 milliseconds               //距离发送下一个 Hello 数据包的时间
        if state UP                                        //接口状态
```

9.3.6　实验 8：配置 MBGP

1. 实验目的

通过本实验可以掌握：
① 启动 BGP 路由进程的方法。

② BGP 进程中通告 IPv6 前缀的方法。
③ 配置 IBGP 和 EBGP 邻居的方法。
④ 配置 BGP 路由更新源的方法。
⑤ 配置 next-hop-self 的方法。

2．实验拓扑

配置 MBGP 实验拓扑如图 9-16 所示。

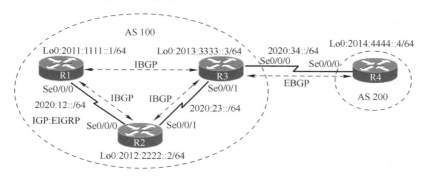

图 9-16　配置 MBGP 实验拓扑

3．实验步骤

本实验中 IBGP 的路由器（R1、R2 和 R3）形成全互连的邻居关系。IBGP 路由器之间运行的 IGP 是 IPv6 EIGRP，使得互相学到环回接口的 IPv6 路由，为 IBGP 使用环回接口作为更新源建立邻居关系提供连通性。

（1）配置路由器 R1

```
R1(config)#ipv6 unicast-routing
R1(config)#ipv6 router eigrp 1                        //配置 IPv6 EIGRP 进程，进程号为 1
R1(config-rtr)#eigrp router-id 1.1.1.1
R1(config)#interface Loopback0
R1(config-if)#ipv6 address 2011:1111::1/64
R1(config-if)#ipv6 eigrp 1
R1(config)#interface Serial0/0/0
R1(config-if)#ipv6 address 2020:12::1/64
R1(config-if)#ipv6 eigrp 1
R1(config)#router bgp 100                             //启动 BGP 进程
R1(config-router)#bgp router-id 1.1.1.1               //配置 BGP 路由器 ID
R1(config-router)#no bgp default ipv4-unicast         //关闭 IPv4 地址族 BGP 路由交换
R1(config-router)#neighbor 2012:2222::2 remote-as 100 //指定邻居及所在的 AS
R1(config-router)#neighbor 2012:2222::2 update-source Loopback0
//指定 BGP 更新源
R1(config-router)#neighbor 2013:3333::3 remote-as 100
R1(config-router)#neighbor 2013:3333::3 update-source Loopback0
R1(config-router)#address-family ipv6                 //进入 IPv6 地址族
R1(config-router-af)#neighbor 2012:2222::2 activate   //激活邻居
```

```
R1(config-router-af)#neighbor 2013:3333::3 activate
R1(config-router-af)#network 2011:1111::/64                //通告 IPv6 前缀
R1(config-router-af)#exit-address-family
```

（2）配置路由器 R2

```
R2(config)#ipv6 unicast-routing
R2(config)#ipv6 router eigrp 1
R2(config-rtr)#eigrp router-id 2.2.2.2
R2(config)#interface Loopback0
R2(config-if)#ipv6 address 2012:2222::2/64
R2(config-if)# ipv6 eigrp 1
R2(config)#interface Serial0/0/0
R2(config-if)#ipv6 address 2020:12::2/64
R2(config-if)# ipv6 eigrp 1
R2(config)#interface Serial0/0/1
R2(config-if)#ipv6 address 2020:23::2/64
R2(config-if)#ipv6 eigrp 1
R2(config)#router bgp 100
R2(config-router)#bgp router-id 2.2.2.2
R2(config-router)#no bgp default ipv4-unicast
R2(config-router)#neighbor 2011:1111::1 remote-as 100
R2(config-router)#neighbor 2011:1111::1 update-source Loopback0
R2(config-router)#neighbor 2013:3333::3 remote-as 100
R2(config-router)#neighbor 2013:3333::3 update-source Loopback0
R2(config-router)#address-family ipv6
R2(config-router-af)#neighbor 2011:1111::1 activate
R2(config-router-af)#neighbor 2013:3333::3 activate
R2(config-router-af)#exit-address-family
```

（3）配置路由器 R3

```
R3(config)#ipv6 unicast-routing
R3(config)#ipv6 router eigrp 1
R3(config-rtr)#eigrp router-id 3.3.3.3
R3(config)#interface Loopback0
R3(config-if)#ipv6 address 2013:3333::3/64
R3(config-if)# ipv6 eigrp 1
R3(config)#interface Serial0/0/0
R3(config-if)#ipv6 address 2020:34::3/64
R3(config)#interface Serial0/0/1
R3(config-if)#ipv6 address 2020:23::3/64
R3(config-if)# ipv6 eigrp 1
R3(config)#router bgp 100
R3(config-router)#bgp router-id 3.3.3.3
R3(config-router)#no bgp default ipv4-unicast
R3(config-router)#neighbor 2011:1111::1 remote-as 100
R3(config-router)#neighbor 2011:1111::1 update-source Loopback0
R3(config-router)#neighbor 2012:2222::2 remote-as 100
R3(config-router)#neighbor 2012:2222::2 update-source Loopback0
R3(config-router)#neighbor 2020:34::4 remote-as 200
```

第 9 章 IPv6

```
R3(config-router)#address-family ipv6
R3(config-router-af)#neighbor 2011:1111::1 activate
R3(config-router-af)#neighbor 2011:1111::1 next-hop-self        //配置下一跳自我
R3(config-router-af)#neighbor 2012:2222::2 activate
R3(config-router-af)#neighbor 2012:2222::2 next-hop-self
R3(config-router-af)#neighbor 2020:34::4 activate
R3(config-router-af)#exit-address-family
```

（4）配置路由器 R4

```
R4(config)#ipv6 unicast-routing
R4(config)#interface Loopback0
R4(config-if)#ipv6 address 2014:4444::4/64
R4(config)#interface Serial0/0/0
R4(config-if)#ipv6 address 2020:34::4/64
R4(config)#router bgp 200
R4(config-router)#bgp router-id 4.4.4.4
R4(config-router)#no bgp default ipv4-unicast
R4(config-router)#neighbor 2020:34::3 remote-as 100
R4(config-router)#address-family ipv6
R4(config-router-af)#neighbor 2020:34::3 activate
R4(config-router-af)#network 2014:4444::/64
R4(config-router-af)#exit-address-family
```

4．实验调试

（1）查看 TCP 连接信息摘要

```
R3#show tcp brief
TCB         Local Address           Foreign Address         (state)
23B1B38C    2020:34::3.179          2020:34::4.51816        ESTAB
3FC76418    2013:3333::3.179        2011:1111::1.58430      ESTAB
3A0BDC14    2013:3333::3.179        2012:2222::2.46484      ESTAB
```

以上输出表明路由器 R1、R2 和 R4 与路由器 R3 的 179 端口建立了 TCP 连接，说明 IPv6 BGP 也使用 TCP（端口 179）作为传输协议。

（2）查看 IPv6 BGP 的摘要信息

```
R3#show ip bgp ipv6 unicast summary
BGP router identifier 3.3.3.3, local AS number 100          //BGP 路由器 ID 及本地 AS
BGP table version is 5, main routing table version 5
//BGP 表的版本号（当 BGP 表变化时该号会逐次加 1）
2 network entries using 336 bytes of memory
2 path entries using 224 bytes of memory
2/2 BGP path/bestpath attribute entries using 320 bytes of memory
1 BGP AS-PATH entries using 24 bytes of memory
0 BGP route-map cache entries using 0 bytes of memory
0 BGP filter-list cache entries using 0 bytes of memory
BGP using 904 total bytes of memory
//以上七行显示了 BGP 使用内存的情况
```

```
BGP activity 2/0 prefixes, 2/0 paths, scan interval 60 secs
//前缀、路径和扫描间隔
Neighbor        V       AS   MsgRcvd MsgSent   TblVer   InQ OutQ Up/Down    State/PfxRcd
2011:1111::1    4      100      10      11        5      0    0  00:03:23         1
2012:2222::2    4      100       8       8        5      0    0  00:03:05         0
2020:34::4      4      200       6       7        5      0    0  00:01:55         1
```

（3）查看 IPv6 BGP 表的信息

```
R3#show ip bgp ipv6 unicast
① R3#show ip bgp ipv6 unicast
BGP table version is 3, local router ID is 3.3.3.3
Status codes: s suppressed, d damped, h history, * valid, > best, i - internal,
              r RIB-failure, S Stale
Origin codes: i - IGP, e - EGP, ? - incomplete
      Network          Next Hop        Metric LocPrf  Weight   Path
r>i  2011:1111::/64   2011:1111::1        0    100      0      i
*>   2014:4444::/64   2020:34::4          0             0    200 i
② R4#show ip bgp ipv6 unicast
BGP table version is 3, local router ID is 4.4.4.4
Status codes: s suppressed, d damped, h history, * valid, > best, i - internal,
              r RIB-failure, S Stale
Origin codes: i - IGP, e - EGP, ? - incomplete
      Network          Next Hop        Metric LocPrf  Weight   Path
*>   2011:1111::/64   2020:34::3                        0    100 i
*>   2014:4444::/64   ::                    0         32768    i
```

IPv6 BGP 表中代码和各 BGP 路由条目的含义请参考 7.2.1 实验 1，这里不再一一解释。

（4）查看 IPv6 BGP 路由条目

以下输出全部省略路由代码部分。

```
① R1#show ipv6 route bgp
B   2014:4444::/64 [200/0]
     via 2013:3333::3
② R2#show ipv6 route bgp
B   2014:4444::/64 [200/0]
     via 2013:3333::3
③ R3#show ipv6 route bgp
B   2014:4444::/64 [20/0]
     via FE80::FA72:EAFF:FEC8:4F98, Serial0/0/0
④ R4#show ipv6 route bgp
B   2011:1111::/64 [20/0]
     via FE80::FA72:EAFF:FE69:18B8, Serial0/0/0
```

以上输出省略路由代码部分，路由表条目中 IPv6 IBGP 的管理距离是 200，而 IPv6 EBGP 的管理距离是 20。

（5）执行 ping 命令

```
R4#ping 2011:1111::1 source 2014:4444::4
Type escape sequence to abort.
```

Sending 5, 100-byte ICMP Echos to 2011:1111::1, timeout is 2 seconds:
Packet sent with a source address of 2014:4444::4
!!!!!
Success rate is 100 percent (5/5), round-trip min/avg/max = 96/117/132 ms

9.4 IPv6 路由重分布和策略路由

9.4.1 实验 9：配置 OSPFv3、IS-IS 和 MBGP 路由重分布

1．实验目的

通过本实验可以掌握：
① 种子度量值的配置。
② 路由重分布参数的含义。
③ MBGP 和 OSPFv3 的重分布方法。
④ 直连路由与 IS-IS 的重分布方法。
⑤ IS-IS 和 MBGP 的重分布方法。

2．实验拓扑

配置 OSPFv3、IS-IS 和 MBGP 重分布实验拓扑如图 9-17 所示。

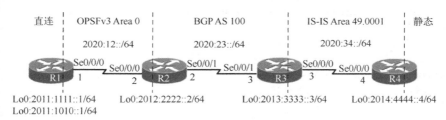

图 9-17　配置 OSPFv3、IS-IS 和 MBGP 重分布实验拓扑

3．实验步骤

（1）配置路由器 R1

```
R1(config)#ipv6 unicast-routing
R1(config)#interface Loopback0
R1(config-if)#ipv6 address 2011:1111::1/64
R1(config)#interface Loopback1
R1(config-if)#ipv6 address 2011:1010::1/64
R1(config)#ipv6 prefix-list A1 seq 5 permit 2011:1010::/64
//定义 IPv6 前缀列表
R1(config)#ipv6 prefix-list A2 seq 5 permit 2011:1111::/64
R1(config)#route-map CONN permit 10
R1(config-route-map)#match ipv6 address prefix-list A1          //匹配前缀列表
```

```
R1(config-route-map)#set metric-type type-1
R1(config-route-map)#set metric 100
R1(config)#route-map CONN permit 20
R1(config-route-map)#match ipv6 address prefix-list A2
R1(config-route-map)#set tag 100
R1(config-route-map)#set metric 200
R1(config)#ipv6 router ospf 1
R1(config-rtr)#router-id 1.1.1.1
R1(config-rtr)#auto-cost reference-bandwidth 1000
R1(config-rtr)#redistribute connected route-map CONN
//重分布直连路由，调用 route-map
R1(config)#interface Serial0/0/0
R1(config-if)#ipv6 address 2020:12::1/64
R1(config-if)#ipv6 ospf 1 area 0
```

（2）配置路由器 R2

```
R2(config)#ipv6 unicast-routing
R2(config)#interface Loopback0
R2(config-if)#ipv6 address 2012:2222::2/64
R2(config)#ipv6 router ospf 1
R2(config-rtr)#router-id 1.1.1.1
R2(config-rtr)#auto-cost reference-bandwidth 1000
R2(config-rtr)#redistribute connected metric 100 metric-type 1
//在将 BGP 路由重分布到 IGP 路由进程时，直连路由不被重分布，即使配置了关键字 include-connected，所以此处还要重分布直连路由
R2(config-rtr)#redistribute bgp 100 metric 100 metric-type 1
R2(config)#interface Serial0/0/0
R2(config-if)#ipv6 address 2020:12::2/64
R2(config-if)#ipv6 ospf 1 area 0
R2(config)#interface Serial0/0/1
R2(config-if)#ipv6 address 2020:23::2/64
R2(config)#router bgp 100
R2(config-router)#bgp router-id 2.2.2.2
R2(config-router)#no bgp default ipv4-unicast
R2(config-router)#neighbor 2020:23::3 remote-as 100
R2(config-router)#address-family ipv6
R2(config-router-af)#neighbor 2020:23::3 activate
R2(config-router-af)#redistribute ospf 1 match external 1 external 2 include-connected
                    //将直连路由、OSPF 外部 OE1 和 OE2 类型路由重分布进 BGP 进程中
R2(config-router-af)#bgp redistribute-internal
//默认情况下，BGP 不会将从 IBGP 学到的路由重分布到 IGP 中，该命令使得 IBGP 路由能够被重分布到 IGP 路由进程中
R2(config-router-af)#network 2012:2222::/64
R2(config-router-af)#exit-address-family
```

（3）配置路由器 R3

```
R3(config)#ipv6 unicast-routing
R3(config)#interface Loopback0
R3(config-if)#ipv6 address 2013:3333::3/64
```

```
R3(config)#interface Serial0/0/1
R3(config-if)#ipv6 address 2020:23::3/64
R3(config)#router isis cisco
R3(config-router)#net 49.0001.3333.3333.3333.00
R3(config-router)#is-type level-2-only
R3(config-router)#address-family ipv6
R3(config-router-af)#redistribute bgp 100
R3(config-router-af)#exit-address-family
R3(config)#interface Serial0/0/0
R3(config-if)#ipv6 address 2020:34::3/64
R3(config-if)#ipv6 router isis cisco
R3(config)#router bgp 100
R3(config-router)#bgp router-id 3.3.3.3
R3(config-router)#no bgp default ipv4-unicast
R3(config-router)#neighbor 2020:23::2 remote-as 100
R3(config-router)#address-family ipv6
R3(config-router-af)#neighbor 2020:23::2 activate
R3(config-router-af)#bgp redistribute-internal
R3(config-router-af)#network 2013:3333::/64
R3(config-router-af)# redistribute isis cisco level-2 include-connected
//重分布 IS-IS 路由，包括将直连路由重分布到 BGP 进程中
R3(config-router-af)#exit-address-family
```

（4）配置路由器 R4

```
R4(config)#ipv6 unicast-routing
R4(config)#ipv6 route 2044:4444::/64 null0
R4(config)#router isis cisco
R4(config-router)#net 49.0001.4444.4444.4444.00
R4(config-router)#is-type level-2-only
R4(config-router)#address-family ipv6
R4(config-router-af)#redistribute static metric 30
//重分布静态路由到 IS-IS 进程中
R4(config-router-af)#exit-address-family
R4(config)#interface Loopback0
R4(config-if)#ipv6 address 2014:4444::4/64
R4(config-if)#ipv6 router isis cisco
R4(config)#interface Serial0/0/0
R4(config-if)#ipv6 address 2020:34::4/64
R4(config-if)#ipv6 router isis cisco
```

4．实验调试

（1）查看 IPv6 路由表

以下输出均省略路由代码部分以及以 C 和 L 开头的路由条目。

```
① R1#show ipv6 route ospf
OE1 2012:2222::/64 [110/747]
     via FE80::FA72:EAFF:FE69:1C78, Serial0/0/0
OE1 2013:3333::/64 [110/747]
```

```
            via FE80::FA72:EAFF:FE69:1C78, Serial0/0/0
OE1  2014:4444::/64 [110/747]
            via FE80::FA72:EAFF:FE69:1C78, Serial0/0/0
OE1  2020:23::/64 [110/747]
            via FE80::FA72:EAFF:FE69:1C78, Serial0/0/0
OE1  2020:34::/64 [110/747]
            via FE80::FA72:EAFF:FE69:1C78, Serial0/0/0
OE1  2044:4444::/64 [110/747]
            via FE80::FA72:EAFF:FE69:1C78, Serial0/0/0
② R2#show ipv6 route
OE1  2011:1010::/64 [110/747]
            via FE80::FA72:EAFF:FED6:F4C8, Serial0/0/0
OE2  2011:1111::/64 [110/200], tag 200
            via FE80::FA72:EAFF:FED6:F4C8, Serial0/0/0
B    2013:3333::/64 [200/0]
        via 2020:23::3
B    2014:4444::/64 [200/20]
        via 2020:23::3
B    2020:34::/64 [200/0]
        via 2020:23::3
B    2044:4444::/64 [200/40]
        via 2020:23::3
③ R3#show ipv6 route
B    2011:1010::/64 [200/747]
        via 2020:23::2
B    2011:1111::/64 [200/200]
        via 2020:23::2
B    2012:2222::/64 [200/0]
        via 2020:23::2
I2   2014:4444::/64 [115/20]
            via FE80::FA72:EAFF:FEC8:4F98, Serial0/0/0
B    2020:12::/64 [200/0]
        via 2020:23::2
I2   2044:4444::/64 [115/**40**]
            via FE80::FA72:EAFF:FEC8:4F98, Serial0/0/0
④ R4#show ipv6 route
I2   2011:1010::/64 [115/10]
            via FE80::FA72:EAFF:FE69:18B8, Serial0/0/0
I2   2011:1111::/64 [115/10]
            via FE80::FA72:EAFF:FE69:18B8, Serial0/0/0
I2   2012:2222::/64 [115/10]
            via FE80::FA72:EAFF:FE69:18B8, Serial0/0/0
I2   2020:12::/64 [115/10]
            via FE80::FA72:EAFF:FE69:18B8, Serial0/0/0
S    2044:4444::/64 [1/0]
        via Null0, directly connected
```

以上①、②、③和④输出表明每台路由器都学到整个网络的路由信息，从而实现了不同路由协议之间 IPv6 路由信息的共享。

（2）查看 IPv6 路由协议相关信息

① R2#**show ipv6 protocols**
IPv6 Routing Protocol is "connected"
IPv6 Routing Protocol is "application"
IPv6 Routing Protocol is "ND"
IPv6 Routing Protocol is "static"
IPv6 Routing Protocol is "**ospf 1**"
  Router ID 2.2.2.2
  Autonomous system boundary router
  Number of areas: 1 normal, 0 stub, 0 nssa
  Interfaces (Area 0):
    Serial0/0/0
  Redistribution:
    Redistributing protocol **connected with metric 100 type 1**
    Redistributing protocol **bgp 100 with metric 100 type 1**
IPv6 Routing Protocol is **"bgp 100"**
  IGP synchronization is disabled
  Redistribution:
    Redistributing protocol **ospf 1 (internal, external 1 & 2,) include-connected**
  Neighbor(s):
    Address   FiltIn FiltOut Weight RoutemapIn RoutemapOut
    2020:23::3
  Distance:

以上输出表明了路由器 R2 上 OSPFv3 和 MBGP 之间双向重分布的情况。

② R3#**show ipv6 protocols**
IPv6 Routing Protocol is "connected"
IPv6 Routing Protocol is "application"
IPv6 Routing Protocol is "ND"
IPv6 Routing Protocol is **"isis cisco"**
  Interfaces:
    Serial0/0/0
  Redistribution:
    Redistributing protocol **bgp 100 at level 2**
IPv6 Routing Protocol is **"bgp 100"**
  IGP synchronization is disabled
  Redistribution:
    Redistributing protocol **isis cisco level 2 include-connected**
  Neighbor(s):
    Address   FiltIn FiltOut Weight RoutemapIn RoutemapOut
    2020:23::2
  Distance:

以上输出表明了路由器 R3 上 IS-IS 和 MBGP 之间双向重分布的情况。

9.4.2 实验 10：配置 IPv6 策略路由

1. 实验目的

通过本实验可以掌握：
① 用 route-map 定义 IPv6 路由策略的方法。
② 在接口下应用路由策略的方法。
③ 基于源 IPv6 地址的策略路由的配置和调试方法。

2. 实验拓扑

配置 IPv6 策略路由实验拓扑如图 9-18 所示。

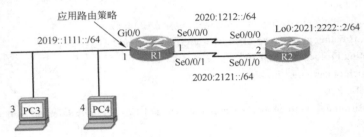

图 9-18　配置 IPv6 策略路由实验拓扑

3. 实验步骤

本实验中，在路由器 R1 的 Gi0/0 接口应用 IPv6 路由策略，为从主机 PC3 来的数据设置出接口 Se0/0/0，为从主机 PC4 来的数据设置出接口 Se0/0/1，所有其他的数据包正常转发，整个网络运行 IPv6 EIGRP 路由协议。

（1）配置路由器 R1

```
R1(config)#ipv6 unicast-routing
R1(config)#ipv6 access-list P1
R1(config-ipv6-acl)#permit ipv6 host 2019:1111::3 any
R1(config)#ipv6 access-list P2
R1(config-ipv6-acl)#permit ipv6 host 2019:1111::4 any
R1(config)#route-map PBR permit 10                          //定义策略
R1(config-route-map)#match ipv6 address P1
R1(config-route-map)#set interface Serial0/0/0
R1(config)#route-map PBR permit 20
R1(config-route-map)#match ipv6 address P2
R1(config-route-map)#set interface Serial0/0/1
R1(config)#ipv6 router eigrp 1
R1(config-rtr)#eigrp router-id 1.1.1.1
R1(config)#interface gigabitEthernet0/0
R1(config-if)#ipv6 address 2019:1111::1/64
R1(config-if)#ipv6 eigrp 1
R1(config-if)#ipv6 policy route-map PBR                      //应用路由策略
```

```
R1(config)#interface Serial0/0/0
R1(config-if)#ipv6 address 2020:1212::1/64
R1(config-if)#ipv6 eigrp 1
R1(config)#interface Serial0/0/1
R1(config-if)#ipv6 address 2020:2121::1/64
R1(config-if)# ipv6 eigrp 1
```

（2）配置路由器 R2

```
R2(config)#ipv6 unicast-routing
R1(config)#ipv6 router eigrp 1
R1(config-rtr)#eigrp router-id    2.2.2.2
R2(config)#interface Loopback0
R2(config-if)#ipv6 address 2021:2222::2/64
R2(config-if)#ipv6 eigrp 1
R2(config)#interface Serial0/0/0
R2(config-if)#ipv6 address 2020:1212::2/64
R2(config-if)#ipv6 eigrp 1
R2(config)#interface Serial0/1/0
R2(config-if)#ipv6 address 2020:2121::2/64
R2(config-if)#ipv6 eigrp 1
```

4．实验调试

（1）在主机 PC3（路由器 R3 模拟）上执行 ping 命令

```
R3#ping 2021:2222::2 repeat 1
```

路由器 R1 上显示的调试信息如下：

```
R1#debug ipv6 policy
01:55:31: IPv6 PBR (CEF): GigabitEthernet0/0, matched src 2019:1111::3 dst 2021:2222::2 protocol 58
01:55:31: IPv6 PBR (CEF): Policy route via Serial0/0/0
```

以上输出信息表明源地址为 2019:1111::3 的主机发送给目的地址为 2021:2222::2 的数据包（协议号 58 表示是 ICMPv6 数据包）在接口 Gi0/0 上匹配路由策略，执行策略路由，设置数据包出接口为 Se0/0/0。

（2）在主机 PC4（路由器 R3 模拟）上执行 ping 命令

```
R3#ping 2021:2222::2 repeat 1
```

路由器 R1 上显示的调试信息如下：

```
02:10:17: IPv6 PBR (CEF): GigabitEthernet0/0, matched src 2019:1111::4 dst 2021:2222::2 protocol 58
02:10:17: IPv6 PBR (CEF): Policy route via Serial0/0/1
```

以上输出信息表明源地址为 2019:1111::4 的主机发送给目的地址为 2021:2222::2 的数据包（协议号 58 表示是 ICMPv6 数据包）在接口 Gi0/0 上匹配路由策略，执行策略路由，设置数据包出接口为 Se0/0/1。

（3）查看接口上应用的路由策略

```
R1#show ipv6 policy
Interface                    Routemap
GigabitEthernet0/0           PBR
```

以上输出信息表明在 Gi0/0 接口上应用了路由策略 PBR。

（4）查看定义的路由策略及路由策略匹配情况

```
R1#show route-map
route-map PBR, permit, sequence 10
  Match clauses:
      ipv6 address P1
  Set clauses:
      interface Serial0/0/0
  Policy routing matches: 20 packets, 1316 bytes      //匹配策略路由的数据包的数量和字节数
route-map PBR, permit, sequence 20
  Match clauses:
      ipv6 address P2
  Set clauses:
      interface Serial0/0/1
  Policy routing matches: 28 packets, 1828 bytes      //匹配策略路由的数据包的数量和字节数
```

9.5 配置 IPv4 向 IPv6 过渡

9.5.1 实验 11：配置手工隧道

1. 实验目的

通过本实验可以掌握：
① 启用 IPv6 路由的方法。
② IPv6 与 IPv4 共存的实现方法。
③ 手工隧道的工作原理和配置。

2. 实验拓扑

配置 IPv6 手工隧道与 GRE 隧道实验拓扑如图 9-19 所示。

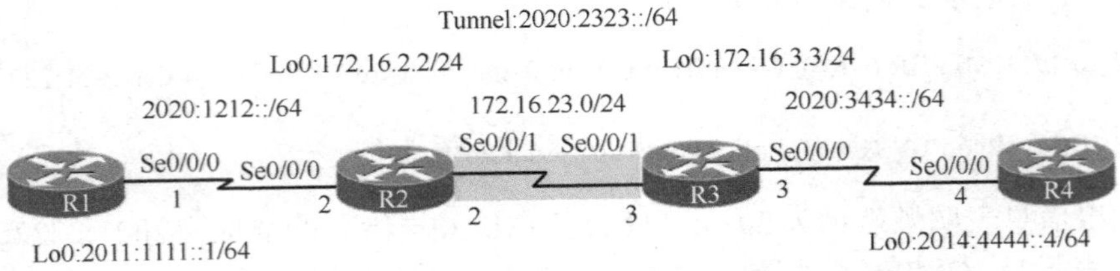

图 9-19 配置 IPv6 手工隧道与 GRE 隧道实验拓扑

第 9 章　IPv6

3. 实验步骤

本实验通过在路由器 R2 和 R3 之间创建手工隧道，把两端的 IPv6 网络连接起来并运行 OSPFv3，实现 IPv4 和 IPv6 网络共存，确保整个 IPv6 网络的互通。路由器 R2 和 R3 之间运行 IPv4 EIGRP 协议，为使用环回接口作为隧道接口的源和目的地提供连通性。

（1）配置路由器 R1

```
R1(config)#ipv6 unicast-routing
R1(config)#ipv6 router ospf 1
R1(config-rtr)#router-id 1.1.1.1
R1(config)#interface Loopback0
R1(config-if)#ipv6 address 2011:1111::1/64
R1(config-if)#ipv6 ospf 1 area 0
R1(config)#interface Serial0/0/0
R1(config-if)#ipv6 address 2020:1212::1/64
R1(config-if)#ipv6 ospf 1 area 0
```

（2）配置路由器 R2

```
R2(config)#ipv6 unicast-routing
R2(config)#interface Loopback0
R2(config-if)#ip address 172.16.2.2 255.255.255.0
R2(config)#interface Serial0/0/1
R2(config-if)#ip address 172.16.23.2 255.255.255.0
R2(config)#router eigrp 1
R2(config-router)#network 172.16.2.2 0.0.0.0
R2(config-router)#network 172.16.23.2 0.0.0.0
R2(config)#ipv6 router ospf 1
R2(config-rtr)#router-id 2.2.2.2
R2(config)#interface Serial0/0/0
R2(config-if)#ipv6 address 2020:1212::2/64
R2(config-if)#ipv6 ospf 1 area 0
R2(config)#interface Tunnel0                          //配置隧道接口
R2(config-if)#ipv6 address 2020:2323::2/64            //配置隧道接口 IPv6 地址
R2(config-if)#ipv6 ospf 1 area 0
R2(config-if)#tunnel source Loopback0                 //配置隧道源接口
R2(config-if)#tunnel destination 172.16.3.3           //配置隧道的目的地址
R2(config-if)#tunnel mode ipv6ip
//配置隧道的模式，ipv6ip 模式用来创建手工隧道
```

（3）配置路由器 R3

```
R3(config)#ipv6 unicast-routing
R3(config)#interface Loopback0
R3(config-if)#ip address 172.16.3.3 255.255.255.0
R3(config)#interface Serial0/0/1
R3(config-if)#ip address 172.16.23.3 255.255.255.0
R3(config)#router eigrp 1
R3(config-router)#network 172.16.3.3 0.0.0.0
R3(config-router)#network 172.16.23.3 0.0.0.0
```

```
R3(config)#ipv6 router ospf 1
R3(config-rtr)#router-id 3.3.3.3
R3(config)#interface Serial0/0/0
R3(config-if)#ipv6 address 2020:3434::3/64
R3(config-if)#ipv6 ospf 1 area 0
R3(config)#interface Tunnel0
R3(config-if)#ipv6 address 2020:2323::3/64
R3(config-if)#ipv6 ospf 1 area 0
R3(config-if)#tunnel source Loopback0
R3(config-if)#tunnel destination 172.16.2.2
R3(config-if)#tunnel mode ipv6ip
```

（4）配置路由器 R4

```
R4(config)#ipv6 unicast-routing
R4(config)#ipv6 router ospf 1
R4(config-rtr)#router-id 4.4.4.4
R4(config)#interface Loopback0
R4(config-if)#ipv6 address 2014:4444::4/64
R4(config-if)#ipv6 ospf 1 area 0
R4(config)#interface Serial0/0/0
R4(config-if)#ipv6 address 2020:3434::4/64
R4(config-if)#ipv6 ospf 1 area 0
```

4. 实验调试

（1）查看隧道建立过程

```
R3#debug tunnel
R3#ping 2020:2323::2 repeat 1
Type escape sequence to abort.
Sending 1, 100-byte ICMP Echos to 2020:2323::2, timeout is 2 seconds:
!
Success rate is 100 percent (1/1), round-trip min/avg/max = 20/20/20 ms
00:16:51: Tunnel0: IPv6/IP encapsulated 172.16.3.3->172.16.2.2 (linktype=79, len=120)
00:16:51: Tunnel0: IPv6/IP to classify 172.16.2.2->172.16.3.3 (tbl=0,"default" len=120 ttl=254 tos=0x0)
ok, oce_rc=0x0
00:16:51: Tunnel0: IPv6/IP (PS) to decaps 172.16.2.2->172.16.3.3 (tbl=0, "default", len=120,ttl=254)
00:16:51: Tunnel0: decapsulated IPv6/IP packet
```

以上输出表明了隧道对出站数据流进行封装，对返回数据包进行解封装的过程，同时可以看到隧道模式为 IPv6/IP。

（2）查看 IP 数据包发送过程

```
R3#debug ip packet detail
R3#ping 2020:2323::2 repeat 1
00:25:47: FIBipv4-packet-proc: route packet from (local) src 172.16.3.3 dst 172.16.2.2
00:25:47: FIBipv4-packet-proc: packet routing succeeded            //数据包路由成功
00:25:47: IP: s=172.16.3.3 (local), d=172.16.2.2 (Serial0/0/1), len 120, sending, proto=41
08:09:47: IP: s=172.16.3.3 (local), d=172.16.2.2 (Serial0/0/1), len 120, sending full packet, proto=41
```

以上输出表明，在隧道模式为 IPv6/IP 封装的 IPv4 数据包中协议字段的值为 41。

（3）查看隧道接口信息

```
R2#show interfaces tunnel 0
Tunnel0 is up, line protocol is up
  Hardware is Tunnel
  MTU 17920 bytes, BW 100 Kbit/sec, DLY 50000 usec,
     reliability 255/255, txload 1/255, rxload 1/255
  Encapsulation TUNNEL, loopback not set
  Keepalive not set
Tunnel linestate evaluation up
Tunnel source 172.16.2.2 (Loopback0), destination 172.16.3.3        //建立隧道的源和目的地
Tunnel Subblocks:
     src-track:
        Tunnel0 source tracking subblock associated with Loopback0
        Set of tunnels with source Loopback0, 1 member (includes iterators), on interface <OK>
  Tunnel protocol/transport IPv6/IP                                 //隧道协议为 IPv6，传输协议为 IPv4
（此处省略部分输出）
```

（4）查看 OSPFv3 路由信息

以下输出均省略路由代码部分。

```
① R1#show ipv6 route ospf
O   2014:4444::4/128 [110/1128]
      via FE80::FA72:EAFF:FE69:1C78, Serial0/0/0
O   2020:2323::/64 [110/1064]
      via FE80::FA72:EAFF:FE69:1C78, Serial0/0/0
O   2020:3434::/64 [110/1128]
      via FE80::FA72:EAFF:FE69:1C78, Serial0/0/0
② R2#show ipv6 route ospf
O   2011:1111::1/128 [110/64]
      via FE80::FA72:EAFF:FED6:F4C8, Serial0/0/0
O   2014:4444::4/128 [110/1064]
      via FE80::AC10:303, Tunnel0
O   2020:3434::/64 [110/1064]
      via FE80::AC10:303, Tunnel0
③ R3#show ipv6 route ospf
O   2011:1111::1/128 [110/1064]
      via FE80::AC10:202, Tunnel0
O   2014:4444::4/128 [110/64]
      via FE80::FA72:EAFF:FEC8:4F98, Serial0/0/0
O   2020:1212::/64 [110/1064]
      via FE80::AC10:202, Tunnel0
④ R4#show ipv6 route ospf
O   2011:1111::1/128 [110/1128]
      via FE80::FA72:EAFF:FE69:18B8, Serial0/0/0
O   2020:1212::/64 [110/1128]
      via FE80::FA72:EAFF:FE69:18B8, Serial0/0/0
O   2020:2323::/64 [110/1064]
      via FE80::FA72:EAFF:FE69:18B8, Serial0/0/0
```

以上输出表明通过手工隧道，实现了两端 IPv6 网络路由信息共享。

（5）采用 ping 命令测试连通性

```
R4#ping 2011:1111::1 source loopback 0
Type escape sequence to abort.
Sending 5, 100-byte ICMP Echos to 2011:1111::1, timeout is 2 seconds:
Packet sent with a source address of 2014:4444::4
!!!!!
Success rate is 100 percent (5/5), round-trip min/avg/max = 44/45/48 ms
```

以上输出验证了跨越隧道的 IPv6 网络的连通性。

9.5.2　实验 12：配置 GRE 隧道

1. 实验目的

通过本实验可以掌握：
① 启用 IPv6 路由的方法。
② IPv6 与 IPv4 共存的实现方法。
③ GRE 隧道的工作原理和配置。

2. 实验拓扑

实验拓扑如图 9-19 所示。

3. 实验步骤

GRE 隧道和手工隧道一样，都是点到点隧道，用于创建永久性连接。本实验的配置和实验 11 基本相同，只是路由器 R2 和 R3 的隧道接口的封装模式为 **gre ip**，这也是隧道接口的默认封装。因此本实验只给出有变化的部分，其他配置和实验 11 完全相同。

（1）配置路由器 R2

```
R2(config)#interface Tunnel0
R2(config-if)#tunnel mode gre ip
//配置隧道的模式，gre ip 模式用来创建 GRE 隧道
```

（2）配置路由器 R3

```
R3(config)#interface Tunnel0
R3(config-if)#tunnel mode gre ip
```

4. 实验调试

（1）查看隧道建立过程

```
R3#debug tunnel
R3#ping 2020:2323::2 repeat 1
Type escape sequence to abort.
```

```
            Sending 1, 100-byte ICMP Echos to 2020:2323::2, timeout is 2 seconds:
            !
            Success rate is 100 percent (1/1), round-trip min/avg/max = 20/20/20 ms
            00:22:04: Tunnel0: GRE/IP encapsulated 172.16.3.3->172.16.2.2 (linktype=79, len=124)
            00:22:04: Tunnel0: GRE/IP (PS) to decaps 172.16.2.2->172.16.3.3 (tbl=0,"default" len=124 ttl=254)
            00:22:04:Pak Decapsulated on Serial0/0/1, ptype 0x86DD, nw start 0xD9E86F0, mac start 0xD9E86D4,
datagram size 100 link type 0x7
            00:22:04: Tunnel0: GRE decapsulated IPV6 packet (linktype=79, len=100)
```

以上输出表明了隧道对出站数据流进行封装，对返回数据包进行解封装的过程，同时可以看到隧道模式为 gre ip。

（2）查看 IP 数据包发送过程

```
        R3#debug ip packet detail
        R3#ping 2020:2323::2 repeat 1
        00:40:28: FIBipv4-packet-proc: route packet from (local) src 172.16.3.3 dst 172.16.2.2
        00:40:28: FIBfwd-proc: packet routed by adj to Serial0/0/1 0.0.0.0
        00:40:28: FIBipv4-packet-proc: packet routing succeeded
        00:40:28: IP: s=172.16.3.3 (local), d=172.16.2.2 (Serial0/0/1), len 124, sending, proto=47
        00:40:28: IP: s=172.16.3.3 (local), d=172.16.2.2 (Serial0/0/1), len 124, sending full packet, proto=47
```

以上输出表明，在隧道模式为 gre ip 封装的 IPv4 数据包中协议字段的值为 47。

（3）查看隧道接口信息

```
        R3#show interfaces tunnel 0
        Tunnel0 is up, line protocol is up
        （此处省略部分输出）
        Tunnel source 172.16.3.3 (Loopback0), destination 172.16.2.2
            Tunnel Subblocks:
                src-track:
                    Tunnel0 source tracking subblock associated with Loopback0
                    Set of tunnels with source Loopback0, 1 member (includes iterators), on interface <OK>
        Tunnel protocol/transport GRE/IP                //隧道协议为 GRE，传输协议为 IPv4
        （此处省略部分输出）
```

（4）查看 OSPFv3 路由信息

以下输出均省略路由代码部分。

```
    ① R1#show ipv6 route ospf
    O   2014:4444::4/128 [110/1128]
            via FE80::FA72:EAFF:FE69:1C78, Serial0/0/0
    O   2020:2323::/64 [110/1064]
            via FE80::FA72:EAFF:FE69:1C78, Serial0/0/0
    O   2020:3434::/64 [110/1128]
            via FE80::FA72:EAFF:FE69:1C78, Serial0/0/0
    ② R4#show ipv6 route ospf
    O   2011:1111::1/128 [110/1128]
            via FE80::FA72:EAFF:FE69:18B8, Serial0/0/0
    O   2020:1212::/64 [110/1128]
            via FE80::FA72:EAFF:FE69:18B8, Serial0/0/0
```

O   2020:2323::/64 [110/1064]
         via FE80::FA72:EAFF:FE69:18B8, Serial0/0/0

以上输出表明通过 GRE 隧道，实现了两端 IPv6 网络的路由信息共享。

9.5.3 实验 13：配置 6to4 隧道

1．实验目的

通过本实验可以掌握：
① 启用 IPv6 路由的方法。
② IPv6 与 IPv4 共存的实现方法。
③ 6to4 隧道的工作原理和配置。
④ 6to4 地址编址规则。
⑤ 6to4 隧道的优点和局限性。

2．实验拓扑

配置 6to4 隧道实验拓扑如图 9-20 所示。

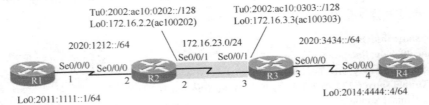

图 9-20 配置 6to4 隧道实验拓扑

3．实验步骤

本实验中，路由器 R1 和 R4 的配置和实验 11 相同，这里仅给出路由器 R2 和 R3 的配置。

（1）配置路由器 R2

本实验只给出隧道接口和路由部分的配置，其余部分和实验 11 相同。

```
R2(config)#interface Tunnel0
R2(config-if)#ipv6 address 2002:AC10:202::/128
//隧道接口的 IPv6 地址由 2002 和转换成十六进制的 IPv4 地址构成
R2(config-if)#tunnel source Loopback0
//只需要配置隧道源，不需要配置隧道目的地址
R2(config-if)#tunnel mode ipv6ip 6to4               //配置隧道的模式
R2(config)#ipv6 route 2020:3434::/64 2002:AC10:303::
R2(config)#ipv6 route 2014:4444::/64 2002:AC10:303::
//静态路由下一跳指向 R3 的隧道接口的 IPv6 地址，该地址内嵌隧道的目的 IPv4 地址
R2(config)#ipv6 route 2002::/16 Tunnel0
//去往 2002 开头的地址，都被送到隧道接口
R2(config)#ipv6 router ospf 1
R2(config-rtr)#redistribute static
```

第 9 章 IPv6

（2）配置路由器 R3

本实验只给出隧道接口和路由部分的配置，其余部分和实验 11 相同。

```
R3(config)#interface Tunnel0
R3(config-if)#ipv6 address 2002:AC10:303::/128
R3(config-if)#tunnel source Loopback0
R3(config-if)#tunnel mode ipv6ip 6to4
R3(config)#ipv6 route 2011:1111::/64 2002:AC10:202::
R3(config)#ipv6 route 2020:1212::/64 2002:AC10:202::
R3(config)#ipv6 route 2002::/16 Tunnel0
R3(config)#ipv6 router ospf 1
R3(config-rtr)#redistribute static
```

4．实验调试

（1）查看隧道建立过程

```
R2#debug tunnel
```

在路由器 R1 上执行 **ping 2014:4444::4 source 2011:1111::1 repeat 1** 命令，路由器 R2 输出的信息如下：

```
18:04:11: Tunnel0: IPv6/IP adjacency fixup, 172.16.2.2->172.16.3.3, tos set to 0x0
18:04:11: Tunnel0: IPv6/IP to classify 172.16.3.3->172.16.2.2 (tbl=0,"default" len=120 ttl=254 tos=0x0) ok, oce_rc=0x0
```

以上输出显示，用于创建隧道的目的地址是从相应静态路由的下一跳 IPv6 地址中自动提取的。

（2）查看隧道接口

```
R2#show interfaces tunnel 0
Tunnel0 is up, line protocol is up
  Hardware is Tunnel
  （此处省略部分输出）
  Tunnel source 172.16.2.2 (Loopback0)
  Tunnel Subblocks:
     src-track:
        Tunnel0 source tracking subblock associated with Loopback0
        Set of tunnels with source Loopback0, 1 member (includes iterators), on interface <OK>
  Tunnel protocol/transport IPv6 6to4            //隧道工作模式
（此处省略部分输出）
```

（3）查看 OSPFv3 路由信息

以下输出均省略路由代码部分。

① R1#show ipv6 route ospf
```
OE2 2002::/16 [110/20]
     via FE80::FA72:EAFF:FE69:1C78, Serial0/0/0
OE2 2014:4444::/64 [110/20]
```

```
                via FE80::FA72:EAFF:FE69:1C78, Serial0/0/0
     OE2 2020:3434::/64 [110/20]
                via FE80::FA72:EAFF:FE69:1C78, Serial0/0/0
② R4#show ipv6 route ospf
     OE2 2002::/16 [110/20]
                via FE80::FA72:EAFF:FE69:18B8, Serial0/0/0
     OE2 2011:1111::/64 [110/20]
                via FE80::FA72:EAFF:FE69:18B8, Serial0/0/0
     OE2 2020:1212::/64 [110/20]
                via FE80::FA72:EAFF:FE69:18B8, Serial0/0/0
```

9.5.4 实验 14：配置 ISATAP 隧道

1．实验目的

通过本实验可以掌握：
① 启用 IPv6 路由的方法。
② IPv6 与 IPv4 共存的实现方法。
③ ISATAP 隧道的工作原理和配置。
④ ISATAP 地址编址规则。

2．实验拓扑

配置 IPv6 EIGRP 实验拓扑如图 9-21 所示。

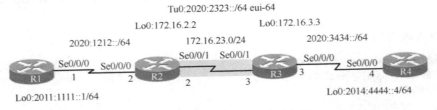

图 9-21　配置 IPv6 EIGRP 实验拓扑

3．实验步骤

本实验中，路由器 R1 和 R4 的配置和实验 11 相同，这里仅给出路由器 R2 和 R3 的配置。

（1）配置路由器 R2

本实验只给出隧道接口和路由部分的配置，其余部分和实验 11 相同。

```
R2(config)#interface Tunnel0
R2(config-if)#ipv6 address 2020:2323::/64 eui-64
R2(config-if)#tunnel source Loopback0
R2(config-if)#tunnel mode ipv6ip isatap              //配置隧道的模式
R2(config)#ipv6 route 2020:3434::/64 Tunnel0 FE80::5EFE:AC10:303
R2(config)#ipv6 route 2014:4444::/64 Tunnel0 FE80::5EFE:AC10:303
//当配置静态路由时，如果下一跳为链路本地地址，则必须指定出接口
```

第 9 章 IPv6

```
R2(config)#ipv6 router ospf 1
R2(config-rtr)#redistribute static
```

（2）配置路由器 R3

本实验只给出隧道接口和路由部分的配置，其余部分和实验 11 相同。

```
R3(config)#interface Tunnel0
R3(config-if)#ipv6 address 2020:2323::/64 eui-64
R3(config-if)#tunnel source Loopback0
R3(config-if)#tunnel mode ipv6ip isatap
R3(config)#ipv6 route 2011:1111::/64 Tunnel0 FE80::5EFE:AC10:202
R3(config)#ipv6 route 2020:1212::/64 Tunnel0 FE80::5EFE:AC10:202
R3(config)#ipv6 router ospf 1
R3(config-rtr)#redistribute static
```

4．实验调试

（1）查看隧道建立过程

```
R2#debug tunnel
```

在路由器 R1 上执行 **ping 2014:4444::4 source 2011:1111::1 repeat 1** 命令，路由器 R2 输出的信息如下：

```
20:01:31: Tunnel0: IPv6/IP adjacency fixup, 172.16.2.2->172.16.3.3, tos set to 0x0
20:01:31: Tunnel0: IPv6/IP to classify 172.16.3.3->172.16.2.2 (tbl=0,"default" len=120 ttl=254 tos=0x0) ok, oce_rc=0x0
```

以上输出显示，用于创建隧道的目的地址是从相应静态路由的下一跳 IPv6 地址中自动提取的。

（2）查看 IPv6 隧道接口信息

```
R2#show ipv6 interface tunnel 0
Tunnel0 is up, line protocol is up
  IPv6 is enabled, link-local address is FE80::5EFE:AC10:202
  No Virtual link-local address(es):
  Global unicast address(es):
    2020:2323::5EFE:AC10:202, subnet is 2020:2323::/64 [EUI]
```
//隧道接口只配置了 IPv6 的前缀，当隧道模式为 isatap 时，路由器使用配置的 IPv6 前缀加上 ISATAP 的 OUI（0000:5EFE）以及用十六进制表示的隧道的源 IPv4 地址构成接口 ID

（此处省略部分输出）

（3）查看隧道接口信息

```
R2#show interfaces tunnel 0
Tunnel0 is up, line protocol is up
  Hardware is Tunnel
```
（此处省略部分输出）
```
  Tunnel source 172.16.2.2 (Loopback0)
  Tunnel Subblocks:
```

src-track:

  Tunnel0 source tracking subblock **associated with Loopback0**
  Set of tunnels with source Loopback0, 1 member (includes iterators), on interface <OK>
  Tunnel protocol/transport **IPv6 ISATAP**　　　　　　　　　　//隧道工作模式
（此处省略部分输出）

（4）查看 OSPFv3 路由信息

（以下输出均省略路由代码部分）
① R1#**show ipv6 route ospf**
OE2 2014:4444::/64 [110/20]
  via FE80::FA72:EAFF:FE69:1C78, Serial0/0/0
OE2 2020:3434::/64 [110/20]
  via FE80::FA72:EAFF:FE69:1C78, Serial0/0/0
② R4#**show ipv6 route ospf**
OE2 2011:1111::/64 [110/20]
  via FE80::FA72:EAFF:FE69:18B8, Serial0/0/0
OE2 2020:1212::/64 [110/20]
  via FE80::FA72:EAFF:FE69:18B8, Serial0/0/0

9.5.5　实验 15：配置 IPv6 静态 NAT-PT

1．实验目的

通过本实验可以掌握：
① NAT-PT 的工作原理和特征。
② NAT-PT 的使用场合。
③ 静态 NAT-PT 的配置。

2．实验拓扑

配置静态 NAT-PT 实验拓扑如图 9-22 所示。

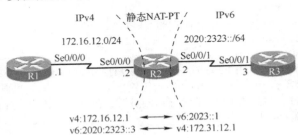

图 9-22　配置静态 NAT-PT 实验拓扑

3．实验步骤

  本实验实现静态 NAT-PT 功能，当从路由器 R1 访问 172.31.12.1 时，路由器 R2 完成协议和地址转换，把 IPv4 地址转换为 IPv6 地址 2020:2323::3；当从路由器 R3 访问 2023::1 时，路由器 R2 完成协议和地址转换，把 IPv6 地址转换为 IPv4 地址 172.16.12.1。只要被转换的地址可达，同时在路由器 R1 或 R3 上有相应的路由条目，就可以通信，尽管实验设计的 IPv4 地址 172.31.12.1 和 IPv6 地址 2023::1 是虚拟的，但是转换后的地址是可达的。

（1）配置路由器 R1

```
R1(config)#ip route 0.0.0.0 0.0.0.0 Serial0/0/0
```

（2）配置路由器 R2

```
R2(config)#ipv6 unicast-routing
R2(config)#interface Serial0/0/0
R2(config-if)#ip address 172.16.12.2 255.255.255.0
R2(config-if)#ipv6 nat
//接口启用 NAT-PT 功能，对前往和来自该接口的数据流进行转换
R2(config)#interface Serial0/0/1
R2(config-if)#ipv6 address 2020:2323::2/64
R2(config-if)#ipv6 nat
R2(config)#ipv6 nat v4v6 source 172.16.12.1 2023::1
//配置 IPv4 到 IPv6 的静态转换条目
R2(config)#ipv6 nat v6v4 source 2020:2323::3 172.31.12.1
//配置 IPv6 到 IPv4 的静态转换条目
R2(config)#ipv6 nat prefix 2023::/96
//配置用于 NAT-PT 转换的 IPv6 前缀，对匹配该前缀的数据包执行转换，前缀长度必须为 96，因为 32 比特的 IPv4 地址被转换成 128 比特的 IPv6 地址，两者长度相差 96 比特
```

（3）配置路由器 R3

```
R3(config)#ipv6 unicast-routing
R3(config)#ipv6 route ::/0 Serial0/0/1
```

4. 实验调试

（1）查看 NAT-PT 表

```
R2#show ipv6 nat translations
Prot   IPv4 source               IPv6 source
       IPv4 destination          IPv6 destination
---    ---                       ---
       172.16.12.1               2023::1
---    172.31.12.1               2020:2323::3
       ---                       ---
```

以上输出表明 NAT-PT 表中包含两条 NAT 静态转换条目。

（2）查看 NAT-PT 的转换过程

```
R2#debug ipv6 nat
R1#ping 172.31.12.1 repeat 1
IPv6 NAT-PT debugging is on
00:52:00: IPv6 NAT:   src (172.16.12.1) -> (2023::1), dst (172.31.12.1) -> (2020:2323::3)
//IPv4 到 IPv6 协议和地址的转换过程
00:52:01: IPv6 NAT: icmp src (2020:2323::3) -> (172.31.12.1), dst (2023::1) -> (172.16.12.1)
```

//IPv6 到 IPv4 协议和地址的转换过程

此时再次查看 NAT-PT 表，内容如下：

```
R2#show ipv6 nat translations
Prot   IPv4 source            IPv6 source
       IPv4 destination       IPv6 destination
---    ---                    ---
       172.16.12.1            2023::1

---    172.31.12.1            2020:2323::3
       172.16.12.1            2023::1

---    172.31.12.1            2020:2323::3
       ---                    ---
```

以上输出表明 NAT-PT 表中包含两条静态转换条目和 ping 操作创建的转换条目（可以通过执行命令 **clear ipv6 nat translation *** 清除该条目）。

（3）查看 IPv6 路由表

此处省略路由代码部分。

```
R2#show ipv6 route connected
C   2020:2323::/64 [0/0]
     via ::, Serial0/0/1
C   2023::/96 [0/0]      //该路由是命令 ipv6 nat prefix 在路由表中创建的
     via ::, NVI0        //NVI0 是 NAT 虚拟接口，R2 收到去往该前缀的数据包就执行 NAT 转换
```

9.5.6　实验 16：配置 IPv6 动态 NAT-PT

1. 实验目的

通过本实验可以掌握：
① NAT-PT 的工作原理和特征。
② NAT-PT 的使用场合。
③ 动态 NAT-PT 的配置。

2. 实验拓扑

配置动态 NAT-PT 实验拓扑如图 9-23 所示。

图 9-23　配置动态 NAT-PT 实验拓扑

3. 实验步骤

本实验分两部分实现：IPv4 地址转换为 IPv6 地址用静态 NAT-PT 实现，IPv6 地址转换为 IPv4 地址用动态 NAT-PT 实现。本实验中，路由器 R1 和 R3 的配置和实验 15 中的相同，这里只给出路由器 R2 的配置。

配置路由器 R2

```
R2(config)#ipv6 unicast-routing
R2(config)#interface Serial0/0/0
R2(config-if)#ip address 172.16.12.2 255.255.255.0
R2(config-if)#ipv6 nat
R2(config)#interface Serial0/0/1
R2(config-if)#ipv6 address 2020:2323::2/64
R2(config-if)#ipv6 nat
R2(config)#ipv6 nat v4v6 source 172.16.12.1 2023::1
//配置 IPv4 到 IPv6 的静态转换条目
R2(config)#ipv6 access-list v6v4
R2(config-ipv6-acl)#permit ipv6 2020:2323::/64 any
//配置 ACL，定义允许动态 NAT-PT 转换的地址
R2(config)#ipv6 nat v6v4 pool POOL 172.31.12.1 172.31.12.10 prefix-length 24
//配置 IPv6 到 IPv4 转换的地址池，地址池的名字为 POOL
R2(config)#ipv6 nat v6v4 source list v6v4 pool POOL
//配置动态 NAT-PT 转换，将地址池和 ACL 关联
R2(config)#ipv6 nat translation timeout 3600
//配置动态转换的全局超时时间，默认为 86400 秒，也可以针对 ICMP、UDP、DNS 和 TCP 等进行单独调整
R2(config)#ipv6 nat max-entries 10000
//限制 NAT-PT 同时处理的转换条目，默认时不限制
R2(config)#ipv6 nat prefix 2023::/96
```

4. 实验调试

（1）查看 NAT-PT 表

```
R2#show ipv6 nat translations
Prot    IPv4 source              IPv6 source
        IPv4 destination         IPv6 destination
---     ---                      ---
        172.16.12.1              2023::1              //静态 NAT 转换条目
R3#ping 2023::1 repeat 1
//由于本实验实现 IPv6 到 IPv4 动态转换，所以测试只能从路由器 R3 发起
R2#show ipv6 nat translations
Prot    IPv4 source              IPv6 source
        IPv4 destination         IPv6 destination
---     ---                      ---
        172.16.12.1              2023::1
icmp    172.31.12.1              2020:2323::3
        172.16.12.1              2023::1
---     172.31.12.1              2020:2323::3
```

以上路由器 R2 的两次输出结果的差异表明静态转换条目一直存在于 NAT-PT 表中，而动态 NAT-PT 转换条目需要有数据流才会创建，而且超过超时时间后会自动从 NAT-PT 表中删除。

（2）配置动态 NAT-PT 过载

路由器 R2 的配置如下：

```
R2(config)#ipv6 nat v6v4 source list v6v4 pool POOL overload
//配置动态 NAT-PT 过载使用 overload 关键字
R3#ping 2023::1 repeat 1
R3#telnet 2023::1 finger
R3#telnet 2023::1
R2#show ipv6 nat translations
Prot   IPv4 source             IPv6 source
       IPv4 destination        IPv6 destination
---    ---                     ---
       172.16.12.1             2023::1
icmp   172.31.12.5,1194        2020:2323::3,1194
       172.16.12.1,1194        2023::1,1194
tcp    172.31.12.5,26401       2020:2323::3,26401
       172.16.12.1,79          2023::1,79
tcp    172.31.12.5,53297       2020:2323::3,53297
       172.16.12.1,23          2023::1,23
```

以上输出显示了三个应用测试之后 NAT-PT 表生成的动态转换条目，包括协议类型、源和目的地址以及端口号，IPv6 到 IPv4 动态转换使用同一个 IPv4 地址 **172.31.12.5**，但是端口号不同。

（3）查看 NAT-PT 转换的统计信息

```
R2#show ipv6 nat statistics
Total active translations: 4 (0 static, 4 dynamic; 3 extended)
//活动的转换条目，4 条动态条目，其中包括 3 个扩展条目
NAT-PT interfaces:
    Serial0/0/0, Serial0/0/1, NVI0              //NAT-PT 的接口
Hits: 0   Misses: 0
Expired translations: 8                         //过期的条目
```

本 章 小 结

IPv6 庞大的地址空间以及对移动性和安全性的支持必将推动 IPv6 的广泛部署。本章介绍了 IPv6 特点、IPv6 地址、IPv6 基本包头格式、IPv6 扩展包头、IPv6 地址类型、IPv6 邻居发现协议、IPv6/IPv4 双栈技术、隧道技术以及 IPv4/IPv6 协议转换技术，并用实验演示和验证了 IPv6 地址配置、IPv6 静态路由配置、RIPng 配置、OSPFv3 配置、IPv6 EIGRP 配置、IPv6 集成 IS-IS 配置、MBGP 配置、IPv6 路由重分布、IPv6 策略路由配置、手工隧道配置、GRE 隧道配置、6to4 隧道配置、ISATAP 隧道配置和 NAT-PT 配置等内容。

参 考 文 献

[1] 梁广民，王隆杰，徐磊，编著. 思科网络实验室 CCNA 实验指南（第 2 版）. 北京：电子工业出版社，2018.
[2] 梁广民，王隆杰，编著. 网络互联技术（第二版）. 北京：高等教育出版社. 2018.
[3] [美]Diane Teare，等著. CCNP ROUTE 300-101 学习指南. YESLAB 工作室，译. 北京：人民邮电出版社，2018.
[4] [美]Jeff Doyle，著. TCP/IP 路由技术（第 1 卷）（第 2 版）英文版. 北京：人民邮电出版社，2017.
[5] [美]Jeff Doyle，著. TCP/IP 路由技术（第 2 卷）（第 2 版）. 夏俊杰，译. 北京：人民邮电出版社，2018.
[6] [美]Kevin Wallace. CCNP ROUTE 300-101 认证考试指南. YESLAB 工作室，译. 北京：人民邮电出版社，2018.
[7] [加]Régis Desmeules，著. Cisco IPv6 网络实现技术（修订版）. 王玲芳，等译. 北京：人民邮电出版社，2018.
[8] [美]Vijay Bollapragada，等著. IPSec VPN 设计. 袁国忠，译. 北京：人民邮电出版社，2012.

反侵权盗版声明

电子工业出版社依法对本作品享有专有出版权。任何未经权利人书面许可，复制、销售或通过信息网络传播本作品的行为；歪曲、篡改、剽窃本作品的行为，均违反《中华人民共和国著作权法》，其行为人应承担相应的民事责任和行政责任，构成犯罪的，将被依法追究刑事责任。

为了维护市场秩序，保护权利人的合法权益，本社将依法查处和打击侵权盗版的单位和个人。欢迎社会各界人士积极举报侵权盗版行为，本社将奖励举报有功人员，并保证举报人的信息不被泄露。

举报电话：（010）88254396；（010）88258888
传　　真：（010）88254397
E-mail：dbqq@phei.com.cn
通信地址：北京市海淀区万寿路 173 信箱
　　　　　电子工业出版社总编办公室
邮　　编：100036